Stochastic Processes

HOLDEN DAY SERIES IN PROBABILITY AND STATISTICS

E. L. Lehmann, Editor

San Francisco

Düsseldorf   Johannesburg   London   Mexico   Panama

São Paulo   Singapore   Sydney   Toronto

Stochastic Processes

EMANUEL PARZEN

Statistical Science Division
State University of New York at Buffalo

 HOLDEN-DAY

11 MP 8079876

Preface

THE THEORY OF STOCHASTIC PROCESSES is generally defined as the "dynamic" part of probability theory, in which one studies a collection of random variables (called a *stochastic process*) from the point of view of their interdependence and limiting behavior. One is observing a stochastic process whenever one examines a process developing in time in a manner controlled by probabilistic laws. Examples of stochastic processes are provided by the path of a particle in Brownian motion, the growth of a population such as a bacterial colony, the fluctuating number of particles emitted by a radioactive source, and the fluctuating output of gasoline in successive runs of an oil-refining mechanism. Stochastic or random processes abound in nature. They occur in medicine, biology, physics, oceanography, economics, and psychology, to name only a few scientific disciplines. If a scientist is to take account of the probabilistic nature of the phenomena with which he is dealing, he should undoubtedly make use of the theory of stochastic processes.

During the past few years I have attempted to develop at Stanford University a first course in stochastic processes, to be taken by students with a knowledge only of the calculus and of the elements of continuous probability theory. This course has three aims:

(i) to give examples of the wide variety of empirical phenomena for which stochastic processes provide mathematical models;

(ii) to provide an introduction to the methods of probability model-building;

(iii) to provide the reader who does not possess an advanced mathematical background with mathematically sound techniques and with a sufficient degree of maturity to enable further study of the theory of stochastic processes.

Written as a text for a course with these aims, this book can be adapted to the needs of students with diverse interests and backgrounds. Many of the chapters can be read independently of one another without loss of continuity. The logical interdependence between the various chapters is indicated in the accompanying diagram.

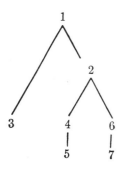

An introductory chapter shows how stochastic processes arise naturally and therefore play a basic role in many scientific fields. Chapter 1 gives precise definitions of the notions of a random variable and a stochastic process and introduces the Wiener process and the Poisson process. In order to study the theory of stochastic processes, one must know how to operate with conditional probabilities and conditional expectations; Chapter 2 attempts to provide the reader with these tools. Chapter 3 discusses some of the basic concepts and techniques of the theory of stochastic processes possessing finite second moments and introduces normal processes and covariance stationary processes.

Chapter 4 treats the properties of the Poisson process and shows that the Poisson process not only arises frequently as a model for the counts of random events but also provides a basic building block with which other useful stochastic processes can be constructed. Renewal counting processes are discussed in Chapter 5. Markov chains (discrete and continuous parameter), random walks, and birth and death processes are treated in Chapters 6 and 7, which also give examples of the wide variety of phenomena to which these stochastic processes may be applied.

Every section contains numerous examples and exercises. Many sections also contain complements; these are extensions of the theory treated in the text, stated in the form of assertions that the reader is asked to prove.

In spite of numerous references to original sources, I have not attempted to indicate in all cases the names of those first responsible for the original developments nor have I attempted to give a complete list of references for further study. I would welcome comments and queries from readers who desire references on any of the topics treated.

The author of a textbook is intellectually indebted to all who have worked in his field. It is impossible to record by name all those whose books and papers have provided ideas for various parts of this book. I have tried to give credit where I could properly assign it. However, the material in this book has been reworked so many times that I may have slighted some who deserve more acknowledgment. To them I offer my apologies together with the assurance that no oversight is intended. I would appreciate receiving any criticisms concerned with keeping the historical record straight, so that these may be incorporated into any revisions of this book.

I would like to thank my colleagues at Stanford University, especially Professors Samuel Karlin and Herbert Solomon, for the constant encouragement they have given me and the stimulating atmosphere they have provided. To my students who have contributed to this book by their comments, I offer my thanks. Particularly valuable assistance has been rendered by Paul Milch. Thanks are due also to Mrs. Betty Jo Prine who supervised the typing. It is a pleasure to acknowledge my gratitude to the Office of Naval Research for the support it has extended for many years to my research work. Finally, I thank my wife for her understanding and encouragement.

EMANUEL PARZEN

London, England
January, 1962

TO CAROL

Table of contents

Role of the theory of stochastic processes

THE SCIENTIST making measurements in his laboratory, the meteorologist attempting to forecast weather, the control systems engineer designing a servomechanism (such as an aircraft autopilot or a thermostatic control), the electrical engineer designing a communications system (such as the radio link between entertainer and audience or the apparatus and cables that transmit messages from one point to another), the economist studying price fluctuations and business cycles, and the neurosurgeon studying brain-wave records—all are encountering problems to which the theory of stochastic processes may be relevant. Before beginning our study of the theory of stochastic processes, let us show why this theory is an essential part of such diverse fields as statistical physics, the theory of population growth, communication and control theory, management science (operations research), and time series analysis. For a survey of other applications of the theory of stochastic processes, especially in astronomy, biology, industry, and medicine, see Bartlett (1962)† and Neyman and Scott (1959).

STATISTICAL PHYSICS

Many parts of the theory of stochastic processes were first developed in connection with the study of fluctuations and noise in physical systems (Einstein [1905], Smoluchowski [1906], Schottky [1918]). Consequently,

†References are listed in detail at the rear of the book.

the theory of stochastic processes can be regarded as the mathematical foundation of statistical physics.

Stochastic processes provide models for physical phenomena such as thermal noise in electric circuits and the Brownian motion of a particle immersed in a liquid or gas.

Brownian motion. When a particle of microscopic size is immersed in a fluid, it is subjected to a great number of random independent impulses owing to collisions with molecules. The resulting vector function $(X(t),$ $Y(t),$ $Z(t))$ representing the position of the particle as a function of time is known as *Brownian motion*.

Thermal noise. Consider a resistor in an electric network. Because of the random motions of the conduction electrons in the resistor, there will occur small random fluctuations in the voltage $X(t)$ across the ends of the resistor. The fluctuating voltage $X(t)$ is called *thermal noise* (its probability law may be shown to depend only on the resistance R and the absolute temperature T of the resistor).

Shot noise. Consider a vacuum diode connected to a resistor. Because the emission of electrons from the heated cathode is not steady, a current $X(t)$ is generated across the resistance which consists of a series of short pulses, each pulse corresponding to the passage of an electron from cathode to anode. The fluctuating current $X(t)$ is called *shot noise*.

STOCHASTIC MODELS
FOR POPULATION GROWTH

The size and composition of a population (whether it consists of living organisms, atoms undergoing fission, or substances decaying radioactively) is constantly fluctuating. Stochastic processes provide a means of describing the mechanisms of such fluctuations (see Bailey [1957], Bartlett [1960], Bharucha-Reid [1960], Harris [1964]).

Some of the biological phenomena for which stochastic processes provide models are (i) the extinction of family surnames, (ii) the consequences of gene mutations and gene recombinations in the theory of evolution, (iii) the spatial distribution of plant and animal communities, (iv) the struggle for existence between two populations that interact or compete, (v) the spread of epidemics, and (vi) the phenomenon of carcinogenesis.

COMMUNICATION
AND CONTROL

A wide variety of problems involving communication and/or control (including such diverse problems as the automatic tracking of moving objects, the reception of radio signals in the presence of natural and artificial disturbances, the reproduction of sound and images, the design of guidance systems, the design of control systems for industrial processes, forecasting, the analysis of economic fluctuations, and the analysis of any kind of record representing observation over time) may be regarded as special cases of the following general problem:

Let T denote a set of points on a time axis such that at each point t in T an observation has been made of a random variable $X(t)$. Given the observations $\{X(t), t \in T\}$, and a quantity Z related to the observation in a manner to be specified, one desires to form in an optimum manner estimates of, and tests of hypotheses about, Z and various functions $h(Z)$. This imprecisely formulated problem provides the general context in which to pose the following usual problems of communication and control.

Prediction (or extrapolation). Having observed the stochastic process $X(t)$ over the interval $s - L \leq t \leq s$, one wishes to predict $X(s + \alpha)$ for any $\alpha > 0$. The interval L of observation may be finite or infinite.

Smoothing. Suppose that the observations

$$\{X(t), s - L \leq t \leq s\}$$

may be regarded as the sum, $X(t) = S(t) + N(t)$, of two stochastic processes (or time series) $S(\cdot)$ and $N(\cdot)$, representing signal and noise, respectively. One desires to estimate the value $S(t)$ of the signal at any time t in $s - L \leq t \leq s$. The terminology *smoothing* derives from the fact that often the noise $N(\cdot)$ consists of very high frequency components compared with the signal $S(\cdot)$; estimating or extracting the signal $S(\cdot)$ can then be regarded as an attempt to pass a smooth curve through a very wiggly record. The problem of predicting $S(s + \alpha)$ for any $\alpha > 0$ is called the problem of smoothing and prediction.

Parameter estimation (signal extraction and detection). Suppose that the observations $\{X(t), 0 \leq t \leq T\}$ may be regarded as a sum, $X(t) = S(t) + N(t)$, where $S(\cdot)$ represents the trajectory (given by $S(t) = x_0 + vt + (a/2)t^2$, say) of a moving object, and $N(\cdot)$ represents errors of measurement. One desires to estimate the velocity v and acceleration a of the object. More generally, one desires to estimate such quantities as $S(t)$ and $(d/dt)S(t)$ at any time t in $0 \leq t \leq T$, under the assumption that the signal $S(\cdot)$ belongs to a known class of functions.

It is clear that the solutions one gives to the foregoing problems depend on the assumptions one makes about the signals and noise received, and also on the criterion one adopts for an optimum solution. It is well established (see Lanning and Battin [1956], Helstrom [1960], Middleton [1960]) that the design of optimum communications and control systems requires recognition of the fact that the signals and noises arising in such systems are best regarded as stochastic processes. Consequently, the first step in the study of modern communication and control systems theory is the study of the theory of stochastic processes. The purpose of such a study is:

(i) to provide a language in which one may state communications and control problems, since the evaluation of communications and control systems must necessarily be in terms of their average behavior over a range of circumstances described probabilistically;

(ii) to provide an insight into the most realistic assumptions to be made concerning the stochastic processes representing the signals and/or noises.

MANAGEMENT
SCIENCE

Stochastic processes provide a method of quantitatively studying and managing business operations and, consequently, play an important role in the modern disciplines of management science and operations research. The two fields in which the theory of stochastic processes has found greatest application are inventory control and waiting-line analysis (see Arrow, Karlin, and Scarf [1958], Syski [1960]).

Inventory control. Two problems of considerable importance to such diverse organizations as retail shops, wholesale distributors, manufacturers, and consumers holding stocks of spare parts are (i) deciding when to place an order for replenishment of their stock of items, and (ii) deciding how large an order to place. Two kinds of uncertainty must be taken into account in making these decisions: (i) uncertainty concerning the number of items that will be demanded during a given time period; (ii) uncertainty concerning the *time-of-delivery lag* that will elapse between the date on which an order is placed and the date on which the items ordered will actually arrive. If it were not for these factors, one could perhaps order new stock in such a manner that it would arrive at the precise instant when it is needed. One would then not have to maintain inventory, which is often very expensive to do. *Inventory control* is concerned with minimizing the cost of maintaining inventories, while at the same time keeping a sufficient stock on hand to meet all contingencies arising from random demand and random time-of-delivery lag of new stock ordered. One approach to the problem of optimal inventory control is to consider inventory policies actually used and to describe the effects of these policies. Given a specific inventory policy, the resultant fluctuating inventory level is a stochastic process.

Queues. A queue (or waiting line) is generated when customers (or servers) arriving at some point to receive (or render) service there must wait to receive (or render) service. The group waiting to receive (or render) service, perhaps including those receiving (or rendering) service, is called the *queue*.

There are many examples of queues. Persons waiting in a railroad station or airport terminal to buy tickets constitute a queue. Planes landing at an airport make up a queue. Ships arriving at a port to load or unload cargo form a queue. Taxis waiting at a taxi stand for passengers are in a queue. Messages transmitted by cable constitute a queue. Mechanics in a plant form a queue at the counters of tool cribs where tools are stored. Machines on a production line which break down and are repaired are in queue for service by mechanics.

In the mathematical theory of queues, waiting lines are classified according to four aspects: (i) the input distribution (the probability law of the times between successive arrivals of customers); (ii) the service time distribution (the probability law of the time it takes to serve a customer); (iii) the number of service channels; (iv) the queue discipline (the manner in which customers are selected to be served; possible policies are "first come, first served," random selection for service, and service according to order of priority). Queueing theory is concerned with the effect that each of these four aspects has on various quantities of interest, such as the length of the queue and the waiting time of a customer for service.

TIME SERIES ANALYSIS

A set of observations arranged chronologically is called a *time series*. Time series are observed in connection with quite diverse phenomena and by a wide variety of researchers. Examples are (i) the economist observing yearly wheat prices, (ii) the geneticist observing daily egg production of a certain new breed of hen, (iii) the meteorologist studying daily rainfall in a given city, (iv) the physicist studying the ambient noise level at a given point in the ocean, (v) the aerodynamicist studying atmospheric turbulence gust velocities, (vi) the electronics engineer studying the internal noise of a radio receiver.

To represent a time series, one proceeds as follows. The set of time points at which measurements are made is called T. In many applications, T is a set of discrete equidistant time points (in which case one writes $T = \{1, 2, \cdots, N\}$, where N is the number of observations) or T is an interval of the real time axis (in which case one writes $T = \{0 \leq t \leq L\}$, where L is the length of the interval). The observation made at time t is denoted by $X(t)$. The set of observations $\{X(t), t \epsilon T\}$ is called a *time series*.

In the development of the theory of stochastic processes, an important role has been played by the study of *economic time series*. Consider the prices of a commodity or corporate stocks traded on an exchange. The record of prices over time may be represented as a fluctuating function $X(t)$. The analysis of such economic time series has been a problem of great interest to economic theorists desiring to explain the dynamics of economic systems and to speculators desiring to forecast prices.

The basic idea (see Wold [1938], Bartlett [1955], Grenander and Rosenblatt [1957], Hannan [1960]) of the statistical theory of analysis of a time series $\{X(t), t \in T\}$ is to regard the time series as an observation made on a family of random variables $\{X(t), t \in T\}$; that is, for each t in T, the observation $X(t)$ is an observed value of a random variable. A family of random variables $\{X(t), t \in T\}$ is called a *stochastic process*. Having made the assumption that the observed time series $\{X(t), t \in T\}$ is an observation (or, in a different terminology, a realization) of a stochastic process $\{X(t), t \in T\}$, the statistical theory of time series analysis attempts to infer from the observed time series the probability law of the stochastic process. The method by which it treats this problem is similar in spirit (although it requires a more complicated analytic technique) to the method by which classical statistical theory treats the problem of inferring the probability law of a random variable X for which one has a finite number of independent observations $X_1, X_2, \cdots, X_n$.

In order to analyze a time series $\{X(t), t \in T\}$, one must first assume a model for $\{X(t), t \in T\}$, which is completely specified except for the values of certain parameters which one proceeds to estimate on the basis of an observed sample. Consequently, the first step in the study of time series analysis is the study of the theory of stochastic processes. The purpose of such a study is

(i) to provide a language in which assumptions may be stated about observed time series;

(ii) to provide an insight into the most realistic and/or mathematically tractable assumptions to be made concerning the stochastic processes that are adopted as models for time series.

Random variables
and stochastic processes

PROBABILITY THEORY is regarded in this book as the study of mathematical models of random phenomena. A random phenomenon is defined as an empirical phenomenon that obeys probabilistic, rather than deterministic, laws.

A random phenomenon that arises through a process (for example, the motion of a particle in Brownian motion, the growth of a population such as a bacterial colony, the fluctuating current in an electric circuit owing to thermal noise or shot noise, or the fluctuating output of gasoline in successive runs of an oil-refining mechanism) which is developing in time in a manner controlled by probabilistic laws is called a *stochastic process*.†

For reasons indicated in the Introduction, from the point of view of the mathematical theory of probability a stochastic process is best defined as a collection $\{X(t), t \epsilon T\}$ of random variables. (The Greek letter ϵ is read "belongs to" or "varying in.") The set T is called the *index set* of the process. No restriction is placed on the nature of T. However, two important cases are when $T = \{0, \pm 1, \pm 2, \cdots\}$ or $T = \{0, 1, 2, \cdots\}$, in which case the stochastic process is said to be a *discrete parameter process*,

---

†The word "stochastic" is of Greek origin; see Hagstroem (1940) for a study of the history of the word. In seventeenth century English, the word "stochastic" had the meaning "to conjecture, to aim at a mark." It is not quite clear how it acquired the meaning it has today of "pertaining to chance." Many writers use the expressions "chance process" or "random process" as synonyms for "stochastic process."

or when $T = \{t: -\infty < t < \infty\}$ or $T = \{t: \geq 0\}$, in which case the stochastic process is said to be a *continuous parameter process*.

This chapter discusses the precise definition of random variables and stochastic processes that will be employed in this book. It also introduces two stochastic processes, the Wiener process and the Poisson process, that play a central role in the theory of stochastic processes.

1-1 RANDOM VARIABLES AND PROBABILITY LAWS

Intuitively, a random variable X is a real-valued quantity which has the property that for every set B of real numbers there exists a probability, denoted by $P[X \text{ is in } B]$, that X is a member of B. Thus X is a *variable* whose values are taken *randomly* (that is, in accord with a probability distribution). In the theory of probability, a random variable is defined as a function on a sample description space. By employing such a definition, one is able to develop a *calculus of random variables* studying the characteristics of random variables generated, by means of various analytic operations, from other random variables.†

In order to give a formal definition of the notion of a random variable we must first introduce the notions of

(i) a sample description space,
(ii) an event,
(iii) a probability function.

The *sample description space* S of a random phenomenon is the space of descriptions of all possible outcomes of the phenomenon.

An *event* is a set of sample descriptions. An event E is said to occur if and only if the observed outcome of the random phenomenon has a sample description in E.

It should be noted that, for technical reasons, one does not usually permit all subsets of S to be events. Rather as the family $\mathfrak{F}$ of events, one adopts a family $\mathfrak{F}$ of subsets of S which has the following properties:

(i) S belongs to $\mathfrak{F}$.

(ii) To $\mathfrak{F}$ belongs the complement E^c of any set E belonging to $\mathfrak{F}$.

(iii) To $\mathfrak{F}$ belongs the union $\bigcup_{n=1}^{\infty} E_n$ of any sequence of sets $E_1, E_2, \cdots$ belonging to $\mathfrak{F}$.

Note that the family of all subsets of S possesses properties (i)–(iii). However, this family is often inconveniently large (in the sense that for

---

†This section constitutes a summary of the main notions of elementary probability theory as developed in Chapters 4–9 of *Modern Probability Theory and Its Applications*, by Emanuel Parzen (New York, Wiley, 1960), hereafter referred to as *Mod Prob*, to which the reader is referred for a more detailed discussion and for examples.

certain sample description spaces S it is impossible to define on the family of all subsets of S a probability function $P[\,\cdot\,]$ satisfying axiom 3 below). For the development of the mathematical theory of probability, it often suffices to take as the family of events the *smallest* family of subsets of S possessing properties (i)–(iii) and also containing as members all the sets in which we expect to be interested. Thus, for example, in the case in which the sample description space S is the real line, one adopts as the family of events the family $\mathcal{B}$ of Borel sets, where $\mathcal{B}$ is defined as the smallest family of sets of real numbers that possesses properties (i)–(iii) and, in addition, contains as members all intervals. (An *interval* is a set of real numbers of the form $\{x: a < x < b\}$, $\{x: a < x \le b\}$, $\{x: a \le x \le b\}$, $\{x: a \le x < b\}$, in which a and b may be finite or infinite numbers.)

To complete the mathematical description of a random phenomenon, one next specifies a probability function $P[\,\cdot\,]$ on the family $\mathcal{F}$ of random events; more precisely, one defines for each event E in $\mathcal{F}$ a number, denoted by $P[E]$ and called the *probability of E* (or the probability that E will occur). Intuitively, $P[E]$ represents the probability that (or relative frequency with which) an observed outcome of the random phenomenon is a member of E.

Regarded as a function on events, $P[\,\cdot\,]$ is assumed to satisfy three axioms:

Axiom 1. $P[E] \ge 0$ for every event E.

Axiom 2. $P[S] = 1$ for the certain event S.

Axiom 3. For any sequence of events $E_1, E_2, \cdots, E_n, \cdots$ which are mutually exclusive (that is, events satisfying the condition that, for any two distinct indices j and k, $E_j E_k = \emptyset$, where $\emptyset$ denotes the impossible event or empty set),

$$P\left[\bigcup_{n=1}^{\infty} E_n\right] = \sum_{n=1}^{\infty} P[E_n].$$

In applied probability theory, sample description spaces are not explicitly employed. Rather, most problems are treated in terms of random variables.

An object X is said to be a *random variable* if (i) it is a real finite valued function defined on a sample description space S on whose family $\mathcal{F}$ of events a probability function $P[\,\cdot\,]$ has been defined, and (ii) for every Borel set B of real numbers the set $\{s: X(s)$ is in $B\}$ is an event in $\mathcal{F}$.

The *probability function* of a random variable X, denoted by $P_X[\,\cdot\,]$, is a function defined for every Borel set B of real numbers by

$$P_X[B] = P[\{s: X(s) \text{ is in } B\}] = P[X \text{ is in } B]. \tag{1.1}$$

In words, $P_X[B]$ is the probability that an observed value of X will be in B.

Two random variables X and Y are said to be *identically distributed* if their probability functions are equal; that is, if $P_X[B] = P_Y[B]$ for all Borel sets B.

The *probability law* of a random variable X is defined as a probability function $P[\cdot]$ that coincides with the probability function $P_X[\cdot]$ of the random variable X. By definition, probability theory is concerned with the statements that can be made about a random variable, when only its probability law is known. Consequently, to describe a random variable, one need only state its probability law.

The probability law of a random variable can always be specified by stating the distribution function $F_X(\cdot)$ of the random variable X, defined for any real number x by

$$F_X(x) = P[X \leq x]. \tag{1.2}$$

A random variable X is called *discrete* if there exists a function, called the *probability mass function* of X and denoted by $p_X(\cdot)$, in terms of which the probability function $P_X[\cdot]$ may be expressed as a sum; for any Borel set B,

$$P_X[B] = P[X \text{ is in } B] = \sum_{\substack{\text{over all } x \text{ in } B \text{ such} \\ \text{that } p_X(x) > 0}} p_X(x). \tag{1.3}$$

It then follows that, for any real number x,

$$p_X(x) = P[X = x]. \tag{1.4}$$

A random variable X is called *continuous* if there exists a function called the *probability density function* of X and denoted by $f_X(\cdot)$, in terms of which $P_X[\cdot]$ may be expressed as an integral†; for any Borel set B,

†We usually assume that the integral in Eq. 1.5 is defined in the sense of Riemann; to ensure that this is the case, we require that the function $f(\cdot)$ be defined and continuous at all but a finite number of points. The integral in Eq. 1.5 is then defined only for events E, which are either intervals or unions of a finite number of non-overlapping intervals. In advanced probability theory the integral in Eq. 1.5 is defined by means of a theory of integration developed in the early 1900's by Henri Lebesgue. Then the function $f(\cdot)$ need only be a Borel function, by which is meant that for any real number c the set $\{x: f(x) < c\}$ is a Borel set. A function that is continuous at all but a finite number of points may be shown to be a Borel function. If B is an interval, or a union of a finite number of non-overlapping intervals, and if $f(\cdot)$ is continuous on B, then the integral of $f(\cdot)$ over B, defined in the sense of Lebesgue, has the same value as the integral of $f(\cdot)$ over B, defined in the sense of Riemann. *In this book the word function (unless otherwise qualified) will mean a Borel function and the word set (of real numbers) will mean a Borel set.*

$$P_X[B] = P[X \text{ is in } B] = \int_B f_X(x)\, dx. \tag{1.5}$$

Since

$$F_X(x) = \int_{-\infty}^{x} f_X(x')\, dx', \, -\infty < x < \infty, \tag{1.6}$$

it follows that

$$f_X(x) = \frac{d}{dx} F_X(x) \tag{1.7}$$

at all real numbers x at which the derivative exists.

The *expectation*, or *mean*, of a random variable X, denoted by $E[X]$, is defined (when it exists) by

$$E[X] = \begin{cases} \displaystyle\int_{-\infty}^{\infty} x\, dF_X(x) \\ \displaystyle\int_{-\infty}^{\infty} x f_X(x)\, dx \\ \displaystyle\sum_{\substack{\text{over all } x \text{ such} \\ \text{that } p_X(x) > 0}} x\, p_X(x) \end{cases} \tag{1.8}$$

depending on whether X is specified by its distribution function,† its probability density function, or its probability mass function. The expectation of X is said to *exist* if the improper integral or infinite series given in Eq. 1.8 converges absolutely (see *Mod Prob*, pp. 203 and 250); in symbols, $E[X]$ exists if and only if $E[\,|X|\,] < \infty$.

The *variance* of X is defined by

$$\text{Var}[X] = E[(X - E[X])^2] = E[X^2] - E^2[X]. \tag{1.9}$$

The *standard deviation* of X is defined by

$$\sigma[X] = \sqrt{\text{Var}[X]}. \tag{1.10}$$

The *moment-generating function* $\psi_X(\cdot)$ of X is defined, for any real number t, by

$$\psi_X(t) = E[e^{tX}]. \tag{1.11}$$

The *characteristic function* $\varphi_X(\cdot)$ of X is defined, for any real number u, by

$$\varphi_X(u) = E[e^{iuX}]. \tag{1.12}$$

†The first integral in Eq. 1.8 is called a Stieltjes integral; for a definition of this integral, see *Mod Prob*, p. 233.

A random variable may not possess a finite mean, variance, or moment-generating function. However, it always possesses a characteristic function. Indeed, there is a one-to-one correspondence between distribution functions and characteristic functions. Consequently, to specify the probability law of a random variable, it suffices to state its characteristic function.

Various "inversion" formulas, giving distribution functions, probability density functions, and probability mass functions explicitly in terms of characteristic functions, may be stated (see *Mod Prob*, Chapter 9). Here we state without proof two inversion formulas:

(i) for any random variable X

$$P[X = x] = \lim_{U \to \infty} \frac{1}{2U} \int_{-U}^{U} e^{-iux} \varphi_X(u)\, du, \; -\infty < x < \infty; \quad (1.13)$$

(ii) if the characteristic function $\varphi_X(\cdot)$ is absolutely integrable in the sense that

$$\int_{-\infty}^{\infty} |\varphi_X(u)|\, du < \infty, \quad (1.14)$$

then X is continuous, and its probability density function is given by

$$f_X(x) = \frac{1}{2\pi} \int_{-\infty}^{\infty} e^{-iux} \varphi_X(u)\, du, \; -\infty < x < \infty. \quad (1.15)$$

One expresses Eq. 1.15 in words by saying that $f_X(\cdot)$ is the *Fourier transform* of $\varphi_X(\cdot)$. Conversely, if X is continuous, then its characteristic function $\varphi_X(\cdot)$ is the Fourier transform of its probability density function:

$$\varphi_X(u) = \int_{-\infty}^{\infty} e^{iux} f_X(x)\, dx, \; -\infty < u < \infty. \quad (1.16)$$

Notice that Eqs. 1.15 and 1.16 are not quite symmetrical formulas; they differ by the factor $(1/2\pi)$ and by a minus sign in the exponent.

Tables 1.1 and 1.2 give some probability laws, their characteristic functions, means, and variances. Table 1.3 gives some examples of random variables that obey these probability laws.

Jointly distributed random variables. Several random variables $X_1, X_2, \cdots, X_n$ are said to be *jointly distributed* if they are defined as functions on the same sample description space. Their joint distribution

TABLE 1.1. **Some frequently encountered discrete probability laws**

Probability law and parameter values	Probability mass function $p_X(x)$		Characteristic function $\varphi_X(u)$	Mean $E[X]$	Variance $\mathrm{Var}[X]$
Binomial $n = 1, 2, \cdots$ $0 \leq p \leq 1$	$\binom{n}{x} p^x q^{n-x},$ $0,$	$x = 0, 1, \cdots, n$ otherwise	$(pe^{iu} + q)^n$ where $q = 1 - p$	np	npq
Poisson $\lambda > 0$	$e^{-\lambda} \dfrac{\lambda^x}{x!},$ $0,$	$x = 0, 1, \cdots$ otherwise	$e^{\lambda(e^{iu}-1)}$	λ	λ
Geometric $0 \leq p \leq 1$	$pq^{x-1},$ $0,$	$x = 1, 2, \cdots$ otherwise	$\dfrac{pe^{iu}}{1 - qe^{iu}}$	$\dfrac{1}{p}$	$\dfrac{q}{p^2}$
Negative binomial $r = 1, 2, \cdots$ $0 \leq p \leq 1$	$\binom{r + x - 1}{x} p^r q^x,$ $0,$	$x = 0, 1, \cdots$ otherwise	$\left(\dfrac{p}{1 - qe^{iu}}\right)^r$	$\dfrac{rq}{p}$	$\dfrac{rq}{p^2}$

function $F_{X_1, X_2, \ldots, X_n}(\cdot, \cdot, \cdots, \cdot)$ is defined for all real numbers $x_1, x_2, \cdots, x_n$ by

$$F_{X_1, X_2, \ldots, X_n}(x_1, x_2, \cdots, x_n) = P[X_1 \leq x_1, X_2 \leq x_2, \cdots, X_n \leq x_n]$$
$$= P[\{s: X_1(s) \leq x_1, X_2(s) \leq x_2, \cdots, X_n(s) \leq x_n\}].$$

Their joint *characteristic function* $\varphi_{X_1, X_2, \ldots, X_n}(\cdot, \cdot, \cdots, \cdot)$ is defined for all real numbers $u_1, u_2, \cdots, u_n$ by

$$\varphi_{X_1, X_2, \ldots, X_n}(u_1, u_2, \cdots, u_n) = E[\exp i(u_1 X_1 + \cdots + u_n X_n)]$$
$$= \int_{-\infty}^{\infty} \cdots \int_{-\infty}^{\infty} \exp i(u_1 x_1 + \cdots + u_n x_n) \, dF_{X_1, \ldots, X_n}(x_1, \cdots, x_n).$$

Independent random variables. Jointly distributed random variables $X_1, X_2, \cdots, X_n$ are said to be *independent* if and only if any of the following equivalent statements is true.

(i) Criterion in terms of probability functions: for all sets $B_1, B_2, \cdots, B_n$ of real numbers

$$P[X_1 \text{ is in } B_1, \cdots, X_n \text{ is in } B_n] = P[X_1 \text{ is in } B_1] \cdots P[X_n \text{ is in } B_n]. \quad (1.17)$$

(ii) Criterion in terms of distribution functions: for all real numbers $x_1, x_2, \cdots, x_n$

$$F_{X_1, \ldots, X_n}(x_1, \cdots, x_n) = F_{X_1}(x_1) \cdots F_{X_n}(x_n). \quad (1.18)$$

TABLE 1.2. Some frequently encountered continuous probability laws

Probability law and parameter values	Probability density function $f_X(x)$	Characteristic function $\varphi_X(u)$	Mean $E[X]$	Variance $\mathrm{Var}[X]$
Uniform over interval a to b	$\dfrac{1}{b-a},\quad a < x < b$ $0,\quad$ otherwise	$\dfrac{e^{iub} - e^{iua}}{iu(b-a)}$	$\dfrac{a+b}{2}$	$\dfrac{(b-a)^2}{12}$
Normal or $N(m,\sigma^2)$ $-\infty < m < \infty$ $\sigma > 0$	$\dfrac{1}{\sigma\sqrt{2\pi}}\exp\left[-\dfrac{1}{2}\left(\dfrac{x-m}{\sigma}\right)^2\right]$	$\exp(ium - \tfrac{1}{2}u^2\sigma^2)$	m	σ^2
Exponential $\lambda > 0$	$\lambda e^{-\lambda x},\quad x > 0$ $0,\quad$ otherwise	$\left(1 - \dfrac{iu}{\lambda}\right)^{-1}$	$\dfrac{1}{\lambda}$	$\dfrac{1}{\lambda^2}$
Gamma† $r > 0$ $\lambda > 0$	$\dfrac{\lambda}{\Gamma(r)}(\lambda x)^{r-1}e^{-\lambda x},\quad x > 0$	$\left(1 - \dfrac{iu}{\lambda}\right)^{-r}$	$\dfrac{r}{\lambda}$	$\dfrac{r}{\lambda^2}$
χ^2 with n degrees of freedom	$\dfrac{1}{2^{n/2}\Gamma\left(\frac{n}{2}\right)}x^{(n/2)-1}e^{-x/2},\quad x > 0$ $0,\quad$ otherwise	$(1 - 2iu)^{-n/2}$	n	$2n$
F with m, n degrees of freedom	$\dfrac{\Gamma\left(\frac{m+n}{2}\right)}{\Gamma\left(\frac{m}{2}\right)\Gamma\left(\frac{n}{2}\right)}\left(\dfrac{m}{n}\right)^{m/2}\dfrac{x^{(m/2)-1}}{\left[1+\dfrac{m}{n}x\right]^{(m+n)/2}},\quad x > 0$ $0,\quad$ otherwise		$\dfrac{n}{n-2}$ if $n > 2$	$\dfrac{2m(m+n-2)}{m(n-2)^2(n-4)}$ if $n > 4$

†For the definition of the gamma function, see p. 62.

TABLE 1.3. **Examples of random variables that obey the probability laws given in Tables 1.1 and 1.2**

Binomial	The number of successes in n independent Bernoulli trials in which the probability of success at each trial is p (a Bernoulli trial is one with two possible outcomes, called *success* and *failure*)
Poisson	The number of occurrences of events of a specified type in a period of time of length 1 when events of this type are occurring randomly at a mean rate λ per unit time (events are said to occur randomly in time if they are occurring in accord with a Poisson process as defined in section 1–2)
Geometric	The number of trials required to obtain the first success in a sequence of independent Bernoulli trials in which the probability of success at each trial is p
Negative binomial	The number of failures encountered in a sequence of independent Bernoulli trials (with probability p of success at each trial) before the rth success
Uniform	The location on a line of a dart tossed in such a way that it always hits between the end-points of the interval a to b and any two sub-intervals (of the interval a to b) of equal length have an equal chance of being hit
Normal	The number of successes in n independent Bernoulli trials (with probability p of success at each trial) approximately obeys a normal probability law with $m = np$ and $\sigma^2 = npq$
Exponential	The waiting time required to observe the first occurrence of an event of a specified type when events of this type are occurring randomly at a mean rate λ per unit time
Gamma	The waiting time required to observe the rth occurrence of an event of a specified type when events of this type are occurring randomly at a mean rate λ per unit time
χ^2	The sum $X_1{}^2 + \cdots + X_n{}^2$ of the squares of n independent random variables, each $N(0,1)$
F	The ratio nU/mV, where U and V are independent random variables, χ^2 distributed with m and n degrees of freedom, respectively

(iii) Criterion in terms of characteristic functions: for all real numbers $u_1, u_2, \cdots, u_n$

$$\varphi_{X_1,\ldots,X_n}(u_1, \cdots, u_n) = \varphi_{X_1}(u_1) \cdots \varphi_{X_n}(u_n). \qquad (1.19)$$

(iv) Criterion in terms of expectations: for all functions $g_1(\,\cdot\,), \cdots, g_n(\,\cdot\,)$ for which all the expectations in Eq. 1.20 exist

$$E[g_1(X_1) \cdots g_n(X_n)] = E[g_1(X_1)] \cdots E[g_n(X_n)]. \qquad (1.20)$$

Uncorrelated random variables. Two jointly distributed random variables X_1 and X_2 with finite second moments are said to be *uncorrelated* if their covariance, defined by†

$$\mathrm{Cov}[X_1,X_2] = E[X_1X_2] - E[X_1]E[X_2] = E[(X_1 - E[X_1])(X_2 - E[X_2])],$$
(1.21)

vanishes; or equivalently if the correlation coefficient, defined by

$$\rho(X_1,X_2) = \frac{\mathrm{Cov}[X_1,X_2]}{\sigma[X_1]\sigma[X_2]},$$
(1.22)

vanishes; or equivalently if

$$\mathrm{Var}[X_1 + X_2] = \mathrm{Var}[X_1] + \mathrm{Var}[X_2].$$
(1.23)

Distribution of the sum $X_1 + X_2$ of two independent random variables X_1 and X_2. It may be shown that if X_1 and X_2 are independent continuous random variables, then $X_1 + X_2$ is a continuous random variable with probability density function given by

$$\begin{aligned}
f_{X_1+X_2}(y) &= \int_{-\infty}^{\infty} f_{X_1}(x)f_{X_2}(y - x)\,dx \\
&= \int_{-\infty}^{\infty} f_{X_1}(y - x)f_{X_2}(x)\,dx.
\end{aligned}$$
(1.24)

If X_1 and X_2 are independent discrete random variables, then $X_1 + X_2$ is a discrete random variable with probability mass function given by

$$\begin{aligned}
p_{X_1+X_2}(y) &= \sum_{x} p_{X_1}(x)p_{X_2}(y - x) \\
&= \sum_{x} p_{X_1}(y - x)p_{X_2}(x).
\end{aligned}$$
(1.25)

Characteristic functions derive their importance from the fact that they provide a general answer to the problem of finding the probability law of the sum of two independent random variables X_1 and X_2. Since the expectation of the product of two independent random variables is equal to the product of their expectations, it follows that

$$E[e^{iu(X_1+X_2)}] = E[e^{iuX_1}e^{iuX_2}] = E[e^{iuX_1}]E[e^{iuX_2}].$$
(1.26)

†Note that, for any random variable X, $\mathrm{Cov}[X,X]$ is well defined and is equal to $\mathrm{Var}[X]$.

Consequently, the characteristic function of a sum of independent random variables is equal to the product of their characteristic functions:

$$\varphi_{X_1+X_2}(u) = \varphi_{X_1}(u)\varphi_{X_2}(u). \tag{1.27}$$

Using Eqs. 1.24, 1.25, and especially 1.27, one may prove results such as the following.

THEOREM 1A   .

Let X_1 and X_2 be independent random variables.

(i) If X_1 is normally distributed with mean m_1 and standard deviation σ_1, and X_2 is normally distributed with mean m_2 and standard deviation σ_2, then $X_1 + X_2$ is normally distributed with mean $m = m_1 + m_2$ and standard deviation $\sigma = \sqrt{\sigma_1{}^2 + \sigma_2{}^2}$.

(ii) If X_1 obeys a binomial probability law with parameters n_1 and p, and X_2 obeys a binomial probability law with parameters n_2 and p, then $X_1 + X_2$ obeys a binomial probability law with parameters $n_1 + n_2$ and p.

(iii) If X_1 is Poisson distributed with mean λ_1, and X_2 is Poisson distributed with mean λ_2, then $X_1 + X_2$ is Poisson distributed with mean $\lambda = \lambda_1 + \lambda_2$.

(iv) If X_1 obeys a gamma probability law with parameters r_1 and λ and X_2 obeys a gamma probability law with parameters r_2 and λ, then $X_1 + X_2$ obeys gamma probability law with parameters $r_1 + r_2$ and λ.

(v) If X_1 obeys a negative binomial law with parameters r_1 and p, and X_2 obeys a negative binomial law with parameters r_2 and p, then $X_1 + X_2$ obeys a negative binomial law with parameters $r_1 + r_2$ and p.

COMPLEMENTS

1A *Moments can be expressed in terms of derivatives of characteristic functions.* Let X_1 and X_2 be two jointly distributed random variables. Show formally that

$$\varphi_{X_1}(u) = \varphi_{X_1, X_2}(u, 0),$$

$$E[X_1] = \frac{1}{i} \frac{d}{du} \varphi_{X_1}(0) = \frac{1}{i} \frac{\partial}{\partial u_1} \varphi_{X_1, X_2}(0,0), \tag{1.28}$$

$$E[X_1{}^2] = -\frac{d^2}{du^2} \varphi_{X_1}(0) = -\frac{\partial^2}{\partial u_1{}^2} \varphi_{X_1, X_2}(0,0), \tag{1.29}$$

$$E[X_1 X_2] = -\frac{\partial^2}{\partial u_1 \, \partial u_2} \varphi_{X_1, X_2}(0,0). \tag{1.30}$$

It may be shown that Eqs. 1.28, 1.29, and 1.30 hold if the moments involved exist.

1B *Moments can be expressed in terms of derivatives of the logarithm of the characteristic function.* Let X_1 and X_2 be jointly distributed random variables. Show formally that

$$iE[X_1] = \frac{d}{du} \log \varphi_{X_1}(0), \tag{1.31}$$

$$i^2 \operatorname{Var}[X_1] = \frac{d^2}{du^2} \log \varphi_{X_1}(0), \tag{1.32}$$

$$i^2 \operatorname{Cov}[X_1, X_2] = \frac{\partial^2}{\partial u_1 \, \partial u_2} \log \varphi_{X_1, X_2}(0,0). \tag{1.33}$$

It may be shown that Eqs. 1.31, 1.32, and 1.33 hold if the moments involved exist.

1C *Two-dimensional generalization of Chebyshev's inequality.*
(i) Let X_1 and X_2 be random variables with means 0, variances 1, and correlation coefficient ρ. Show that

$$E[\max (X_1^2, X_2^2)] \le 1 + \sqrt{1 - \rho^2}.$$

Hint. $2 \max (X_1^2, X_2^2) = |X_1^2 - X_2^2| + X_1^2 + X_2^2$,
$$E^2[|X_1^2 - X_2^2|] \le E[|X_1 - X_2|^2] E[|X_1 + X_2|^2].$$

(ii) Using problem (i), show that for any pair of random variables with correlation coefficient ρ, and for any $\lambda > 0$

$$P[|X_1 - E[X_1]| \ge \lambda \sigma[X_1] \quad \text{or} \quad |X_2 - E[X_2]| \ge \lambda \sigma[X_2]]$$
$$\le \frac{1}{\lambda^2} \{1 + \sqrt{1 - \rho^2}\}.$$

(iii) Using problem (ii), derive the usual form of Chebyshev's inequality: for any random variable X with finite variance and any $\lambda > 0$

$$P[|X - E[X]| \ge \lambda \sigma[X]] \le \frac{1}{\lambda^2}.$$

For higher dimension generalizations of Chebyshev's inequality, see Olkin and Pratt (1958).

1D *A case of the Law of Large Numbers.* The (Weak) Law of Large Numbers states that if $\{X_n\}$ is a sequence of independent random variables identically distributed as a random variable X whose mean is finite, then, as $n \to \infty$,

$$M_n = \frac{1}{n} \sum_{k=1}^{n} X_k \to E[X] \quad \text{in probability}$$

in the sense that, for every $\epsilon > 0$,

$$P[|M_n - E[X]| > \epsilon] \to 0 \quad \text{as} \quad n \to \infty.$$

A partial extension of this result is given by the following exercise.

Let U be uniformly distributed in the interval $-\pi$ to π. For $k = 1$, $2, \cdots$, define $X_k = \cos kU$. For $n = 1, 2, \cdots$, define

$$S_n = X_1 + X_2 + \cdots + X_n$$

(i) Find $E[X_k]$, $\mathrm{Var}[X_k]$, $E[S_n]$, $\mathrm{Var}[S_n]$. *Hint.* Are $X_1, X_2, \cdots$ uncorrelated? (ii) Let $\epsilon > 0$. Find

$$\lim_{n \to \infty} P[\, | \frac{1}{n} S_n | < \epsilon].$$

1E *A case of the Central Limit Theorem.* The (classical) Central Limit Theorem states that if $\{X_n\}$ is a sequence of independent random variables identically distributed as a random variable X with finite mean and variance, and if $S_n = X_1 + X_2 + \cdots + X_n$, then as $n \to \infty$

$$S_n^* = \frac{S_n - E[S_n]}{\sigma[S_n]} \to N(0,1) \quad \text{in distribution}$$

in the sense that for every real number x

$$\lim_{n \to \infty} P[S_n^* \leq x] = \frac{1}{\sqrt{2\pi}} \int_{-\infty}^{x} e^{-(1/2)y^2} \, dy.$$

To understand the meaning of this assertion, consider the following example. Let X_1, X_2, X_3, X_4 be independent random variables, uniformly distributed on the unit interval. Let $S_2 = X_1 + X_2$, $S_4 = X_1 + X_2 + X_3 + X_4$,

$$S_2^* = \frac{S_2 - E[S_2]}{\sigma[S_2]}, \quad S_4^* = \frac{S_4 - E[S_4]}{\sigma[S_4]}.$$

(i) Show that

$$f_{S_2^*}(x) = \frac{1 - | x/\sqrt{6} |}{\sqrt{6}}, \quad | x | \leq \sqrt{6}$$

$$= 0, \text{ otherwise}$$

$$f_{S_4^*}(y) = \frac{(2/3) - 4 | y/2\sqrt{3} |^2 + 4 | y/2\sqrt{3} |^3}{\sqrt{3}}, \quad | y | \leq \sqrt{3}$$

$$= \frac{\{2 - | y/\sqrt{3} | \}^3}{6\sqrt{3}}, \quad \sqrt{3} \leq | y | \leq 2\sqrt{3}$$

$$= 0, \text{ otherwise}.$$

(ii) Let Z be normally distributed with mean 0 and variance 1. On one graph, plot the probability density functions of S_2^*, S_4^*, and Z. Would you say that S_4^* and Z have approximately the same probability law? Approximately evaluate $P[1 < S_4 < 3]$.

EXERCISES

1.1 A fish bowl contains 3 plump goldfish and 7 scrawny goldfish. Felix, the cat, hungrily scoops 3 fish out of the bowl. They are all plump goldfish, and Felix prepares to eat them. Just then the cat's master, a mean probabilist, happens by and says, "Felix, if you can repeat this performance 2 times out of 3, I'll let you eat the goldfish. If you can't, then you'll be fed your usual meal of chopped-up snails!" What is the probability that Felix gets fish for supper?

1.2 It is estimated that the probability of detecting a moderate attack of tuberculosis using an x-ray photograph of the chest is 0.6. In a city with 60,000 inhabitants, a mass x-ray survey is planned so as to detect all the people with tuberculosis. Two x-ray photographs will be taken of each individual, and he or she will be judged a suspect if at least one of these photographs is found "positive." Suppose that in the city there are 2000 persons with moderate attacks of tuberculosis. Let X denote the number of them who, as a result of the survey, will be judged "suspects." Find the mean and variance of X.

1.3 A young man and a young lady plan to meet between 5 and 6 P.M., each agreeing not to wait more than 10 minutes for the other. Find the probability that they will meet if they arrive independently at random times between 5 and 6 P.M.

1.4 A man with n keys wants to open his door. He tries the keys independently and at random. Let N_n be the number of trials required to open the door. Find $E[N_n]$ and $\text{Var}[N_n]$ if (i) unsuccessful keys are not eliminated from further selections; (ii) if they are. Assume that exactly one of the keys can open the door.

1.5 In firing a missile at a given target, suppose that the vertical and horizontal components of the deviation of the impact from the center of the target can be regarded as normally and independently distributed about the center of the target with standard deviation 3 miles. Suppose that one is aiming at a circular target of diameter 6 miles. What is the probability of the missile hitting the target? What is the smallest number of missiles that must be fired so that the probability is at least 0.99 of getting at least one hit?

1.6 Let X and Y have joint probability density function

$$f_{X,Y}(x,y) = \frac{1}{\pi} \quad \text{if} \quad x^2 + y^2 \leq 1,$$
$$= 0, \text{ otherwise.}$$

Are X and Y (i) uncorrelated; (ii) independent?

1.7 Show that the sum of n independent random variables, each obeying a gamma distribution with parameters r and λ, has the same distribution as the sum of rn independent exponentially distributed random variables with mean $1/\lambda$.

In Exercises 1.8–1.13, find $P[1 < X \leq 4]$ if (i) $X = X_1 + X_2$, (ii) $X = \min(X_1, X_2)$ in which X_1 and X_2 are independent identically distributed random variables possessing the probability law described. *Hint.* Show that $P[a < \min(X_1, X_2) \leq b] = P^2[X_1 > a] - P^2[X_1 > b]$.

1.8 X_1 and X_2 are normal with means 1 and variances 2.

1.9 X_1 and X_2 are Poisson distributed with means 2.

1.10 X_1 and X_2 obey the binomial probability law with mean 1 and variance 0.8.

1.11 X_1 and X_2 obey the exponential probability law with mean 1.5.

1.12 X_1 and X_2 obey the geometric probability law with mean 2.

1.13 X_1 and X_2 are uniformly distributed over the interval 0 to 3.

1.14   Let X_1, X_2, X_3, X_4 be independent $N(0,1)$ random variables. Let $a_1, a_2 > 0$. Let $R_1 = a_1\sqrt{X_1{}^2 + X_2{}^2}$, $R_2 = a_2\sqrt{X_3{}^2 + X_4{}^2}$. Show that
$$P[R_1 > R_2] = a_1{}^2/(a_1{}^2 + a_2{}^2).$$

In Exercises 1.15–1.17, let U be a random variable uniformly distributed on the interval 0 to 1. Let $g(u)$ be a non-decreasing function, defined for $0 \le u \le 1$. Find $g(u)$ if the random variable $g(U)$ has the distribution given. *Hint.* If

$$F(y) = P[g(U) \le y] = P[U \le g^{-1}(y)] = g^{-1}(y)$$

then $g(y) = F^{-1}(y)$. The inverse function $F^{-1}(y)$ is defined as follows: $F^{-1}(y)$ is equal to the smallest value of x so that $F(x) \le y$ (compare *Mod Prob*, p. 313).

1.15 $g(U)$ is Cauchy distributed; that is, $f_{g(U)}(y) = \{\pi(1 + y^2)\}^{-1}$.

1.16 $g(U)$ is exponentially distributed, with mean $1/\mu$.

1.17 $g(U)$ is normally distributed, with mean m and variance σ^2.

1.18   Let X be a random variable with characteristic function

$$\varphi_X(u) = \exp\{\mu(e^{\lambda(e^{iu}-1)} - 1)\}.$$

where λ and μ are positive constants. (i) Show that

$$P[X = 0] = e^{-\mu(1 - e^{-\lambda})}.$$

Hint. Use Eq. 1.13. (ii) Find $E[X]$ and $\text{Var}[X]$.

In Exercises 1.19–1.21, find the mean and variance of $\cos \pi X$, where X has the distribution given.

1.19 X is $N(m, \sigma^2)$.

1.20 X is Poisson distributed, with mean λ.

1.21 X is uniformly distributed on the interval -1 to 1.

1-2 DESCRIBING THE PROBABILITY LAW OF A STOCHASTIC PROCESS

In this section various methods of describing a stochastic process are discussed. First, recall the formal definition given at the beginning of this chapter: a stochastic process is a family of random variables $\{X(t), t \in T\}$ indexed by a parameter t varying in an index set T.

It seems plausible that a stochastic process $\{X(t), t \in T\}$, defined on an infinite index set T, can for practical purposes be adequately represented by some finite number of ordinates. Consequently, one way of describing a stochastic process $\{X(t), t \in T\}$ is to specify the joint probability law of the n random variables $X(t_1), \cdots, X(t_n)$ for all integers n and n points $t_1, t_2, \cdots, t_n$ in T. To specify the joint probability law of the n random variables $X(t_1), \cdots, X(t_n)$, one may specify either (i) the joint distribution function, given for all real numbers $x_1, \cdots, x_n$ by

$$F_{X(t_1),\ldots,X(t_n)}(x_1, \cdots, x_n) = P[X(t_1) \leq x_1, X(t_2) \leq x_2, \cdots, X(t_n) \leq x_n]$$
(2.1)

or (ii) the joint characteristic function, given for all real numbers $u_1, \cdots, u_n$ by

$$\varphi_{X(t_1),\ldots,X(t_n)}(u_1, \cdots, u_n)$$
(2.2)
$$= E[\exp i(u_1 X(t_1) + \cdots + u_n X(t_n))]$$
$$= \int_{-\infty}^{\infty} \cdots \int_{-\infty}^{\infty} \exp i(u_1 x_1 + \cdots + u_n x_n) \, dF_{X(t_1),\ldots,X(t_n)}(x_1, \cdots, x_n).$$

The distribution function in Eq. 2.1 and the characteristic function in Eq. 2.2 are said to be n-dimensional, since they represent the joint probability law of n random variables.

EXAMPLE 2A

A sequence of independent random variables $X_1, X_2, \cdots, X_n, \cdots$ constitutes a (discrete parameter) stochastic process whose index set is $T = \{1, 2, \cdots\}$. To specify the joint probability law of the process, it suffices to specify the individual characteristic functions

$$\{\varphi_{X_n}(u), n = 1, 2, \cdots\},$$

since

$$\varphi_{X_1,\ldots,X}(u_1, \cdots, u_n) = \varphi_{X_1}(u_1) \cdots \varphi_{X_n}(u_n).$$
(2.3)

The sequence $\{S_n\}$ of consecutive sums $S_n = X_1 + X_2 + \cdots + X_n$ of independent random variables $\{X_n\}$ constitutes a (discrete parameter) stochastic process. To specify the joint probability law of the process, it again suffices to specify the individual characteristic functions

$$\{\varphi_{X_n}(u), n = 1, 2, \cdots\},$$

since

$$\varphi_{S_1, \ldots, S_n}(u_1, \cdots, u_n) = \varphi_{X_1}(u_1 + u_2 + \cdots + u_n) \tag{2.4}$$
$$\times \varphi_{X_2}(u_2 + \cdots + u_n) \cdots \varphi_{X_{n-1}}(u_{n-1} + u_n)\varphi_{X_n}(u_n).$$

To prove Eq. 2.4, one uses the easily verified fact that

$$\sum_{k=1}^{n} u_k S_k = u_n(S_n - S_{n-1}) + (u_{n-1} + u_n)(S_{n-1} - S_{n-2}) \tag{2.5}$$
$$+ \cdots + (u_2 + \cdots + u_n)(S_2 - S_1) + (u_1 + \cdots + u_n)S_1.$$

The sequence $\{S_n\}$ of consecutive sums of independent random variables is known as a *random walk*, since one may interpret S_n as the displacement from its starting position of a point (or particle) that is performing a random walk on a straight line by taking at the kth step a random displacement X_k. The reader will find it instructive to plot the sample paths of some random walks by drawing samples from a table of random numbers.

Another way of describing a stochastic process is to give a formula for the value $X(t)$ of the process at each point t in terms of a family of random variables whose probability law is known.

EXAMPLE 2B

Consider the continuous parameter stochastic process $\{X(t), t \geq 0\}$ defined by

$$X(t) = A \cos \omega t + B \sin \omega t, \tag{2.6}$$

where the frequency ω is a known positive constant, and A and B are independent normally distributed random variables with means 0 and variances σ^2. Sufficient information is then at hand to find the probability of any assertion concerning the stochastic process; for example, let us find, for any constant c,

$$P\left[\int_0^{2\pi/\omega} X^2(t) \, dt > c\right].$$

Now, letting $L = 2\pi/\omega$,

$$\int_0^L X^2(t) \, dt = A^2 \int_0^L \cos^2 \omega t \, dt + 2AB \int_0^L \cos \omega t \sin \omega t \, dt + B^2 \int_0^L \sin^2 \omega t \, dt$$
$$= \frac{L}{2} (A^2 + B^2).$$

Therefore,

$$P\left[\int_0^{2\pi/\omega} X^2(t) \, dt > c\right] = P\left[A^2 + B^2 > \frac{c\omega}{\pi}\right] = \int_{c\omega/\pi}^{\infty} \frac{1}{2\sigma^2} \exp\left\{-\frac{y}{2\sigma^2}\right\} \, dy$$
$$= \exp\left\{-\frac{c\omega}{2\pi\sigma^2}\right\},$$

since $(A^2 + B^2)/\sigma^2$ has a χ^2 distribution with 2 degrees of freedom (see *Mod Prob*, p. 321).

It should be pointed out that a stochastic process $\{X(t), t \in T\}$ is in reality a function of two arguments $\{X(t,s), t \in T, s \in S\}$. For a fixed value of t, $X(t, \cdot)$ is a function on the probability space S, or equivalently $X(t, \cdot)$ is a random variable. On the other hand, for fixed s in S, $X(\cdot, s)$ is a function of t that represents a possible observation on the stochastic process $\{X(t), t \in T\}$. The function $X(\cdot, s)$ is called a *realization*, or *sample function*, of the process. When considering a stochastic process one should always attempt to form a picture of what a typical sample function of the process will look like (Figures 1.1 and 1.2).

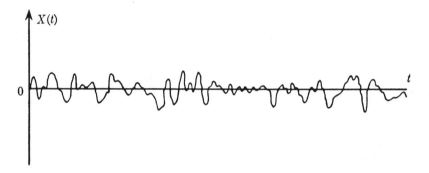

Fig. 1.1. A typical noise record. The noise arising in a resistor as modified by an amplifier may be displayed as a finite sample of a stochastic process $\{X(t), t \geq 0\}$.

EXAMPLE 2C

The stochastic process $\{X(t), t \geq 0\}$ defined by Eq. 2.6 may be written

$$X(t) = R \cos(\omega t - \theta) \tag{2.7}$$

where $R = \sqrt{A^2 + B^2}$, $\theta = \tan^{-1}(B/A)$. In view of Eq. 2.7, a sample function of this process is clearly a sine wave of amplitude R and phase θ, as is diagrammed in Figure 1.2.

COMPLEMENTS

2A Two stochastic processes $\{X(t), t \geq 0\}$ and $\{Y(t), t \geq 0\}$ are said to be *identically distributed* if they have the same family of finite dimensional probability laws. Show that the following two processes are identically distributed.

$$X(t) = A \cos \omega t + B \sin \omega t, \tag{2.8}$$

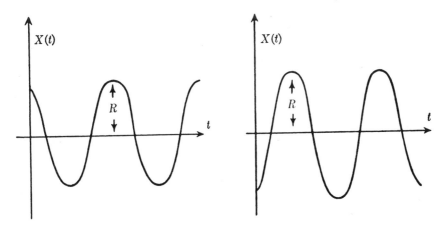

Fig. 1.2. Typical sample functions of the stochastic process $X(t) = R \cos (\omega t + \theta)$.

where ω is a known positive constant, and A and B are independent normally distributed random variables with means 0 and variances σ^2;

$$Y(t) = R \cos(\omega t + \theta), \qquad (2.9)$$

where ω is a known positive constant, R and θ are independent random variables, θ is uniformly distributed on 0 to 2π, and R is Rayleigh distributed with probability density function for $r > 0$

$$f_R(r) = \frac{r}{\sigma^2} \exp\left[-\frac{1}{2}\left(\frac{r}{\sigma}\right)^2\right].$$

2B Two stochastic processes $\{X(t), t \geq 0\}$ and $\{Y(t), t \geq 0\}$ are said to be *independent* if, for any integer n and n time points $t_1, \cdots, t_n$, the random vectors

$$(X(t_1), \cdots, X(t_n)) \quad \text{and} \quad (Y(t_1), \cdots, Y(t_n))$$

are independent in the sense that for all (Borel) sets A and B of n-tuples of real numbers

$$P[(X(t_1), \cdots, X(t_n)) \,\epsilon\, A \quad \text{and} \quad (Y(t_1), \cdots, Y(t_n)) \,\epsilon\, B]$$
$$= P[(X(t_1), \cdots, X(t_n)) \,\epsilon\, A] P[(Y(t_1), \cdots, Y(t_n)) \,\epsilon\, B].$$

Show that the processes defined by Eqs. 2.8 and 2.9 are independent if the random variables A, B, R, and θ are independent.

2C Show that if one specifies the n-dimensional probability laws of a stochastic process (that is, the set of all n-dimensional distributions) one thereby specifies all lower dimensional probability laws of the process. *Hint.* Show that for any set $\{t_1, \cdots, t_n\}$ and $m < n$,

$$\varphi_{X(t_1),\ldots,X(t_m)}(u_1, \cdots, u_m) = \varphi_{X(t_1),\ldots,X(t_n)}(u_1, \cdots, u_m, 0, \cdots, 0).$$

EXERCISES

In Exercises 2.1–2.3, consider a sequence of independent random variables $X_1, X_2, \cdots$ identically distributed as a random variable X. Let $S_n = X_1 + X_2 + \cdots + X_n$. Under the assumption given in each exercise concerning the probability law of X, find for any integer n,

(i) the joint characteristic function of $X_1, \cdots, X_n$;

(ii) the joint characteristic function of $S_1, \cdots, S_n$;

(iii) the joint characteristic function of $Y_1, \cdots, Y_n$, where $Y_k = X_{k+1} - X_k$.

2.1 X is $N(0, \sigma^2)$.

2.2 X is Poisson distributed with mean λ.

2.3 X is exponentially distributed with mean $1/\lambda$.

In Exercises 2.4 − 2.8 a stochastic process $\{X(t), t \geq 0\}$ is defined by an explicit formula in terms of two random variables A and B, which are independent normally distributed random variables with mean 0 and variance σ^2. Find the mean of the following random variables:

(i) $\max\limits_{0 \leq t \leq 1} X(t)$,

(ii) $\max\limits_{0 \leq t \leq 1} |X(t)|$,

(iii) $\int_0^1 X(t)\, dt$,

(iv) $\int_0^1 X^2(t)\, dt$.

2.4 $X(t) = At + B$.

2.5 $X(t) = t^2 + At + B$ [omit part (ii)].

2.6 $X(t) = e^{At}$.

2.7 $X(t) = A \cos \pi t$.

2.8 $X(t) = A$ for $0 < t < \frac{1}{2}$,
$\quad\quad = B$ for $t > \frac{1}{2}$.

1-3   THE WIENER PROCESS AND THE POISSON PROCESS

Two stochastic processes, the Wiener process and the Poisson process, play a central role in the theory of stochastic processes. These processes are valuable, not only as models for many important phenomena, but also as building blocks with which other useful stochastic processes can be constructed. In order to define these processes, we must introduce the notion of a stochastic process with *independent increments*.

A continuous parameter stochastic process $\{X(t), 0 \leq t < \infty\}$ is said to have *independent increments* if $X(0) = 0$ and, for all choices of indices $t_0 < t_1 < \cdots < t_n$, the n random variables

$$X(t_1) - X(t_0), \cdots, X(t_n) - X(t_{n-1}) \tag{3.1}$$

are independent. The process is said to have *stationary independent increments* if, in addition, $X(t_2 + h) - X(t_1 + h)$ has the same distribution as $X(t_2) - X(t_1)$ for all choice of indices t_1 and t_2, and every $h > 0$.

It is easily verified, for a stochastic process with independent increments, that

$$\varphi_{X(t_1),\ldots,X(t_n)}(u_1, \cdots, u_n)$$
$$= \varphi_{X(t_1)}(u_1 + \cdots + u_n) \prod_{k=2}^{n} \varphi_{X(t_k)-X(t_{k-1})}(u_k + \cdots + u_n). \tag{3.2}$$

Consequently, from a knowledge of the individual probability laws of $X(t)$ and $X(t) - X(s)$, for all non-negative s and t, one can deduce the joint probability law of any n random variables $X(t_1), \cdots, X(t_n)$.

The Wiener process. In the theory of stochastic processes and its applications, a fundamental role is played by the Wiener process. Among other applications, the Wiener process provides a model for Brownian motion and thermal noise in electric circuits. A particle (of, say, diameter 10^{-4} cm) immersed in a liquid or gas exhibits ceaseless irregular motions, which are discernible under the microscope. The motion of such a particle is called *Brownian motion*, after the English botanist Robert Brown, who discovered the phenomenon in 1827 (just after the introduction of microscopes with achromatic objectives; see Perrin [1916], p. 84). The same phenomenon is also exhibited in striking fashion by smoke particles suspended in air.

The explanation of the phenomenon of Brownian motion was one of the major successes of statistical mechanics and kinetic theory. In 1905, Einstein showed that Brownian motion could be explained by assuming that the particles are subject to the continual bombardment of the molecules of the surrounding medium. His pioneering work was generalized, extended, and experimentally verified by various physicists. (For a history of the theory of Brownian motion, see the papers collected in Wax [1954] and Einstein [1956].)

Let $X(t)$ denote the displacement (from its starting point) after time t of a particle in Brownian motion. By definition $X(0) = 0$. A Brownian particle undergoes perpetual motion due to continual "impacts" upon the particle due to the force field of its surrounding medium. The displacement of the particle over a time interval (s,t) which is long compared to the time between impacts can be regarded as the sum of a large number of small displacements. Consequently by the central limit theorem it seems reasonable to assume that $X(t) - X(s)$ is normally distributed. Next, it is reasonable to assume that the probability distribution of the displacement $X(t) - X(s)$ should be the same as that of $X(t + h)$ $- X(s + h)$, for any $h > 0$, since it is supposed that the medium is in equilibrium and the probability law of a particle's displacement over an interval should depend only on the length $t - s$ of the interval, and not on the time at which we begin observation.

We next assume that the motion of the particle is due entirely to very frequent irregular molecular impacts. Mathematically, this is interpreted to say that the stochastic process $X(t)$ has independent increments; the displacements over non-overlapping intervals are independent since the number and size of impacts in non-overlapping intervals are independent.

We are thus led to define the notion of a *Wiener process* (after N. Wiener who was among the first to consider mathematical Brownian motion [1923], [1930]). The Wiener process is also called the Wiener-Lévy process, and the Brownian motion process.

A stochastic process $\{X(t), t \geq 0\}$ is said to be a Wiener process if

(i) $\{X(t), t \geq 0\}$ has stationary independent increments,
(ii) For every $t > 0$, $X(t)$ is normally distributed,
(iii) For all $t > 0$, $E[X(t)] = 0$,
(iv) $X(0) = 0$.

Since $X(t)$ has independent increments, and $X(0) = 0$, to state the probability law of the stochastic process $X(t)$ it suffices to state the probability law of the increment $X(t) - X(s)$ for any $s < t$. Since $X(t) - X(s)$ is normal its probability law is determined by its mean and variance. One easily verifies that $E[X(s) - X(t)] = 0$. Therefore,

$$\varphi_{X(t)-X(s)}(u) = \exp\{-\tfrac{1}{2} u^2 \, \mathrm{Var}[X(t) - X(s)]\} \tag{3.3}$$

From the fact that $X(t)$ has stationary independent increments it may be shown (see Exercise 3.15) that there is some positive constant, denoted by σ^2, such that for $t \geq s \geq 0$

$$\mathrm{Var}[X(t) - X(s)] = \sigma^2 \, | \, t - s \, |. \tag{3.4}$$

The probability law of a Wiener process is thus determined by axioms (i)–(iv) up to a parameter σ^2. This parameter is an empirical characteristic of the process which must be determined from observations. In the case that the Wiener process is a model for Brownian motion, σ^2 is the mean square displacement of the particle per unit time. It was shown by Einstein in 1905 that

$$\sigma^2 = \frac{4RT}{Nf} \tag{3.5}$$

where R is the universal gas constant, N the Avogadro number, T the absolute temperature, and f the friction coefficient of the surrounding medium. Relation 3.5 made possible the determination of the Avogadro number from Brownian motion experiments, an achievement for which Perrin was awarded the Nobel prize in 1926 (see Perrin [1916]).

The Wiener process originated as a model for Brownian motion (in the work of Wiener [1923]) and for price movements in stock and commodity markets (in the work of Bachelier [1900]; for a recent review, see Osborne [1959]). More recently, the Wiener process has found application in quantum mechanics (see Kac [1951, 1959], Montroll [1952], Wiener [1953]). Another important application of the Wiener process is concerned with the asymptotic distribution of goodness of fit tests for distribution functions (see Anderson and Darling [1952]); this application is discussed in Section 3–5.

The Poisson process. Consider random events such as (1) the arrival of an electron emitted from the cathode of a vacuum tube at the anode, (2) the arrival of α-rays, emitted from a radioactive source, at a Geiger counter, (3) the arrival of customers for service, where the service is to be rendered by such diverse mechanisms as the cashiers or salesmen of a department store, the stock clerk of a factory, the runways of an airport, the cargo-handling facilities of a port, the maintenance man of a machine shop, and the trunk lines of a telephone exchange, or (4) the occurrence of accidents, errors, breakdowns, and other similar calamities.

One may describe these events by a *counting function* $N(t)$, defined for all $t > 0$, which represents the number of events that have occurred during the time period from 0 to t (more precisely in the interval $(0,t]$, open at 0 and closed at t); a typical function $N(\cdot)$ is graphed in Figure 1.3. By time 0 is meant the time at which we begin to observe whether or not the specified random event is occurring. For each time t, the value $N(t)$ is an observed value of a random variable. The family of random variables $\{N(t), t \geq 0\}$ constitutes a stochastic process. For each random variable $N(t)$, the only possible values are the integers $0, 1, 2, \cdots$. A stochastic process $\{N(t), t \geq 0\}$ whose random variables can assume as values only the integers $0, 1, 2, \cdots$ is called an integer-valued process. A particularly important integer-valued process is the Poisson process.

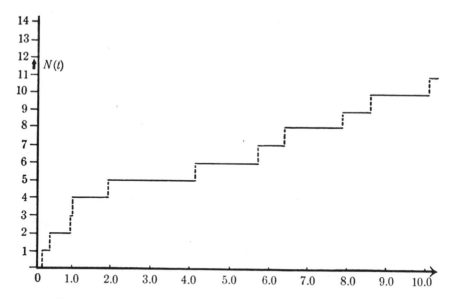

Fig. 1.3. Typical sample function of a Poisson process $\{N(t), t \geq 0\}$ with mean rate $\nu = 1$ per unit time.

An integer-valued process $\{N(t), t \geq 0\}$ is said to be a *Poisson process, with mean rate (or intensity) ν*, if the following assumptions are fulfilled:

(i) $\{N(t), t \geq 0\}$ has stationary independent increments;

(ii) for any times s and t such that $s < t$ the number $N(t) - N(s)$ of counts in the interval s to t is Poisson distributed, with mean $\nu(t - s)$. Consequently, for $k = 0, 1, 2, \cdots$

$$P[N(t) - N(s) = k] = e^{-\nu(t-s)} \frac{\{\nu(t-s)\}^k}{k!}, \tag{3.6}$$

$$E[N(t) - N(s)] = \nu(t - s), \quad \text{Var}[N(t) - N(s)] = \nu(t - s). \tag{3.7}$$

From Eq. 3.7 one sees that the parameter ν represents the mean rate of occurrence per unit time of the events being counted.

There are several very general sets of assumptions which lead to the Poisson process. These are stated in Chapter 4. It will be seen that the conditions which an integer-valued stochastic process must satisfy in order to be a Poisson process may be expected to arise often.

Events whose counting function $N(\cdot)$ is a Poisson process will be said to be occurring in accord with a Poisson process at a mean rate ν or to be of Poisson type with intensity ν.

EXAMPLE 3A

Radioactive decay. All modern theories of radioactive decay involve the assumption that in an assembly of nuclei of a given element all the nuclei are identical, independent, and have the same probability of decaying in unit time. It then follows that emissions of radioactive radiations from a radioactive source constitute events whose counting process is a Poisson process. An example of an experimental verification of the fact that the Poisson process adequately describes radioactive disintegration is given in Evans (1955), p. 777 ff.

EXAMPLE 3B

Shot noise in electron tubes. The sensitivity attainable with electronic amplifiers and apparatus is inherently limited by the spontaneous current fluctuations present in such devices, usually called *noise*. One source of noise in vacuum tubes is shot noise, which is due to the random emission of electrons from the heated cathode. Suppose that the potential difference between the cathode and the anode is so great that all electrons emitted by the cathode have such high velocities that there is no accumulation of electrons between the cathode and the anode (and thus no space charge). Emission of electrons from the cathode may be shown to be events of Poisson type (see Davenport and Root [1958], pp. 112–119). Consequently, the number of electrons emitted from the cathode in a time interval of length t obeys a Poisson probability law with parameter νt, in which ν is the mean rate of emission of electrons from the cathode.

EXAMPLE 3C

Machine failures. Consider an item (such as a vacuum tube or a machine) which is used until it fails and is then repaired (and consequently renewed) or replaced by a new item of similar type. The lifetime (length of time such an item serves before it fails) is assumed to be a random variable T. The lifetimes $T_1, T_2, \cdots, T_n, \cdots$ of the successive items put into service are assumed to be independent random variables, identically distributed as T. For $t > 0$, let $N(t)$ be the number of items that have failed in the interval 0 to t. If the lifetime of each item is exponentially distributed, it may be shown that $\{N(t), t \geq 0\}$ is a Poisson process. Indeed it is shown in Chapters 4 and 5 that $\{N(t), t \geq 0\}$ is a Poisson process if and only if each item has an exponentially distributed lifetime.

Other examples of events of Poisson type, and methods of determining whether a given process is Poisson, may be found in Davis (1952).

Poisson processes in space. In developing theories of the distribution of galaxies in the stellar system or of the distribution of centers of populations (animals, epidemics, and so on) it is convenient to regard the centers of the galaxies or the populations as points distributed in space.

Consider an array of points distributed in a space S, where S is a Euclidean space of dimension $d \geq 1$. For each region R in S, let $N(R)$ denote the number of points (finite or infinite) contained in the region R. The array of points is said to be distributed according to a stochastic mechanism if for every (measurable) region R in S, $N(R)$ is a random variable. The array of points is said to be of *Poisson type with density* ν or to be distributed in accord with a Poisson process with density ν if the following assumptions are fulfilled:

(i) the numbers of points in non-overlapping regions are independent random variables; more precisely, for any integer n and n non-overlapping (disjoint) regions $R_1, R_2, \cdots, R_n$, the random variables $N(R_1), \cdots, N(R_n)$ are independent;

(ii) for any region R of finite volume, $N(R)$ is Poisson distributed with mean $\nu V(R)$, where $V(R)$ denotes the (d-dimensional Euclidean) volume of the region R.

EXAMPLE 3D

Spatial distribution of plants and animals (ecology). The distributions in space of animals and plants are described by Poisson processes and generalized Poisson processes (defined in Section 4–2). The reason for this is well described by Skellam (1952): "Consider a wide expanse of exposed open ground of a uniform character such as would be provided by the muddy bed of a recently drained shallow lake, and consider the disposition of the independently dispersed windborne seeds of one of the species [of plants] which will colonize the area. That the number occurring in a quadrat square marked on the surface is a Poisson variate is seen from the fact that there are many such seeds each with an extremely small chance of falling into the quadrat." If seeds are distributed on a plane in accord with a Poisson process at a mean rate ν per unit area, then the number of seeds in a region of area A has a Poisson distribution with mean νA.

To illustrate the usefulness of the Poisson process in space, we obtain the following result.

THEOREM 3A

Distribution of the nearest neighbor in a Poisson distribution of particles in space. (i) If particles are distributed on a plane in accord with a Poisson process at a mean rate ν per unit area, and T is the distance between an individual particle and its nearest neighbor, then $\pi \nu T^2$ is exponentially distributed with mean 1. (ii) If particles are distributed in three-dimensional space in accord with a Poisson process at a mean rate ν per unit volume, and if T is the distance between an individual particle and its nearest neighbor, then $(4/3)\pi\nu T^3$ is exponentially distributed with mean 1; consequently, for $t > 0$,

$$f_T(t) = 4\pi\nu t^2 \exp\{-\tfrac{4}{3}\pi\nu t^3\} . \tag{3.8}$$

Remark. Eq. 3.8 plays an important role in astrophysics in the analysis of the force acting on a star which is a member of a stellar system (see Chandrasekhar [1943], p. 72).

Proof. We explicitly prove only assertion (i). First note that $P[T^2 > y/\pi\nu]$ is equal to the probability that a circle of radius $\{y/\pi\nu\}^{1/2}$ contains zero particles. Now the probability that a circle of area A contains zero particles is given by $e^{-\nu A}$. Since a circle of radius $\{y/\pi\nu\}^{1/2}$ has area y/ν, it follows that

$$P[\pi\nu T^2 > y] = e^{-\nu},$$

and the desired conclusion is at hand.

EXERCISES

Solve Exercises 3.1 to 3.9 by defining a stochastic process $\{N(t), t \geq 0\}$ with stationary independent increments in terms of which one may formulate the problem.

3.1 If customers arrive at the rate of 2 a minute, what is the probability that the number of customers arriving in 2 minutes is (i) exactly 3, (ii) 3 or less, (iii) 3 or more, (iv) more than 3, (v) less than 3?

3.2 If customers arrive at the rate of 2 a minute, what is the probability that in each of two non-overlapping 2–minute periods, the number of customers arriving is (i) exactly 3, (ii) 3 or less, (iii) 3 or more?

3.3 If customers arrive at the rate of 2 a minute, find (i) the mean number of customers in a 5–minute period, (ii) the variance of the number of customers in a 5–minute period, (iii) the probability that there will be at least 1 customer in a 5–minute period.

3.4 Certain parts essential to the running of a machine fail in accord with a Poisson process at a rate of 1 every 5 weeks. If there are 2 spare parts in inventory, and if a new supply will arrive in 9 weeks, what is the probability that during the next 9 weeks, due to a lack of spare parts, production will be stopped for a week or more?

3.5 What is the probability that a stock of 4 items of a certain product will last less than a day if sales of the item constitute (i) a Poisson process with an average daily volume of 4 units, (ii) a Poisson process whose average daily volume is a random variable equal to 3, 4, 5 with probabilities 0.25, 0.50, and 0.25 respectively?

3.6 The customers of a certain newsboy arrive in accord with a Poisson probability law at a rate of 1 customer per minute. What is the probability that 5 minutes have elapsed since (i) his last customer arrived, (ii) his next to last customer arrived?

3.7 The customers of a newsboy arrive in accordance with a Poisson process, at a mean arrival rate of 2 a minute. (i) Find the conditional probability that there will be no customers within the next 2 minutes, given that there were 1 or more customers within the last 2 minutes. (ii) The newsboy often makes the following bet. He will pay his opponent $1 if his next customer does not arrive within a minute; otherwise the opponent must pay him $1. What is the expected value of the newsboy's winnings on this bet?

3.8 A radioactive source is observed during 4 non-overlapping time intervals of 6 seconds each. The number of particles emitted during each period is counted. If the particles emitted obey a Poisson probability law, at a rate of 0.5 particles emitted per second, find the probability that (i) in each of the 4 time intervals, 3 or more particles are counted, (ii) in at least 1 of the 4 time intervals, 3 or more particles are counted.

3.9 Consider the number of suicides in a certain city in which suicides occur at an average rate of 2 per week. (i) Find the probability that in the town there will be 6 or more suicides in a week. (ii) What is the expected number of weeks in a year in which 6 or more suicides will be reported in this town? (iii) Would you find it surprising that during 1 year there were at least 2 weeks in which 6 or more suicides were reported? (iv) If only every other suicide is reported, what is the probability that in a week 2 or more suicides are reported?

3.10 Prove assertion (ii) of Theorem 3A.

3.11 Find the probability density function of the distribution of the nearest neighbor in a Poisson distribution of particles in d-dimensional Euclidean space.

3.12 Show that the mean distance $E[T]$ between a particle and its nearest neighbor in a collection of particles randomly distributed in 3-dimensional space is given by

$$E[T] = \left(\frac{3}{4\pi}\right)^{1/3} \Gamma\left(\frac{4}{3}\right) \nu^{-1/3} = 0.554 \nu^{-1/3}.$$

3.13 Consider events happening on the interval $-\infty$ to ∞ in accord with a Poisson process with density ν. For each time $t > 0$, let U_t be the time from the most recent event happening before t to t, and let V_t be the time from t to the first event happening after t. Find the probability density functions of V_t and $U_t + V_t$.

3.14 Let $\{T_n\}$ be a sequence of independent random variables exponentially distributed with mean $1/\nu$. For $t > 0$, define $N(t)$ as the integer satisfying

$$T_1 + T_2 + \cdots + T_{N(t)} \leq t < T_1 + T_2 + \cdots + T_{N(t)+1}.$$

Show that $\{N(t), t \geq 0\}$ is a Poisson process with mean rate ν.
Note. This assertion is proved in Theorem 2B of Chapter 5.

3.15 Prove that Eq. 3.4 holds. *Hint.* The function $f(t) = \text{Var}[X(t)]$ satisfies the functional equation

$$f(t_1 + t_2) = f(t_1) + f(t_2)$$

discussed in Section 4–1.

3.16 Let $\{X(t), t \geq 0\}$ be a Poisson process. Show that

$$E\left[\left|\frac{X(t)}{t} - \nu\right|^2\right] = \frac{\nu}{t} \to 0 \quad \text{as} \quad t \to \infty,$$

In view of this result would you consider $X(t)/t$ to be a satisfactory estimate of ν on the basis of having observed the process $X(t)$ over the interval 0 to t?

3.17 *The sum of two independent Poisson processes is a Poisson process:* Let $\{N_1(t), t \geq 0\}$ and $\{N_2(t), t \geq 0\}$ be two independent Poisson processes with mean rates ν_1 and ν_2 respectively; for example, $N_1(t)$ may represent the number of light bulbs replaced in an office building, and $N_2(t)$ may represent the number of light bulbs replaced in a second office building. Define $N(t) = N_1(t) + N_2(t)$. Show that $\{N(t), t \geq 0\}$ is a Poisson process with mean rate $\nu = \nu_1 + \nu_2$.

In Exercises 3.18 to 3.21, consider a stochastic process $\{X(t), t \geq 0\}$, defined by

$$X(t) = X_1(t) - X_2(t), \, t \geq 0$$

where $\{X_1(t), t \geq 0\}$ and $\{X_2(t), t \geq 0\}$ are independent Poisson processes with mean rates ν_1 and ν_2 respectively. For example, $X_1(t)$ may represent the number of taxis waiting (at time t at a certain rail terminal) for passengers, while $X_2(t)$ represents the number of potential passengers waiting for taxis at the terminal. Then $X(t) = X_1(t) - X_2(t)$ represents the excess (which may be negative) of taxis over passengers.

3.18 Show that $\{X(t), t \geq 0\}$ has stationary independent increments.

3.19 Show that $\{X(t), t \geq 0\}$ is not a Poisson process.

3.20 Find $P[X(t) - X(s) = k]$ for $t > s \geq 0$ and $k = 0, \pm 1, \pm 2, \cdots$.

3.21 Find $\lim_{t \to \infty} P[\, |X(t)| < c]$ for any real number c. Conclude that, as time increases, the excess of taxis over passengers becomes arbitrarily large in absolute value.

1-4 TWO-VALUED PROCESSES

In probability theory, random variables which take only two values, such as success and failure, play an important role. Similarly, a very important class of stochastic processes are those which take only two values, which may be taken to be real numbers A and B. Such a stochastic process will be called a two-valued (or dot-dash) process. A typical sample function of a two-valued process is graphed in Figure 1.4.

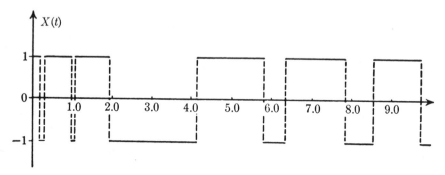

Fig. 1.4. Typical sample function of a random telegraph signal $\{X(t), t \geq 0\}$.

A two-valued process $\{X(t), t \geq 0\}$ whose possible values are 1 and -1 will be called a one-minus-one process. It is clear that if $X(t)$ is a one-minus-one process, then

$$Y(t) = \frac{B+A}{2} + \left\{\frac{B-A}{2}\right\} X(t) \qquad (4.1)$$

is a two-valued process whose possible values are A and B.

One way in which one may represent a one-minus-one process $\{X(t), t \geq 0\}$ is as follows. Define $N(0) = 0$. For $t > 0$ let $N(t)$ be the number of times in the interval $(0,t]$ that the one-minus-one process $X(\cdot)$ has changed its value. We call $\{N(t), t \geq 0\}$ the counting process of the one-minus-one process. It is immediate that one may write

$$X(t) = X(0)(-1)^{N(t)}, \qquad (4.2)$$

in which $X(0)$ is the initial value of the one-minus-one process.

If $\{N(t), t \geq 0\}$ is a Poisson process, we call $\{X(t), t \geq 0\}$ a *random telegraph signal*. More precisely, a stochastic process $\{X(t), t \geq 0\}$ is called **a random telegraph signal** if (i) its values are 1 and -1 successively, (ii) the times at which the values change are distributed according to a Poisson process $\{N(t), t \geq 0\}$ at mean rate ν and (iii) the initial value $X(0)$ is a random variable, independent of the Poisson process $\{N(t), t \geq 0\}$, such that

$$P[X(0) = 1] = P[X(0) = -1] = \tfrac{1}{2}. \qquad (4.3)$$

It is relatively easy to construct a physical device to generate a random telegraph signal; a block diagram of such a device is given in Figure 1.5. Random telegraph signals play an important role in the construction of random signal generators (see Wonham and Fuller [1958]).

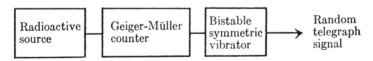

Fig. 1.5. Block diagram of a random telegraph signal generator.

THEOREM 4A

Let $\{N(t), t \geq 0\}$ be an integer-valued stochastic process with independent increments. For $t > 0$, let

$$X(t) = X(0)(-1)^{N(t)} \tag{4.4}$$

where the initial value $X(0)$ is chosen independently of the process $\{N(t), t \geq 0\}$, and is equally probable to be 1 or -1. Then $\{X(t), t \geq 0\}$ is a one-minus-one process with two-dimensional characteristic function

$$\varphi_{X(t_1),X(t_2)}(u_1, u_2) = \cos u_1 \cos u_2 - K(t_1, t_2) \sin u_1 \sin u_2, \tag{4.5}$$

where for $t_1 < t_2$ we define

$$\begin{aligned} K(t_1, t_2) &= P[N(t_2) - N(t_1) \text{ is even}] - P[N(t_2) - N(t_1) \text{ is odd}] \\ &= \varphi_{N(t_2)-N(t_1)}(\pi) \end{aligned} \tag{4.6}$$

Remark. Since $e^{\pi i} = -1$, it follows that for any integer-valued random variable N

$$\begin{aligned} P[N \text{ is even}] - P[N \text{ is odd}] &= \sum_{n=0}^{\infty} (-1)^n P[N = n] \\ &= \sum_{n=0}^{\infty} e^{i\pi n} P[N = n] = E[e^{i\pi N}] = \varphi_N(\pi). \end{aligned} \tag{4.7}$$

For a Poisson process $\{N(t), t \geq 0\}$ with intensity ν

$$\varphi_{N(t_2)-N(t_1)}(u) = e^{\nu(t_2-t_1)(e^{iu}-1)},$$

so that

$$K(t_1, t_2) = \varphi_{N(t_2)-N(t_1)}(\pi) = e^{-2\nu(t_2-t_1)}. \tag{4.8}$$

Proof: We first note that (for $t_1 < t_2$)

$P[X(t_1) = 1]$
$$= P[X(0) = 1]P[N(t_1) \text{ is even}] + P[X(0) = -1]P[N(t_1) \text{ is odd}]$$
$$= \tfrac{1}{2}\{P[N(t_1) \text{ is even}] + P[N(t_1) \text{ is odd}]\} = \tfrac{1}{2},$$

Similarly, $P[X(t_1) = -1] = 1/2$, so that $X(t_1)$ is equally likely to be 1 or -1. Therefore,

$p_{1,1} = P[X(t_1) = 1, X(t_2) = 1] = P[X(t_1) = 1 \text{ and } N(t_2) - N(t_1) \text{ is even}]$
$$= \tfrac{1}{2}P[N(t_2) - N(t_1) \text{ is even}].$$

Similarly

$$p_{-1,-1} = P[X(t_1) = -1, X(t_2) = -1] = \tfrac{1}{2}P[N(t_2) - N(t_1) \text{ is even}],$$
$$p_{1,-1} = P[X(t_1) = 1, X(t_2) = -1] = \tfrac{1}{2}P[N(t_2) - N(t_1) \text{ is odd}],$$
$$p_{-1,1} = P[X(t_1) = -1, X(t_2) = 1] = \tfrac{1}{2}P[N(t_2) - N(t_1) \text{ is odd}].$$

Now

$$\varphi_{X(t_1),X(t_2)}(u_1,u_2) = e^{i(u_1+u_2)}p_{1,1} + e^{-i(u_1+u_2)}p_{-1,-1}$$
$$+ e^{i(u_1-u_2)}p_{1,-1} + e^{-i(u_1-u_2)}p_{-1,1}$$
$$= \cos(u_1 + u_2)P[N(t_2) - N(t_1) \text{ is even}]$$
$$+ \cos(u_1 - u_2)P[N(t_2) - N(t_1) \text{ is odd}].$$

Since $\cos(u_1 \pm u_2) = \cos u_1 \cos u_2 \mp \sin u_1 \sin u_2$, the desired conclusion is at hand.

Another way in which one can represent a two-valued process is in terms of the time between changes of values. Let $\{X(t), t \geq 0\}$ be a zero-one process (that is, a stochastic process which takes only the values 0 and 1).

EXAMPLE 4A

A model for system reliability. Consider a system which can be in one of two states, "on" or "off." If it is "on" it serves for a random time before breakdown. If it is "off," it is off for a random time before being repaired. Let $X(t)$ be equal to 1 or 0 depending on whether the system is "on" or "off" at time t.

EXAMPLE 4B

A model for semi-conductor noise produced by electron "traps." An electron flowing in a semi-conductor is said to be either in a free state or a trapped state. The electron is assumed to be free for a random time which is exponentially distributed with mean $1/\mu$, and is then trapped for a random time which is exponentially distributed with mean $1/\lambda$. Define $X(t)$ to be 1 or 0 depending on whether the electron is free or trapped at time t. (For another version of this model for semi-conductor noise, see Machlup [1954].)

In order to study the properties of the zero-one process described in Examples 4A and 4B some assumptions need be made about the behavior of the times between changes of value. The following assumptions are sometimes reasonable:

(i) the times between value changes are independent random variables;

(ii) the time it takes to change from 0 to 1 is distributed as a random variable U, and the time it takes to change from 1 to 0 is distributed as a random variable V.

In most applications of zero-one processes, the following properties seem to be of most interest:

(a) the probabilities

$$P[X(t) = 0] \quad \text{and} \quad P[X(t) = 1] \tag{4.9}$$

that $X(t)$ will at time t be in each of its possible states,

(b) the probability law of the fraction of time during the interval 0 to t, denoted $\beta(t)$, that the process has the value 1. By introducing the notion of a stochastic integral (see Section 3–3) one may represent $\beta(t)$ as the integral

$$\beta(t) = \frac{1}{t} \int_0^t X(t') \, dt'. \tag{4.10}$$

For many applications, it suffices to know the behavior of the zero-one process $\{X(t), t \geq 0\}$ after it has been operating for a long time (that is, for large values of t). The limit theorems of probability theory then provide very simple answers.

THEOREM 4B

Let $\{X(t), t \geq 0\}$ be a zero-one process satisfying assumptions (i) and (ii).

If U and V have finite means, and if $U + V$ has a continuous distribution, then

$$\lim_{t \to \infty} P[X(t) = 0] = \frac{E[U]}{E[U] + E[V]},$$

$$\lim_{t \to \infty} P[X(t) = 1] = \frac{E[V]}{E[U] + E[V]}. \tag{4.11}$$

If U and V have finite variances, then the fraction $\beta(t)$ of the interval 0 to t that the process has the value 1 is asymptotically normally distributed in the sense that for every real number x

$$\lim_{t \to \infty} P\left[\sqrt{t}\, \frac{(\beta(t) - m)}{\sigma} \le x\right] = \frac{1}{\sqrt{2\pi}} \int_{-\infty}^{x} e^{-(1/2)v^2}\, dy, \qquad (4.12)$$

where

$$m = \frac{E[V]}{E[U] + E[V]},$$
$$\sigma^2 = \frac{E^2[V]\,\mathrm{Var}[U] + E^2[U]\,\mathrm{Var}[V]}{\{E[U] + E[V]\}^3}. \qquad (4.13)$$

A proof of Eq. 4.11 is sketched in Section 5–3. The proof of Eq. 4.12 is beyond the scope of the book (see Rényi [1957]). A proof of a related theorem is given in Section 6–10.

The zero crossing problem. Two-valued processes can arise in yet a third way. Let $\{Z(t), t \ge 0\}$ be a stochastic process (such as the Wiener process). One may be interested in determining the probability law of the number of zeros of the stochastic process in an interval a to b. If one defines

$$X(t) = 1 \qquad \text{if } Z(t) \ge 0$$
$$= -1 \qquad \text{if } Z(t) < 0, \qquad (4.14)$$

then the number of times $X(t)$ changes its value in the interval a to b is equal to the number of zeros of $Z(t)$ in this interval. The problem of finding the probability law of the number of zeros of a stochastic process, or of the number of changes of value of a two-valued process, is essentially unsolved. (For references to some known results, see Longuet-Higgins [1958], Mac-Fadden [1958], and Slepian [1961].)

In introducing the notion of a two-valued process, our aim has been twofold. On the one hand, we have desired to indicate some of the questions which are of interest. On the other hand, we have tried to give an idea of the kinds of answers that are available, even if their proofs are beyond the scope of this book. It is hoped that by following an approach of this kind, this book will provide the reader with enough fundamentals for competent professional application, and with a sufficient degree of maturity to enable further study, of the theory of stochastic processes.

Conditional probability and conditional expectation

IN THE THEORY of stochastic processes, a position of importance is occupied by the notions of (i) the conditional probability

$$P[A \mid X = x]$$

of an event A given (the value of) a random variable X, (ii) the conditional distribution function

$$F_{Y|X}(y \mid x) = P[Y \leq y \mid X = x]$$

of a random variable Y, given (the value of) a random variable X, and (iii) the conditional expectation

$$E[Y \mid X = x]$$

of a random variable Y, given (the value of) a random variable X.

Rigorous treatments of these notions require rather sophisticated mathematics and are beyond the scope of this book. Nevertheless, in order to study the theory of stochastic processes one must know how to operate with conditional probabilities and conditional expectations. This chapter attempts to provide the reader with these tools.

2-1 CONDITIONING BY A DISCRETE RANDOM VARIABLE

The conditional probability $P[A \mid B]$ of the event A, given the event B, is defined as follows:

$$P[A \mid B] = \frac{P[AB]}{P[B]} \quad \text{if } P[B] > 0,$$

$$= \text{undefined if } P[B] = 0. \tag{1.1}$$

In view of Eq. 1.1 a natural way to define the conditional probability $P[A \mid X = x]$ of an event A, given the event $[X = x]$ that the observed value of the random variable X is equal to x, might be

$$P[A \mid X = x] = \frac{P[A \text{ occurs and } X = x]}{P[X = x]} \quad \text{if } P[X = x] > 0, \tag{1.2}$$

while $P[A \mid X = x]$ is undefined if $P[X = x] = 0$. This definition is clearly unsatisfactory if X is a continuous random variable, since then $P[X = x]$ is zero for all x. However if X is discrete, the definition of $P[A \mid X = x]$ given by Eq. 1.2 is satisfactory, because the set C of values of x for which $P[X = x] > 0$ has probability one of containing the observed value of X, and we wish to define $P[A \mid X = x]$ only for real numbers x which could actually arise as observed values of x.

The conditional distribution function $F_{Y|X}(\cdot \mid \cdot)$ of a random variable Y, given a discrete random variable X, is also well defined by

$$F_{Y|X}(y \mid x) = P[Y \le y \mid X = x] = \frac{P[Y \le y \text{ and } X = x]}{P[X = x]} \tag{1.3}$$

for all values of y and values of x such that $P[X = x] > 0$.

The conditional mean of Y, given that $X = x$, is defined as the mean of the conditional distribution function $F_{Y|X}(\cdot \mid x)$. In symbols, one may write the conditional mean as a Stieltjes integral:

$$E[Y \mid X = x] = \int_{-\infty}^{\infty} y \, dF_{Y|X}(y \mid x)$$

$$= \lim_{n \to \infty} \sum_{k=-\infty}^{\infty} \frac{k}{2^n} P\left[\frac{k}{2^n} < Y \le \frac{k+1}{2^n} \mid X = x\right].$$

If X and Y are jointly discrete, one can define the *conditional probability mass function* of Y, given X, by

$$p_{Y|X}(y \mid x) = P[Y = y \mid X = x] = \frac{p_{X,Y}(x,y)}{p_X(x)} \tag{1.4}$$

if x is such that $p_X(x) > 0$. The conditional probability mass function has the desired relation to the conditional distribution function: for x such that $p_X(x) > 0$

$$F_{Y|X}(y \mid x) = \sum_{\text{over } y' \leq y} p_{Y|X}(y' \mid x). \tag{1.5}$$

To prove Eq. 1.5 one uses the fact that

$$P[Y \leq y \quad \text{and} \quad X = x] = \sum_{\text{over } y' \leq y} p_{X,Y}(y',x).$$

In view of Eqs. 1.3 and 1.5, in the case of jointly discrete random variables X and Y, the conditional expectation of Y, given X, may be written (for x such that $p_X(x) > 0$)

$$E[Y \mid X = x] = \sum_{\text{over } y} y\, p_{Y|X}(y \mid x). \tag{1.6}$$

Before defining the notions of conditional distributions and conditional expectations for non-discrete random variables, let us show how these notions are used.

The notion of conditional expectation has two main uses. On the one hand, it is important because it provides one possible answer to the problem of prediction and in general plays an important part in time series analysis and statistical decision theory. On the other hand, it is important because it provides a means of analyzing dependent random variables in a series of steps. The usefulness of conditional expectation derives from the three basic properties given in Theorem 1A (which we state in general although we prove them only in the case of jointly discrete random variables).

THEOREM 1A

Three fundamental properties of conditional expectation. Let X and Y be jointly distributed random variables and $g(x,y)$ a function of two variables. Suppose that $E[Y]$ is finite and that X is discrete. Let $E[Y \mid X = x]$ denote the conditional expectation of Y, given that $X = x$ (this notion is defined in the case of jointly discrete random variables by Eq. 1.6 and in the general case is defined in Section 2–2). (i) Then

$$E[Y] = \sum_{\text{over } x} E[Y \mid X = x] p_X(x); \tag{1.7}$$

in words, from a knowledge of the conditional mean of Y given X, one can obtain the (unconditional) mean of Y. (ii) For all x such that $p_X(x) > 0$

$$E[Y \mid X = x] = E[Y] \text{ if } X \text{ and } Y \text{ are independent}; \tag{1.8}$$

in words, the conditional expectation $E[Y \mid X = x]$ does not depend on x and is equal to the unconditional mean $E[Y]$ if X and Y are independent. (iii) For all x such that $p_X(x) > 0$,

$$E[g(X,Y) \mid X = x] = E[g(x,Y) \mid X = x] \qquad (1.9)$$

for any function $g(\,\cdot\,,\,\cdot\,)$ such that the expectation $E[g(X,Y)]$ exists.

To express Eq. 1.9 in words, let $U = g(X,Y)$ and $V = g(x,Y)$. Now U is a random variable which is a function of X and Y, while V is (for a given real number x) a function only of Y. What Eq. 1.9 says is that the conditional mean of U, given that $X = x$, is equal to the conditional mean of V, given that $X = x$.

Proof. We prove Eqs. 1.7–1.9 only for jointly discrete random variables X and Y. To prove Eq. 1.7, one writes

$$
\begin{aligned}
E[Y] &= \sum_{\text{over } x} \sum_{\text{over } y} y \, p_{X,Y}(x,y) \\
&= \sum_{\text{over } x} p_X(x) \sum_{\text{over } y} y \, p_{Y\mid X}(y \mid x) \qquad (1.10) \\
&= \sum_{\text{over } x} p_X(x) E[Y \mid X = x].
\end{aligned}
$$

That Eq. 1.8 holds follows from Eq. 1.6 and the easily verified fact that for all x such that $p_X(x) > 0$

$$p_{Y\mid X}(y \mid x) = p_Y(y) \text{ if } X \text{ and } Y \text{ are independent.} \qquad (1.11)$$

To prove Eq. 1.9, note that

$$P[U = u, X = x] = \sum_{\substack{\text{over } y \text{ such} \\ \text{that } g(x,y)=u}} p_{X,Y}(x,y), \qquad (1.12)$$

$$P[V = v, X = x] = \sum_{\substack{\text{over } y \text{ such} \\ \text{that } g(x,y)=v}} p_{X,Y}(x,y). \qquad (1.13)$$

Consequently, for all real numbers u, and x such that $p_X(x) > 0$,

$$p_{U\mid X}(u \mid x) = p_{V\mid X}(u \mid x), \qquad (1.14)$$

$$E[U \mid X = x] = \sum_{\text{over } u} u \, p_{U\mid X}(u \mid x) = \sum_{\text{over } u} u \, p_{V\mid X}(u \mid x) = E[V \mid X = x], \qquad (1.15)$$

and the proof of Eq. 1.9 is now complete.

It should be noted that the notion of conditional probability is a special case of the notion of conditional expectation. Given an event A, which is a subset of a sample description space S, the indicator function I_A of the event A is defined on S by

$$I_A(s) = 1 \text{ if } s \text{ belongs to } A, \tag{1.16}$$
$$= 0 \text{ if } s \text{ does not belong to } A.$$

It is easily verified that

$$E[I_A \mid X = x] = P[\{I_A = 1\} \mid X = x] = P[A \mid X = x]. \tag{1.17}$$

Consequently, from Theorem 1A we obtain immediately the following frequently used results.

THEOREM 1B

Three fundamental properties of conditional probability. For any event A, discrete random variables X and Y, function g, and set B

$$P[A] = \sum_{\text{over } x} P[A \mid X = x] p_X(x) \tag{1.18}$$

$$P[Y \text{ is in } B \mid X = x] = P[Y \text{ is in } B] \text{ if } X \text{ and } Y \text{ are independent} \tag{1.19}$$

$$P[g(X,Y) \text{ is in } B \mid X = x] = P[g(x,Y) \text{ is in } B \mid X = x]. \tag{1.20}$$

Examples 1A and 1B illustrate the use of Theorems 1A and 1B, and at the same time illustrate the use of functional equations, such as difference equations, to solve probability problems.

EXAMPLE 1A

Consider n independent tosses of a coin which has probability p of falling heads. Let S_n be the number of tosses in which the coin falls heads. Find the probability P_n that S_n is even.

Solution. One approach to this problem would be to find the probability law of the random variable S_n, which is clearly binomial:

$$P[S_n = k] = \binom{n}{k} p^k q^{n-k}, \qquad k = 0, 1, \cdots, n,$$

in which $q = 1 - p$. Consequently,

$$P_n = \binom{n}{0} q^n + \binom{n}{2} p^2 q^{n-2} + \cdots$$
$$= \tfrac{1}{2}\{(p + q)^n + (q - p)^n\}$$
$$= \tfrac{1}{2}\{1 + (q - p)^n\}.$$

Another approach to finding P_n is to obtain a difference equation which P_n satisfies. First let us give an informal argument. The event that

n tosses of a coin result in an even number of heads can occur in one of two ways; on the first toss there is a tail and on the remaining $(n-1)$ trials there are an even number of heads, or on the first toss there is a head and on the remaining $(n-1)$ tosses there are an odd number of heads. Consequently, for $n = 2, 3, \cdots$

$$P_n = qP_{n-1} + p(1 - P_{n-1}) = (q - p)P_{n-1} + p, \qquad (1.21)$$

while $P_1 = q$. By Complement 1A, the solution of this difference equation is given by, for $n = 1, 2, \cdots$,

$$P_n = \left(q - \frac{p}{1 - (q - p)}\right)(q - p)^{n-1} + \frac{p}{1 - (q - p)}, \qquad (1.22)$$
$$= \tfrac{1}{2}(1 + (q - p)^n).$$

We now give a rigorous derivation of Eq. 1.21. For $j = 1, 2, \cdots, n$, let X_j be equal to 1 or 0 according as the outcome of the jth toss is heads or tails. By Eq. 1.18

$$P_n = P[S_n \text{ even}] = P[S_n \text{ even} \mid X_1 = 0]P[X_1 = 0]$$
$$+ P[S_n \text{ even} \mid X_1 = 1]P[X_1 = 1]. \qquad (1.23)$$

Next, by Eqs. 1.19 and 1.20

$$P[S_n \text{ even} \mid X_1 = 0] = P\left[\sum_{j=2}^{n} X_j \text{ even} \mid X_1 = 0\right] = P\left[\sum_{j=2}^{n} X_j \text{ even}\right]$$
$$= P[S_{n-1} \text{ even}] = P_{n-1}. \qquad (1.24)$$

Similarly,

$$P[S_n \text{ even} \mid X_1 = 1] = P[S_{n-1} \text{ odd}] = 1 - P_{n-1}. \qquad (1.25)$$

From Eqs. 1.23–1.25 one obtains Eq. 1.21.

EXAMPLE 1B

Consider n independent trials of an experiment which has probability p of success. Let S_n be the number of successes in the n trials. Find its mean $E[S_n]$.

Solution. Since the number S_n of successes in n Bernoulli trials is known to obey a binomial probability law, its mean is given by

$$E[S_n] = \sum_{k=0}^{n} k \binom{n}{k} p^k q^{n-k} = np.$$

However, let us show how one may find $E[S_n]$ by use of difference equations.

We first argue informally as follows. Let $m_n = E[S_n]$ be the expected number of successes in n Bernoulli trials. Then the expected number of successes in the $(n-1)$ trials following the first trial is equal to m_{n-1}. Now if the first trial is a success then $m_n = 1 + m_{n-1}$, while if the first trial is a failure then $m_n = m_{n-1}$. Therefore, for $n = 2, 3, \cdots$,

$$m_n = p(1 + m_{n-1}) + qm_{n-1} = (p+q)m_{n-1} + p = m_{n-1} + p. \quad (1.26)$$

Since $m_1 = p$, it follows that $m_n = m_{n-k} + kp = m_1 + (n-1)p = np$.

One may rigorously derive the difference equation 1.26 as follows. In terms of conditional means we may write, by Eq. 1.7.

$$E[S_n] = E[S_n \mid X_1 = 0]P[X_1 = 0] + E[S_n \mid X_1 = 1]P[X_1 = 1]. \quad (1.27)$$

One may now infer Eq. 1.26, using the facts that

$$
\begin{aligned}
E[S_n \mid X_1 = 0] &= E[X_1 + X_2 + \cdots + X_n \mid X_1 = 0] \\
&= E[X_2] + \cdots + E[X_n] = E[S_{n-1}], \\
E[S_n \mid X_1 = 1] &= E[X_1 + X_2 + \cdots + X_n \mid X_1 = 1] \\
&= 1 + E[X_2] + \cdots + E[X_n] = 1 + E[S_{n-1}].
\end{aligned}
$$

To illustrate the use of ideas of conditional probability in the theory of stochastic processes, we consider the properties of a stochastic process obtained from a Poisson process by random selection.

Preservation of the Poisson process under random selection. An important aspect of the Poisson process is that it is preserved under random selection. Given events of Poisson type, suppose one does not count all of the events occurring. Rather each event occurring is counted with a probability p. It is shown in Theorem 1C that the events counted are still of Poisson type.

THEOREM 1C

For $t \geq 0$, let $N(t)$ denote the number of events of a given type occurring in the interval 0 to t. If $N(\cdot)$ is a Poisson process with mean rate ν, if each event occurring has probability p of being recorded and if the recording of one event is independent of the recording of other events and is also independent of the process $\{N(t), t \geq 0\}$, and if $M(t)$ denotes the number of events recorded in the interval 0 to t, then $\{M(t), t \geq 0\}$ is a Poisson process with parameter $\mu = \nu p$.

Proof. It is clear that $M(\cdot)$ is a process with independent increments. Consequently, to prove the theorem it suffices to prove that, for any times $t > s \geq 0$ and integer k,

$$P[M(t) - M(s) = k] = e^{-\nu p(t-s)} \frac{[\nu p(t-s)]^k}{k!}.$$

Now, given that n events have occurred in the interval s to t, the number recorded satisfies a binomial probability law:

$$P[M(t) - M(s) = k \mid N(t) - N(s) = n] = \binom{n}{k} p^k q^{n-k}. \tag{1.28}$$

Since

$$P[N(t) - N(s) = n] = e^{-\nu(t-s)} \frac{[\nu(t-s)]^n}{n!},$$

it follows that

$$\begin{aligned}
P[M(t) - M(s) = k] &= \sum_{n=k}^{\infty} \binom{n}{k} p^k q^{n-k} e^{-\nu(t-s)} \frac{[\nu(t-s)]^n}{n!} \\
&= \frac{e^{-\nu(t-s)}}{k!} [p\nu(t-s)]^k \sum_{n=k}^{\infty} \frac{[q\nu(t-s)]^{n-k}}{(n-k)!} \\
&= \frac{[p\nu(t-s)]^k}{k!} e^{-\nu(t-s)} e^{(1-p)\nu(t-s)}.
\end{aligned}$$

The proof of Theorem 1C is now complete.

EXAMPLE 1C

A defective counter. Suppose that particles arrive at a Geiger counter in accord with a Poisson process at a mean rate of ν per unit time. The Geiger counter activates a recording mechanism containing a defective relay which operates correctly with probability p. Consequently, each particle arriving at the counter has probability p of being recorded. For $t \geq 0$, let $M(t)$ be the number of particles recorded in the interval 0 to t. It follows that $M(\cdot)$ is a Poisson counting process, with mean rate $\mu = \nu p$.

EXAMPLE 1D

Selective customers. Suppose that customers pass by a shop in accord with a Poisson process at mean rate ν. If each customer has probability p of entering the shop, customers enter the shop in accord with a Poisson process with mean rate νp.

EXAMPLE 1E

Ecology. Suppose that each seed described in Example 3D of Chapter 1 has only probability p of germination. The surviving seeds continue to be distributed in accord with a Poisson process.

In applying Theorem 1C, the essential condition to be verified is Eq. 1.28. Indeed the proper definition of random selection is that Eq. 1 28 holds. An example of a selection process which seems to be random but does not satisfy Eq. 1.28 is given in Example 1E of Chapter 5.

COMPLEMENT

1A *Solution of a difference equation.* Show that if a sequence of numbers $p_1, p_2, \cdots$ satisfies the difference equation

$$p_n = ap_{n-1}, \quad n = 2, 3, \cdots, \tag{1.29}$$

in which a is a given constant, then

$$p_n = p_1 a^{n-1}, \quad n = 1, 2, \cdots.$$

Consequently, show that the difference equation

$$p_n = ap_{n-1} + b, \quad n = 2, 3, \cdots,$$

in which a and b are given constants, has as its solution

$$
\begin{aligned}
p_n &= \left(p_1 - \frac{b}{1-a}\right)a^{n-1} + \frac{b}{1-a} \quad && \text{if} \quad a \neq 1, \\
&= (n-1)b + p_1 \quad && \text{if} \quad a = 1.
\end{aligned}
$$

Hint. Show that $p_n' = p_n - \dfrac{b}{1-a}$ satisfies Eq. 1.29.

EXERCISES

1.1 If X and Y are independent Poisson random variables, show that the conditional distribution of X, given $X + Y$, is binomial.

1.2 If X is Poisson distributed with mean λ, and the conditional distribution of Y, given that $X = n$, is binomial with parameters n and p, show that Y is Poisson distributed with mean λp.

1.3 *The binomial distribution and the Poisson process,* I. For any Poisson process $\{N(t), t \geq 0\}$ show that, for $s < t$,

$$P[N(s) = k \mid N(t) = n] = \binom{n}{k}\left(\frac{s}{t}\right)^k\left(1 - \frac{s}{t}\right)^{n-k}.$$

1.4 *The binomial distribution and the Poisson process,* II. Show that if $\{N_1(t), t \geq 0\}$ and $\{N_2(t), t \geq 0\}$ are independent Poisson processes, with respective intensities ν_1 and ν_2, then for $0 \leq k \leq n$

$$P[N_1(t) = k \mid N_1(t) + N_2(t) = n] = \binom{n}{k}p^k q^{n-k},$$

in which

$$p = \frac{1}{1 + (\nu_2/\nu_1)}, \quad q = 1 - p.$$

1.5 Do Exercise 3.8 of Chapter 1 under the assumption that the counting mechanism is defective, and each particle arriving at the counter has only probability 2/3 of being recorded.

1.6 Each pulse arriving at a Geiger counter has only probability 1/3 of being recorded. Suppose that pulses arrive in accord with a Poisson process at a mean rate of 6 per second. If Z is the number of pulses recorded in half a minute, find $E[Z]$, $\text{Var}[Z]$, $P[Z \geq 2]$.

1.7 Persons pass by a certain restaurant in accord with a Poisson process at a rate of 1,000 per hour. Suppose each person has probability 0.01 of entering the restaurant. Let Z be the number of patrons entering the restaurant in a 10-minute period. Find $E[Z]$, $\text{Var}[Z]$, $P[Z \geq 2]$.

Waiting-time problems. In Examples 1A and 1B we were considering probabilities defined on a finite number of trials. Many problems (for example, those concerning waiting times) involve probabilities defined on a possibly infinite number of trials. Difference equations represent a particularly powerful method of treating problems of this type.

1.8 Show that the expected number of trials it takes to achieve r successes, in a series of independent repeated Bernoulli trials with probability p of success at each trial, is equal to r/p, for any integer $r = 1, 2, \cdots$. *Hint.* Let S_r be the waiting time to the rth success. Let $m_r = E[S_r]$. Using the fact that

$$m_r = p(1 + m_{r-1}) + q(1 + m_r)$$

find a difference equation that m_r satisfies for $r = 2, 3, \cdots$. Since $m_1 = 1/p$, it follows that $m_r = r/p$. To prove that $m_1 = 1/p$, show that m_1 satisfies the equation $m_1 = p + q(m_1 + 1) = 1 + qm_1$.

1.9 *Thief of Baghdad Problem.* The Thief of Baghdad has been placed in a dungeon with three doors. One of the doors leads into a tunnel which returns him to the dungeon after one day's travel through the tunnel. Another door leads to a similar tunnel (called the Long Tunnel) whose traversal requires 3 days rather than 1 day. The third door leads to freedom. Assume that the Thief is equally likely to choose each door (that is, each time he chooses a door he does not know what lies beyond.) Find the mean number of days the Thief will be imprisoned from the moment he first chooses a door to the moment he chooses the door leading to freedom.

1.10 Three players (denoted a, b, and c) take turns at playing a fair game according to the following rules. At the start a and b play, while c is out. The winner of the match between a and b plays c. The winner of the second match then plays the loser of the first match. The game continues in this way until a player wins twice in succession, thus becoming the winner of the game. Let A, B, and C denote respectively the events that a, b, or c is the winner of the game. (i) Find $P[A]$, $P[B]$, $P[C]$. (ii) Find the mean duration of the game.

2-2 CONDITIONING BY A CONTINUOUS RANDOM VARIABLE

If X and Y have joint probability density function $f_{X,Y}(x,y)$, the conditional probability density function $f_{Y|X}(\cdot\,|\,\cdot)$ of Y given X is defined, for all y and for all x such that $f_X(x) > 0$, by

$$f_{Y|X}(y \mid x) = \frac{f_{X,Y}(x,y)}{f_X(x)} = \frac{f_{X,Y}(x,y)}{\int_{-\infty}^{\infty} f_{X,Y}(x,y)\,dy}. \qquad (2.1)$$

Since the set $C = \{x : f_X(x) > 0\}$ has probability one of containing the observed value of X, we regard Eq. 2.1 as a satisfactory definition, since almost all observed values of X will lie in the set C, and we wish to define $f_{Y|X}(y \mid x)$ only at points x that could actually arise as observed values of X.

The conditional expectation, given X, of a random variable Y is defined by (for all x such that $f_X(x) > 0$)

$$E[Y \mid X = x] = \int_{-\infty}^{\infty} y\, f_{Y|X}(y \mid x)\,dy. \qquad (2.2)$$

That Eqs. 2.1 and 2.2 seem reasonable definitions for jointly continuous random variables is clear by comparing these definitions with Eqs. 1.4 and 1.6 respectively. To justify these definitions more fully we must first study the properties which the notions of conditional probability and expectation should have.

In order to define the notion of the conditional probability $P[A \mid X = x]$ of an event A given that the random variable X has an observed value equal to x, it suffices, in view of Eq. 1.17, to define the notion of the conditional expectation $E[Y \mid X = x]$ of a random variable Y, given that $X = x$. Now, two central properties which $E[Y \mid X = x]$ should have are as follows:

(i) If one takes the expectation of $E[Y \mid X = x]$ with respect to the probability law of X one should obtain $E[Y]$; in symbols,

$$E[Y] = \int_{-\infty}^{\infty} E[Y \mid X = x]\,dF_X(x). \qquad (2.3)$$

(ii) The conditional expectation $E[g(X)Y \mid X = x]$ of a random variable which is the product of Y and a function $g(X)$ of X should satisfy

$$E[g(X)Y \mid X = x] = g(x)E[Y \mid X = x] \qquad (2.4)$$

for any random variable $g(X)$ which is a function of X and such that $E[g(X)Y]$ is finite.

Combining Eqs. 2.3 and 2.4, it follows that the notion of conditional expectation should have the property that for suitable functions $g(X)$

$$E[g(X)Y] = \int_{-\infty}^{\infty} g(x)E[Y \mid X = x] \, dF_X(x). \tag{2.5}$$

Let us now introduce the notation $E[Y \mid X]$ to denote the function of X whose value, when $X = x$, is defined to be equal to $E[Y \mid X = x]$. We may then write Eq. 2.5 in the form

$$E[g(X)Y] = E[g(X)E[Y \mid X]] \tag{2.6}$$

for any random variable $g(X)$ which is a function of X and such that $E[g(X)Y]$ is finite.

It may be shown that Eq. 2.6 provides a general method of defining conditional expectation. Given a random variable X, and a random variable Y with finite mean it may be shown, using the so-called Radon-Nikodym Theorem (see Doob [1953], p. 17, or Loève [1960], p. 341), that there exists a unique† function of X, denoted $E[Y \mid X]$, satisfying Eq. 2.6 for all functions $g(\cdot)$ for which the expectation on the left side of Eq. 2.6 exists. In other words, in advanced probability theory the notion of conditional expectation (and consequently the notion of conditional probability) is not defined explicitly but is defined implicitly as the function (of X) satisfying Eq. 2.6. The justification for this procedure is that it may be shown (see Theorem 3A) that the notion of conditional expectation thus defined possesses all the properties which intuition demands that it should.

To illustrate the use of the axiomatic definition of conditional expectation given by Eq. 2.6, let us prove that for jointly continuous random variables X and Y one may evaluate the conditional expectation by Eq. 2.2. On the one hand,

$$E[Yg(X)] = \int_{-\infty}^{\infty} \int_{-\infty}^{\infty} yg(x) f_{X,Y}(x,y) \, dx \, dy. \tag{2.7}$$

On the other hand, let $E[Y \mid X]$ denote the function of X defined by Eq. 2.2. Then

$$E[E[Y \mid X]g(X)] = \int_{-\infty}^{\infty} \int_{-\infty}^{\infty} E[Y \mid X = x]g(x) f_{X,Y}(x,y) \, dx \, dy$$
$$= \int_{-\infty}^{\infty} E[Y \mid X = x]g(x) f_X(x) \, dx, \tag{2.8}$$

---

†The conditional expectation is unique in the sense that two functions (of X) which possess the properties of conditional expectation can differ only on a set of probability zero, and therefore are equal with probability one. For the purposes of applied probability theory, and consequently, in this book, two random variables which have probability one of being equal may be considered to be identical. A proof of the uniqueness of $E[Y \mid X]$ is sketched in Section 2-3.

since $\int_{-\infty}^{\infty} f_{X,Y}(x,y)\, dy = f_X(x)$. From Eqs. 2.2 and 2.8 it follows that

$$E[E[Y \mid X]g(X)] = \int_{-\infty}^{\infty} \left\{ \int_{-\infty}^{\infty} y\, f_{Y|X}(y \mid x)\, dy \right\} g(x)\, f_X(x)\, dx. \quad (2.9)$$

In writing Eq. 2.9 we are making use of the fundamental fact that if $\varphi_1(x)$ and $\varphi_2(x)$ are two functions of x such that

$$P[X \text{ belongs to } \{x: \varphi_1(x) = \varphi_2(x)\}] = 1, \quad (2.10)$$

then

$$E[\varphi_1(X)] = \int_{-\infty}^{\infty} \varphi_1(x)\, dF_X(x) = \int_{-\infty}^{\infty} \varphi_2(x)\, dF_X(x) = E[\varphi_2(X)]. \quad (2.11)$$

From Eqs. 2.9 and 2.1 it follows that

$$E[E[Y \mid X]g(X)] = \int_{-\infty}^{\infty} \int_{-\infty}^{\infty} yg(x)\, f_{X,Y}(x,y)\, dx\, dy. \quad (2.12)$$

Comparing Eqs. 2.12 and 2.7, one sees that the function of X defined by the right-hand side of Eq. 2.2 satisfies Eq. 2.6. Since this function satisfies the implicit definition of conditional expectation, we are justified in asserting that, in the case of jointly continuous random variables, $E[Y \mid X]$ as defined by Eq. 2.6 is equal to the right-hand side of Eq. 2.2. Similarly one may show that if $\varphi(Y)$ is a random variable with finite mean which is a function of Y, and if X and Y are jointly continuous, then (for all x such that $f_X(x) > 0$)

$$E[\varphi(Y) \mid X = x] = \int_{-\infty}^{\infty} \varphi(y)\, f_{Y|X}(y \mid x)\, dy. \quad (2.13)$$

EXAMPLE 2A

Jointly normally distributed random variables. Two jointly distributed random variables X_1 and X_2 are said to be jointly normal if their joint characteristic function is given by, for any real numbers u_1 and u_2,

$$\varphi_{X_1,X_2}(u_1,u_2) = \exp[i(u_1 m_1 + u_2 m_2) - \tfrac{1}{2}(u_1^2 \sigma_1^2 + u_2^2 \sigma_2^2 + 2u_1 u_2 K_{12})], \quad (2.14)$$

where

$$\begin{aligned}
m_1 &= E[X_1], \; m_2 = E[X_2], \\
\sigma_1^2 &= \mathrm{Var}[X_1], \; \sigma_2^2 = \mathrm{Var}[X_2], \\
K_{12} &= \mathrm{Cov}[X_1,X_2] = \sigma_1 \sigma_2 \rho,
\end{aligned} \quad (2.15)$$

and ρ is the correlation coefficient between X_1 and X_2.

If $|\rho| < 1$, the random variables X_1 and X_2 possess a joint probability density function given by, for any real numbers x_1 and x_2,

$$f_{X_1, X_2}(x_1, x_2) = \frac{1}{2\pi\sigma_1\sigma_2\sqrt{1-\rho^2}} \tag{2.16}$$
$$\times \exp\left\{\frac{-1}{2(1-\rho^2)}\left[\left(\frac{x_1-m_1}{\sigma_1}\right)^2 + \left(\frac{x_2-m_2}{\sigma_2}\right)^2 - 2\rho\left(\frac{x_1-m_1}{\sigma_2}\right)\left(\frac{x_2-m_2}{\sigma_2}\right)\right]\right\}.$$

By algebraically rearranging the exponent in Eq. 2.16 it may be shown that

$$f_{X_1, X_2}(x_1, x_2) = \frac{1}{\sigma_1}\varphi\left(\frac{x_1-m_1}{\sigma_1}\right)\frac{1}{\sigma_2\sqrt{1-\rho^2}}\varphi\left(\frac{x_2-m_2-\left(\frac{\sigma_2}{\sigma_1}\right)\rho(x_1-m_1)}{\sigma_2\sqrt{1-\rho^2}}\right),$$
$$\tag{2.17}$$

where

$$\varphi(y) = \frac{1}{\sqrt{2\pi}}e^{-(1/2)y^2} \tag{2.18}$$

is the normal density function. From Eq. 2.17 it follows that

$$f_{X_1}(x_1) = \frac{1}{\sigma_1}\varphi\left(\frac{x_1-m_1}{\sigma_1}\right), \tag{2.19}$$

$$f_{X_2|X_1}(x_2 \mid x_1) = \frac{f_{X_1, X_2}(x_1, x_2)}{f_{X_1}(x_1)} = \frac{1}{\sigma_2\sqrt{1-\rho^2}}\varphi\left(\frac{x_2-m_2-\left(\frac{\sigma_2}{\sigma_1}\right)\rho(x_1-m_1)}{\sigma_2\sqrt{1-\rho^2}}\right).$$
$$\tag{2.20}$$

In words, Eq. 2.20 may be expressed as follows: if X_1 and X_2 are jointly normally distributed, then the conditional probability law of X_2, given X_1, is the normal probability law with mean $m_2 + \rho\frac{\sigma_2}{\sigma_1}(x_1 - m_1)$ and standard deviation $\sigma_2\sqrt{1-\rho^2}$. In symbols,

$$E[X_2 \mid X_1 = x_1] = m_2 + \rho\frac{\sigma_2}{\sigma_1}(x_1 - m_1), \tag{2.21}$$

$$E[(X_2 - E[X_2 \mid X_1])^2 \mid X_1 = x_1] = \sigma_2^2(1-\rho^2), \tag{2.22}$$

where $m_1, m_2, \sigma_1, \sigma_2$, and ρ are defined by Eq. 2.15. Although Eqs. 2.21 and 2.22 were derived here under the assumption that $|\rho| < 1$, it may be shown that they hold also in the case that $|\rho| = 1$.

Conditional variance, moments, and characteristic function. Many random variables arise as a random mixture of other random variables. In the study of such random variables, the notions of conditional variance, conditional moments, and conditional characteristic function play a central role.

The conditional variance of Y, given X, is denoted by $\mathrm{Var}[Y \mid X]$, and is defined by

$$\mathrm{Var}[Y \mid X] = E[(Y - E[Y \mid X])^2 \mid X], \tag{2.23}$$

assuming that $E[Y^2] < \infty$. In the case of jointly continuous random variables,

$$\mathrm{Var}[Y \mid X = x] = \int_{-\infty}^{\infty} (y - E[Y \mid X = x])^2 f_{Y \mid X}(y \mid x) \, dy. \tag{2.24}$$

There is an important formula expressing the unconditional variance $\mathrm{Var}[Y]$ in terms of the conditional variance: if $E[Y^2] < \infty$, then

$$\mathrm{Var}[Y] = E[\mathrm{Var}[Y \mid X]] + \mathrm{Var}[E[Y \mid X]]. \tag{2.25}$$

In words, *the variance is equal to the mean of the conditional variance plus the variance of the conditional mean.*

To prove Eq. 2.25 we make use of the basic fact: if $E[Y] < \infty$, then

$$E[Y] = E[E[Y \mid X]]. \tag{2.26}$$

To prove Eq. 2.26 let $g(X) = 1$ in Eq. 2.6. From Eq. 2.26 it follows that

$$\mathrm{Var}[Y] = E[\mid Y - E[Y] \mid^2] = E[E[\mid Y - E[Y] \mid^2 \mid X]]. \tag{2.27}$$

Now, for any random variable Z and any constant a,

$$E[\mid Z - a \mid^2] = E[\mid Z - E[Z] \mid^2] + \{E[Z] - a\}^2. \tag{2.28}$$

In the same way that one proves Eq. 2.28 one may show that

$$E[\mid Y - E[Y] \mid^2 \mid X] = E[\mid Y - E[Y \mid X] \mid^2 \mid X] + \{E[Y \mid X] - E[Y]\}^2 \tag{2.29}$$

Upon taking the expectation of both sides of Eq. 2.29 one obtains Eq. 2.25.

EXAMPLE 2B

The sum of a random number of independent random variables. Let N be the number of female insects in a certain region, let X be the number of eggs laid by an insect, and let Y be the number of eggs in the region. Let us find the mean and variance of Y.

Now Y can be represented as follows:

$$Y = X_1 + X_2 + \cdots + X_N, \qquad (2.30)$$

where X_i is the number of eggs laid by the ith insect. We assume that the random variables $X_1, X_2, \cdots$ are not only independent of one another but also are independent of the random variable N representing the number of egg-laying insects. We assume further that the random variables X_i are identically distributed as a random variable X with finite mean and variance.

To find the mean and variance of Y, we must find the conditional mean and variance of Y, given N. Given that $N = n$, Y is distributed as the sum of n independent random variables with common mean $E[X]$ and common variance $\text{Var}[X]$. Consequently,

$$E[Y \mid N = n] = nE[X], \quad \text{Var}[Y \mid N = n] = n\,\text{Var}[X]. \qquad (2.31)$$

We next rewrite Eq. 2.31 in a somewhat more abstract form:

$$E[Y \mid N] = N\,E[X], \quad \text{Var}[Y \mid N] = N\,\text{Var}[X]. \qquad (2.32)$$

From Eq. 2.32 it follows that

$$E[E[Y \mid N]] = E[N]E[X], \quad E[\text{Var}[Y \mid N]] = E[N]\text{Var}[X]. \qquad (2.33)$$

From Eqs. 2.26, 2.25, and 2.33 we finally obtain expressions for the mean and variance of Y:

$$E[Y] = E[N]E[X], \qquad (2.34)$$
$$\text{Var}[Y] = E[N]\text{Var}[X] + \text{Var}[N]E^2[X]. \qquad (2.35)$$

It is well known that the moments and central moments of a random variable Y can be simply expressed in terms of its characteristic function

$$\varphi_Y(u) = E[e^{iuY}]. \qquad (2.36)$$

It may be verified that conditional moments and central moments may be similarly expressed in terms of the *conditional characteristic function*

$$\varphi_{Y \mid X}(u \mid x) = E[e^{iuY} \mid X = x]. \qquad (2.37)$$

In the case of jointly continuous random variables,

$$\varphi_{Y \mid X}(u \mid x) = \int_{-\infty}^{\infty} e^{iuy} f_{Y \mid X}(y \mid x)\, dy. \qquad (2.38)$$

From a knowledge of the conditional characteristic function we may obtain the unconditional characteristic function:

$$\varphi_Y(u) = E[E[e^{iuY} \mid X]] = \int_{-\infty}^{\infty} \varphi_{Y|X}(u \mid x)\, dF_X(x). \tag{2.39}$$

EXAMPLE 2C

Compound Poisson distributions. Greenwood and Yule (1920) found that the negative binomial distribution gave a better fit than did the Poisson distribution to accidents in munitions factories in England during the First World War. They showed that this phenomenon could be explained by assuming that the number of accidents per worker is Poisson distributed with a mean characterizing that worker, and that the different means occurring among the workers could be considered the values of a random variable λ.

More precisely, let X be a Poisson distributed random variable with mean λ. Then

$$E[e^{iuX}] = e^{\lambda(e^{iu}-1)}. \tag{2.40}$$

Suppose, however, that the mean λ is chosen in accord with a probability distribution $F(\lambda)$; for example, if X is the number of accidents per worker in a factory, the mean number λ of accidents per worker may vary from worker to worker. The assumption (2.40) would then be written [if λ is regarded as the observed value of a random variable Λ with distribution function $F(\lambda)$]

$$E[e^{iuX} \mid \Lambda = \lambda] = e^{\lambda(e^{iu}-1)}. \tag{2.41}$$

Consequently, the characteristic function of X is given by

$$
\begin{aligned}
\varphi_X(u) &= \int_{-\infty}^{\infty} E[e^{iuX} \mid \Lambda = \lambda]\, dF(\lambda) \\
&= \int_{-\infty}^{\infty} e^{\lambda(e^{iu}-1)}\, dF(\lambda) \\
&= \varphi_\Lambda(\{e^{iu}-1\}/i),
\end{aligned}
\tag{2.42}
$$

where $\varphi_\Lambda(\cdot)$ is the characteristic function of Λ.

The name *compound Poisson distribution* is given to a probability law, with characteristic function of the form of Eq. 2.42, which arises when one observes a random mixture of events of Poisson type; for diverse examples of compound Poisson distributions, see Feller (1943) and Satterthwaite (1942).

We here consider the important case in which Λ is assumed to obey a gamma distribution with parameters r and λ_0; in symbols,

$$\varphi_\Lambda(u) = \left(1 - \frac{iu}{\lambda_0}\right)^{-r}. \tag{2.43}$$

Then

$$\varphi_X(u) = \left(1 - \frac{\{e^{iu} - 1\}}{\lambda_0}\right)^{-r} = \left(\frac{\lambda_0}{1 + \lambda_0 - e^{iu}}\right)^r$$

$$= \left(\frac{p}{1 - qe^{iu}}\right)^r, \quad \text{where} \quad p = \frac{\lambda_0}{1 + \lambda_0}, \quad q = 1 - p. \tag{2.44}$$

In words, X possesses a negative binomial distribution with parameters r and p. The mean and variance of X are given by

$$E[X] = \frac{rq}{p} = \frac{r}{\lambda_0}, \quad \text{Var}[X] = \frac{rq}{p^2} = \frac{r(1 + \lambda_0)}{\lambda_0^2}. \tag{2.45}$$

The notion of conditional expectation may be extended to several random variables. In particular, let us consider n random variables $X_1, \cdots, X_n$, and a random variable Y with finite mean. The conditional expectation of Y, given the values of $X_1, \cdots, X_n$ is denoted by $E[Y \mid X_1, \cdots, X_n]$ and is defined to be the unique random variable with finite mean which is a function of $X_1, \cdots, X_n$ and satisfies, for every random variable $g(X_1, \cdots, X_n)$ which is a bounded function of $X_1, \cdots, X_n$,

$$E[E[Y \mid X_1, \cdots, X_n] g(X_1, \cdots, X_n)] = E[Y g(X_1, \cdots, X_n)]. \tag{2.46}$$

COMPLEMENT

2A In the same way that the conditional variance is defined, one can define conditional moments. In particular, let us consider the third and fourth central moments.

The third and fourth central moments of a random variable Y, respectively denoted by $\mu_3[Y]$ and $\mu_4[Y]$, are defined by

$$\mu_3[Y] = E[(Y - E[Y])^3], \quad \mu_4[Y] = E[(Y - E[Y])^4].$$

Similarly, we define

$$\mu_3[Y \mid X] = E[(Y - E[Y \mid X])^3 \mid X],$$
$$\mu_4[Y \mid X] = E[(Y - E[Y \mid X])^4 \mid X].$$

Show that

$$\mu_3[Y] = E[\mu_3[Y \mid X]] + \mu_3[E[Y \mid X]],$$
$$\mu_4[Y] = E[\mu_4[Y \mid X]] + 6E[\text{Var}[Y \mid X]] \,\text{Var}[E[Y \mid X]] + \mu_4[E[Y \mid X]].$$

EXERCISES

2.1 Show that Eq. 2.13 holds.

2.2 *Branching, cascade, or multiplicative process.* A population of individuals gives rise to a new population. Assume that the probability that an individual gives rise to k new individuals ("offspring") is p_k for $k = 0, 1, \cdots$, and the number of individuals arising from different individuals are independent random variables. The new population forms a first generation, which in turn reproduces a second generation, and so on. For $n = 0, 1, \cdots$ let X_n be the size of the nth generation. Note that

$$X_{n+1} = \sum_{j=1}^{X_n} Z_j(n),$$

where $Z_j(n)$ is the number of individuals in the $(n + 1)$st generation arising from the jth individual in the nth generation. Suppose that the number of offspring of an individual has finite mean μ and variance σ^2,

$$\mu = \sum_{m=0}^{\infty} m p_m < \infty, \, \sigma^2 = \sum_{m=0}^{\infty} (m - \mu)^2 p_m < \infty.$$

Show that

$$m_n = E[X_n \mid X_0 = 1] = \mu^n$$

$$\sigma_n^2 = \mathrm{Var}[X_n \mid X_0 = 1] = \begin{cases} \sigma^2 \mu^{n-1} \dfrac{\mu^n - 1}{\mu - 1} & \text{if } \mu \neq 1 \\ n\sigma^2 & \text{if } \mu = 1. \end{cases}$$

Hint. Show that $m_{n+1} = \mu m_n$, $\sigma_{n+1}^2 = \mu \sigma_n^2 + \mu^{2n} \sigma^2$.

2.3   The number of animals that a certain hunter traps in a day is a random variable N with mean m and variance σ^2. The hunter is interested in the number of silver foxes that he catches. Let Y be the number of silver foxes trapped. Find the mean and variance of Y, under the following assumptions: the event that a trapped animal ˙s a silver fox has probability p and is independent of whether any other trapped animal is a silver fox, and is also independent of the number N of animals which were trapped.

2.4   The number of accidents occurring in a factory in a week is a random variable with mean μ and variance σ^2. The numbers of individuals injured in single accidents are independently distributed, each with mean ν and variance τ^2. Find the mean and variance of the number of individuals injured in a week.

2.5   Let X_1 and X_2 be jointly normally distributed random variables (representing the observed amplitudes of a noise voltage recorded a known time interval apart). Assume that their joint probability density function is given by Eq. 2.16 with $m_1 = 1$, $m_2 = 2$, $\sigma_1 = 1$, $\sigma_2 = 4$, $\rho = 0.4$. Find $P[X_2 > 1 \mid X_1 = 1]$.

2.6   Let X_1 and X_2 be jointly normally distributed random variables, representing the daily sales (in thousands of units) of a certain product in a certain store on two successive days. Assume that the joint probability density function of X_1 and X_2 is given by Eq. 2.16, with $m_1 = m_2 = 3$, $\sigma_1 = \sigma_2 = 1$, $\rho = 0.8$. Find K so that

(i) $P[X_2 > K] = 0.05$,

(ii) $P[X_2 > K \mid X_1 = 2] = 0.05$.

Suppose the store desires to have on hand on a given day enough units of the product so that with probability 0.95 it can supply all demands for the product on the day. How large should its inventory be on a given morning if (iii) yesterday's sales were 2,000 units, (iv) yesterday's sales are not known?

2.7 Suppose that X is normal with mean 0 and variance 1, and the conditional distribution of Y, given that $X = x$, is normal with mean ρx and variance $1 - \rho^2$. Find the conditional distribution of X, given Y.

2.8 Suppose μ is normal with mean m and variance τ^2, and the conditional distribution of X, given that $\mu = \mu_0$, is normal with mean μ_0 and variance σ^2 (where σ^2 is a constant). Show that (i) the unconditional distribution of X is normal with mean m and variance $\sigma^2 + \tau^2$, (ii) the conditional distribution of μ, given that $X = x$, is normal with mean and variance satisfying

$$E[\mu \mid X = x] = \frac{x\tau^2 + m\sigma^2}{\tau^2 + \sigma^2},$$

$$\mathrm{Var}[\mu \mid X = x] = \frac{\tau^2\sigma^2}{\tau^2 + \sigma^2}.$$

In Exercises 2.9 to 2.12, find
(i) the conditional probability density function of X given Y,
(ii) $E[Y \mid X]$,
assuming that X and Y have the joint probability density function given.

2.9 $f_{X,Y}(x,y) = 6xy(2 - x - y)$ if $0 \leq x, y \leq 1$,
            $= 0$,   otherwise.

2.10 $f_{X,Y}(x,y) = 4y(x - y)e^{-(x+y)}$ if $0 \leq x < \infty, 0 \leq y \leq x$,
             $= 0$,   otherwise.

2.11 $f_{X,Y}(x,y) = \tfrac{1}{8}(y^2 - x^2)e^{-y}$ if $0 \leq y < \infty, \mid x \mid \leq y$,
             $= 0$,   otherwise.

2.12 $f_{X,Y}(x,y) = \dfrac{\sqrt{3}}{2\pi} e^{-(x^2 + xy + y^2)}$ $-\infty < x, y < \infty$.

2.13 *Compound exponential distributions.* Consider the decay of particles in a cloud chamber (or, similarly, the breakdown of equipment or the occurrence of accidents). Assume that the time X it takes a particle to decay is a random variable obeying an exponential probability law with parameter y. However, it is not assumed that the value of y is the same for all particles. Rather, it is assumed that there are particles of different types (or equipment of different types or individuals of different accident proneness). More specifically, it is assumed that for a particle randomly selected from the cloud chamber the parameter y is a particular value of a random variable Y obeying a gamma probability law with a probability density function

$$f_Y(y) = \frac{\beta^\alpha}{\Gamma(\alpha)} \, y^{\alpha-1} e^{-\beta y}, \qquad \text{for} \quad y > 0, \qquad (2.47)$$

In which the parameters α and β are positive constants characterizing the experimental conditions under which the particles are observed. The assumption that the time X it takes a particle to decay obeys an exponential law is now expressed as an assumption on the conditional probability law of X given Y:

$$f_{X|Y}(x \mid y) = ye^{-xy} \quad \text{for} \quad x > 0.$$

Find the individual probability law of the time X (of a particle selected at random) to decay.

2.14 A certain firm finds that the quantity X it sells of a certain item is gamma distributed with density function

$$f_X(x \mid \mu) = \frac{4}{\mu^2} \, xe^{-2x/\mu}, \; x > 0$$

and with mean $E[X] = \mu$. The parameter μ is itself a random variable whose reciprocal $1/\mu$ obeys a gamma probability law with probability density function given by Eq. 2.47. Find the probability density function of X.

2.15 A random sample of size 8 of a random variable X is a set of observed values of random variables $X_1, X_2, \cdots, X_8$ which are independent and identically distributed as X. Assuming that X possesses a continuous probability law, find the probability that each of the last 3 observations (namely X_6, X_7, and X_8) is larger than all of the first 5 observations; more precisely find

$$P[\min (X_6, X_7, X_8) > \max (X_1, \cdots, X_5)].$$

Show that this probability is independent of the functional form of the probability density function of the observed random variables.

Hint. Let $U = \min (X_6, X_7, X_8)$, $V = \max (X_1, \cdots, X_5)$. Show that

$$P[U > V] = \int_{-\infty}^{\infty} \{1 - F_U(v)\} f_V(v) \, dv$$
$$= 5 \int_{-\infty}^{\infty} \{1 - F_X(v)\}^3 \{F_X(v)\}^4 f_X(v) \, dv$$
$$= 5 \int_0^1 (1 - y)^3 y^4 \, dy.$$

Evaluate the integral using the beta function defined for $m > -1$ and $n > -1$ by

$$B(m + 1, n + 1) = \int_0^1 y^m (1 - y)^n \, dy$$
$$= \frac{\Gamma(m + 1)\Gamma(n + 1)}{\Gamma(m + n + 2)}$$

and the gamma function, defined for $n > -1$ by

$$\Gamma(n+1) = \int_0^\infty x^n e^{-x}\, dx$$
$$= n! \qquad \text{if } n = 0, 1, 2, \cdots.$$

2.16   Let $X_1, \cdots, X_n, X_{n+1}, \cdots, X_{n+m}$ be independent identically distributed continuous random variables. Let $V = \max (X_1, \cdots, X_n)$. Find

(i) $P[X_{n+1} \geq V]$,
(ii) $P[\min (X_{n+1}, \cdots, X_{n+m}) \geq V]$.

2.17   Let $T_1, \cdots, T_n$ be independent exponentially distributed random variables with means $1/\lambda_1, \cdots, 1/\lambda_n$ respectively. Show that for $j = 1, \cdots, n$

$$P[T_j = \min (T_1, \cdots, T_n)] = \lambda_j / (\lambda_1 + \cdots + \lambda_n).$$

2-3   PROPERTIES OF CONDITIONAL EXPECTATION

In this section we show that the conditional expectation $E[Y \mid X]$ of Y given X, defined as the unique function of X satisfying Eq. 2.6, enjoys exactly the same properties as does the expectation of a random variable. (This section may be omitted in a first reading of the book.)

THEOREM 3A

Properties of conditional expectation. Let X be a random variable, c a real number, U, V, Y, and $\varphi(X)$ random variables with finite means. Then

$$E[Y \mid X] \text{ is unique,} \tag{3.0}$$
$$E[Y \mid X] = E[Y] \text{ if } X \text{ and } Y \text{ are independent,} \tag{3.1}$$
$$E[c \mid X] = c, \tag{3.2}$$
$$E[\varphi(X) \mid X] = \varphi(X), \tag{3.3}$$
$$E[cY \mid X] = cE[Y \mid X], \tag{3.4}$$
$$E[\varphi(X)Y \mid X] = \varphi(X)E[Y \mid X], \tag{3.5}$$
$$E[U + V \mid X] = E[U \mid X] + E[V \mid X], \tag{3.6}$$
$$0 \leq Y \text{ implies } 0 \leq E[Y \mid X], \tag{3.7}$$
$$U \leq Y \leq V \text{ implies } E[U \mid X] \leq E[Y \mid X] \leq E[V \mid X], \tag{3.8}$$
$$| E[Y \mid X] |^r \leq \{E[(\mid Y \mid) \mid X]\}^r \leq E[\mid Y \mid^r \mid X] \text{ for } r \geq 1. \tag{3.9}$$

Remark. Strictly speaking the assertions made in Eqs. 3.0–3.9 only hold with probability one. However, in this book we shall ignore the distinction between statements that hold without qualification, and those that hold except for a set of circumstances which has probability zero of occurring.

Proof. To prove Eq. 3.0 we show that there is at most one function of X satisfying Eq. 2.6. Suppose there existed two functions $h_1(X)$ and $h_2(X)$ such that

$$E[g(X)Y] = E[g(X)h_1(X)] \tag{3.10}$$
$$E[g(X)Y] = E[g(X)h_2(X)]$$

for every random variable $g(X)$ which is a bounded function of X. Let

$$A_1 = \{x: h_1(x) - h_2(x) > 0\}, \quad A_2 = \{x: h_1(x) - h_2(x) < 0\}.$$

We desire to show that

$$P[h_1(X) \neq h_2(X)] = P[X \epsilon A_1] + P[X \epsilon A_2] = 0. \tag{3.11}$$

We first give a heuristic proof of Eq. 3.11. From Eq. 3.10 it follows that for suitable random variables $g(X)$

$$E[g(X)\{h_1(X) - h_2(X)\}] = 0. \tag{3.12}$$

Now if $g(X) = h_1(X) - h_2(X)$ belonged to the class of random variables for which Eq. 3.12 holds, it would follow that

$$E[\,|\,h_1(X) - h_2(X)\,|^2] = 0, \tag{3.13}$$

which implies that

$$P[\,|\,h_1(X) - h_2(X)\,| = 0] = 1,$$

which is the desired conclusion. In the case that we cannot assume that Eq. 3.12 holds for $g(X) = h_1(X) - h_2(X)$ we proceed as follows. The functions $I_{A_1}(X)$ and $I_{A_2}(X)$ are bounded and therefore are admissible values of $g(X)$ in Eq. 3.12. Therefore,

$$E[I_{A_1}(X)\{h_1(X) - h_2(X)\}] = 0, \tag{3.14}$$
$$E[I_{A_2}(X)\{h_1(X) - h_2(X)\}] = 0.$$

On the other hand

$$I_{A_1}(X)\{h_1(X) - h_2(X)\} \quad \text{and} \quad I_{A_2}(X)\{h_2(X) - h_1(X)\}$$

are non-negative functions. Therefore Eq. 3.14 implies that

$$0 = P[I_{A_1}(X)\{h_1(X) - h_2(X)\} \neq 0] \geq P[X \epsilon A_1], \tag{3.15}$$
$$0 = P[I_{A_2}(X)\{h_1(X) - h_2(X)\} \neq 0] \geq P[X \epsilon A_2].$$

From Eq. 3.15 one deduces Eq. 3.11. The proof of Eq. 3.11 is now complete.

To prove Eq. 3.1, note that by independence of X and Y

$$E[Yg(X)] = E[Y]E[g(X)] = E[E[Y]g(X)].$$

Since $E[Y]$ satisfies the defining property, Eq. 2.6, of conditional expectation, Eq. 3.1 holds. One proves Eq. 3.2 in a similar manner; Eq. 2.6 clearly holds if one lets $E[Y \mid X] = c$ and $Y = c$.

We next prove Eq. 3.5, which implies Eqs. 3.3 and 3.4. Clearly,

$$E[\{\varphi(X)E[Y \mid X]\}g(X)] = E[E[Y \mid X]\{\varphi(X)g(X)\}] = E[Y\{\varphi(X)g(X)\}]$$
$$= E[\{Y\varphi(X)\}g(X)].$$

In view of the defining property, Eq. 2.6, of conditional expectation, the proof of Eq. 3.5 is complete. One proves Eq. 3.6 in a similar manner:

$$E[\{E[U \mid X] + E[V \mid X]\}g(X)] = E[E[U \mid X]g(X)] + E[E[V \mid X]g(X)]$$
$$= E[Ug(X)] + E[Vg(X)]$$
$$= E[\{U + V\}g(X)].$$

To prove Eq. 3.7 note that if $Y \geq 0$, then for any non-negative function $g(X)$

$$0 \leq E[g(X)Y] = E[g(X)E[Y \mid X]]. \tag{3.16}$$

Suppose now that the set A of values x satisfying $E[Y \mid X = x] < 0$ has positive probability. Let $g(x) = 1$ or 0 depending on whether or not x belongs to A. Then $g(X)E[Y \mid X = x]$ is non-positive, and is actually negative on a set of positive probability. Consequently, $E[g(X)E[Y \mid X]] < 0$, in contradiction to Eq. 3.16. The proof of Eq. 3.7 and therefore of Eq. 3.8 is complete.

We next prove Eq. 3.9. From Eq. 3.8, with $U = - \mid Y \mid$ and $V = \mid Y \mid$, it follows that

$$\mid E[Y \mid X] \mid \leq E[(\mid Y \mid) \mid X]. \tag{3.17}$$

Consequently, the first inequality is proved. To prove the second inequality let us show that if Z is a non-negative random variable, and if $r \geq 1$, then

$$\{E[Z \mid X]\}^r \leq E[Z^r \mid X]. \tag{3.18}$$

To prove (3.18) note that (by Taylor's theorem; compare *Mod Prob*, p. 434)

$$z^r - z_0{}^r \geq (z - z_0)rz_0{}^{r-1}$$

for $r \geq 1$ and non-negative numbers z and z_0. Therefore,

$$Z^r - \{E[Z \mid X]\}^r \geq (Z - E[Z \mid X])r\{E[Z \mid X]\}^{r-1}. \qquad (3.19)$$

Take the conditional expectation given X of both sides of Eq. 3.19. Then, using Eqs. 3.3, 3.5, and 3.6

$$E[Z^r \mid X] - \{E[Z \mid X]\}^r \geq r\{E[Z \mid X]\}^{r-1}E[(Z - E[Z \mid X]) \mid X] = 0$$

from which Eq. 3.18 follows. The proof of Theorem 3A is now complete.

EXERCISES

3.1 Let $\{X(t), t \geq 0\}$ be a stochastic process with independent increments and finite mean value function $m(t) = E[X(t)]$. Show that for any times $0 < t_1 < \cdots < t_n < t_{n+1}$

$$E[X(t_{n+1}) \mid X(t_1), \cdots, X(t_n)] = X(t_n) + m(t_{n+1}) - m(t_n).$$

3.2 *Martingales.* A stochastic process $\{X(t), t \geq 0\}$ with finite means is said to be a (continuous parameter) *martingale* if for any set of times

$$t_1 < t_2 < \cdots < t_n < t_{n+1},$$
$$E[X(t_{n+1}) \mid X(t_1), \cdots, X(t_n)] = X(t_n);$$

in words, the conditional expectation of $X(t_{n+1})$, given the values $X(t_1), \cdots, X(t_n)$ is equal to the most recently observed value $X(t_n)$. Show that the Wiener process is a martingale. (For a complete treatment of martingale theory, see Doob [1953]).

3.3 *Martingales* (discrete parameter). A stochastic process $\{X_n, n = 1, 2, \cdots\}$ with finite means is said to be a (discrete parameter) martingale if for any integer n

$$E[X_{n+1} \mid X_1, \cdots, X_n] = X_n.$$

(i) Show that the sequence $\{S_n\}$ of consecutive sums of independent random variables with zero means is a martingale.
(ii) Show that if $\{X_n\}$ is a martingale then for any integers

$$m_1 < m_2 < \cdots < m_n < m_{n+1},$$
$$E[X_{m_{n+1}} \mid X_{m_1}, \cdots, X_{m_n}] = X_{m_n}.$$

Normal processes
and covariance stationary
processes

ONE APPROACH to the problem of developing mathematical models for empirical phenomena evolving in accord with probabilistic laws is to characterize such phenomena in terms of the behavior of their first and second moments. This approach has found important applications in statistical communications and control theory and in time series analysis.

This chapter discusses some of the basic concepts and techniques of the theory of stochastic processes which possess finite second moments. Among such processes, two types are most important for applications: normal processes and covariance stationary processes. Why this is the case is shown.

3-1 THE MEAN VALUE FUNCTION AND THE COVARIANCE KERNEL OF A STOCHASTIC PROCESS

From a knowledge of the mean and variance of a random variable, one cannot in general determine its probability law, unless the functional form of the probability law is known up to several unspecified parameters which are simply related to the mean and variance. For example, if X is a random variable obeying a normal distribution, then a knowledge of its mean and variance yields a knowledge of all probabilities concerning X; while if X is Poisson distributed, then its probability law is determined by its mean. In the general case in which the functional form of the probability law of the random variable is unknown, its mean and variance still serve partially to summarize the probability law since by means of various inequalities, such as Chebyshev's inequality, one can form crude estimates of various features of the probability law.

The role played for a single random variable by its mean and variance is played for a stochastic process by its mean value function and its covariance kernel.

Let $\{X(t), t \in T\}$ be a stochastic process with finite second moments. Its *mean value function*, denoted by $m(t)$, is defined for all t in T by

$$m(t) = E[X(t)], \tag{1.1}$$

and its *covariance kernel*, denoted by $K(s,t)$, is defined for all s and t in T by

$$K(s,t) = \text{Cov}[X(s), X(t)]. \tag{1.2}$$

EXAMPLE 1A

Many stochastic processes arise as functions of a finite number of random variables. For example, suppose $X(t)$ represents the position of a particle in motion with a constant velocity. One may assume that $X(t)$ is of the form

$$X(t) = X_0 + Vt, \tag{1.3}$$

where X_0 and V are random variables, representing the initial position and velocity respectively. The mean value function and covariance kernel of $\{X(t), t \geq 0\}$ are given by

$$m(t) = E[X(t)] = E[X_0] + tE[V], \tag{1.4}$$
$$K(s,t) = \text{Cov}[X(s), X(t)] = \text{Var}[X_0] + (s+t)\,\text{Cov}[X_0,V] + st\,\text{Var}[V].$$

One sees that in order to obtain the mean value function and covariance kernel of $\{X(t), t \geq 0\}$, one does not need to know the joint probability law of X_0 and V, but only their means, variances, and covariance.

EXAMPLE 1B

Mean value function and covariance kernel of the Wiener and Poisson processes. Let $\{X(t), t \geq 0\}$ be the Wiener process with parameter σ^2. By Eqs. 3.3 and 3.4 of Chapter 1 it follows that for all $t \geq 0$

$$m(t) = E[X(t)] = 0,$$
$$\text{Var}[X(t)] = \sigma^2 t. \tag{1.5}$$

We next compute the covariance kernel $K(s,t)$ for $s < t$:

$$\begin{aligned}
\text{Cov}[X(s), X(t)] &= \text{Cov}[X(s), X(t) - X(s) + X(s)] \\
&= \text{Cov}[X(s), X(t) - X(s)] + \text{Cov}[X(s), X(s)] \\
&= \text{Var}[X(s)] = \sigma^2 s,
\end{aligned} \tag{1.6}$$

since $X(s)$ and $X(t) - X(s)$ are independent, and therefore have zero covariance. Consequently, the covariance kernel of the Wiener process with parameter σ^2 is given by

$$K(s,t) = \sigma^2 \min(s,t) \quad \text{for all } s, t \geq 0. \tag{1.7}$$

Similarly, one may show that if $\{N(t), t \geq 0\}$ is the Poisson process with intensity ν, then the mean value function and covariance kernel are respectively given by

$$m(t) = E[N(t)] = \nu t, \tag{1.8}$$
$$K(s,t) = \text{Cov}[N(s), N(t)] = \nu \min(s,t). \tag{1.9}$$

Indeed, from (1.6) it follows that for any stochastic process $\{X(t), t \geq 0\}$ with independent increments,

$$\text{Cov}[X(s), X(t)] = \text{Var}[X(\min\{s,t\})]. \tag{1.10}$$

The importance of the mean value function and the covariance kernel derives from two facts:

(i) it is usually much easier to find the mean value function and the covariance kernel of a stochastic process than it is to find its complete probability law;

(ii) nevertheless, many important questions about a stochastic process can be answered on a basis of a knowledge only of its mean value function and covariance kernel.

In Section 3–3 we show how the mean value function and covariance kernel may be used to study the behavior of sample averages over a stochastic process, and of integrals and derivatives of stochastic processes.

EXAMPLE 1C

The increment process of a Poisson process. Let $\{N(t), t \geq 0\}$ be a Poisson process of intensity ν, and let L be a positive constant. A new stochastic process $\{X(t), t \geq 0\}$ can be defined by

$$X(t) = N(t + L) - N(t). \tag{1.11}$$

For example, if $N(t)$ represents the number of events of a certain kind occurring in the interval 0 to t, then $X(t)$ represents the number of events occurring in a time interval of length L beginning at t. While in principle one could determine the joint probability law of $X(t_1), \cdots, X(t_n)$ for any choice of n time points $t_1, \cdots, t_n$, it is usually more convenient to begin one's study of a stochastic process by computing its mean value function and covariance kernel.

For the process $\{X(t), t \geq 0\}$ defined by Eq. 1.11, the mean value function is

$$m(t) = E[X(t)] = E[N(t + L) - N(t)] = \nu L. \tag{1.12}$$

We next compute the covariance kernel $K(s,t) = \text{Cov}[X(s), X(t)]$. We may assume that $s \leq t$. We next distinguish two cases: (i) $t \leq s + L$ and (ii) $t > s + L$. In case (ii), $X(s)$ and $X(t)$ are independent random variables and consequently have zero covariance. In case (i), we write

$$\begin{aligned} K(s,t) &= \text{Cov}[N(s + L) - N(s), N(t + L) - N(t)] \\ &= \text{Cov}[N(s + L) - N(t) + N(t) - N(s), N(t + L) - N(t)] \\ &= \text{Cov}[N(s + L) - N(t), N(t + L) - N(t)], \end{aligned} \tag{1.13}$$

since $N(t) - N(s)$ and $N(t + L) - N(t)$ have zero covariance. Next, writing $N(t + L) - N(t) = N(t + L) - N(s + L) + N(s + L) - N(t)$, one infers from Eq. 1.13 that

$$K(s,t) = \text{Var}[N(s + L) - N(t)] = \nu\{s + L - t\} = \nu\{L - (t - s)\}. \tag{1.14}$$

We have thus established that the covariance kernel of the process $\{X(t), t \geq 0\}$ defined by Eq. 1.11 is (for all $s, t \geq 0$)

$$\begin{aligned} K(s,t) &= \nu\{L - |t - s|\} && \text{if } |t - s| \leq L, \\ &= 0 && \text{if } |t - s| > L. \end{aligned} \tag{1.15}$$

3-2   STATIONARY AND EVOLUTIONARY PROCESSES

It has been found useful in the theory of stochastic processes to divide stochastic processes into two broad classes: stationary and evolutionary. Intuitively a stationary process is one whose distribution remains the same as time progresses, because the random mechanism producing the process is not changing as time progresses. An evolutionary process is one which is not stationary.

A Poisson process $\{N(t), t \geq 0\}$ is evolutionary since the distribution of $N(t)$ is functionally dependent on t. On the other hand the process $\{X(t), t \geq 0\}$ defined by Eq. 1.11 appears as if it might be stationary, because the distribution of the number of events occurring in a time interval of fixed length L does not functionally depend on the time t at which the interval begins.

In order to give a more formal definition of stationarity we introduce the notion of a linear index set.

An index set T is said to be a *linear* index set if it has the property that the sum $t + h$, of any two members t and h of T, also belongs to T. Examples of such index sets are $T = \{1, 2, \cdots\}$, $T = \{0, \pm 1, \pm 2, \cdots\}$, $T = \{t: t \geq 0\}$, and $T = \{t: -\infty < t < \infty\}$.

A stochastic process $\{X(t), t \, \epsilon \, T\}$, whose index set T is linear, is said to be

(i) *strictly stationary of order* k, where k is a given positive integer, if for any k points $t_1, \cdots, t_k$ in T, and any h in T, the k-dimensional random vectors

$$(X(t_1), \cdots, X(t_k)) \quad \text{and} \quad (X(t_1 + h), \cdots, X(t_k + h))$$

are identically distributed,

(ii) *strictly stationary* if for any integer k it is strictly stationary of order k.

To prove that a stochastic process is strictly stationary requires the verification of a very large number of conditions. For example, consider the stochastic process $\{X(t), t \geq 0\}$ defined by Eq. 1.11. One easily verifies that it is strictly stationary of order 1. With a little more effort one could prove that it is strictly stationary of order 2. Indeed, the process can be shown to be strictly stationary. However, in this book we will not discuss the theory of strictly stationary processes since it involves rather sophisticated mathematics.

There is another notion of stationarity, called covariance stationarity, whose theory is much easier to develop, and which suffices for many of the practical applications.

A stochastic process $\{X(t), t \, \epsilon \, T\}$ is said to be *covariance stationary*† if it possesses finite second moments, if its index set T is linear, and if its covariance kernel $K(s,t)$ is a function only of the absolute difference $|s - t|$, in the sense that there exists a function $R(v)$ such that for all s and t in T

$$K(s,t) = R(s - t); \tag{2.1}$$

or, more precisely, $R(v)$ has the property that for every t and v in T

$$\text{Cov}[X(t), X(t + v)] = R(v). \tag{2.2}$$

---

†The term "covariance stationary" is used in this book for the terms "weakly stationary," "stationary in the wide sense," or "second order stationary" used by other writers (such as Doob [1953], p. 95, or Loève [1960], p. 482).

We call $R(v)$ the covariance function of the covariance stationary time series $\{X(t), t \, \epsilon \, T\}$.

In view of Eq. 1.15, the stochastic process $\{X(t), t \geq 0\}$, considered in Example 1C, is covariance stationary with covariance function (see Figure 3.1)

$$R(v) = v \, \{L - |v|\} \qquad \text{if } |v| \leq L, \qquad (2.3)$$
$$\quad\;\; = 0 \qquad\qquad\;\; \text{if } |v| > L.$$

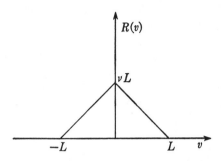

Fig. 3.1. Graph of the covariance function $R(v)$ given by Eq. 2.3.

It should be noted that in order for a stochastic process with finite second moments to be covariance stationary it is not necessary that its mean value function be a constant. Thus if $\{X(t), t > 0\}$ is the stochastic process considered in Example 1C, the stochastic process

$$Y(t) = \cos\!\left(\frac{2\pi}{L}\, t\right) + X(t) \qquad (2.4)$$

is covariance stationary even though its mean value function

$$E[Y(t)] = vL + \cos\!\left(\frac{2\pi}{L}\, t\right) \qquad (2.5)$$

is functionally dependent on t.

The meaning and uses of stationarity. The assumptions involved in assuming that a stochastic process is strictly stationary are best illustrated by the following quotation from Neyman and Scott (1959):

> The universe, as we see it, is a single realization of a four-dimensional stochastic process (three spatial coordinates and one time coordinate). The vaguely expressed assumption that "the distribution and motion of matter in sufficiently large spatial regions are, by and large, intrinsically much the same . . ." corresponds to a conceptually rigorous hypothesis that the stochastic process in question is stationary in the three spatial coordinates.

Operationally, the same assumption of stationarity may be expressed by postulating that in every region of space there exists a particular chance mechanism, the same for all regions, governing the distribution of matter and its motions.

It should be noted that the fact that a stochastic process is stationary does not mean that a typical sample function of the process has the same appearance at all points of time. Suppose that at some time in the history of the world an explosion takes place in some very isolated normally quiet forest. Let $X(t)$ represent the sound intensity at time t in the forest. Then $\{X(t), -\infty < t < \infty\}$ is a strictly stationary process if the time at which the explosion occurs is chosen uniformly on the real line.

The notion of a stationary stochastic process derives its importance from the fact that the *ergodic theorem* was first proved for stationary processes and from the fact that the notion of the *spectrum* was first defined for a stationary stochastic process. In this section, we briefly discuss what is meant by an ergodic theorem. In Section 3–6, we discuss the spectrum. It is emphasized that while these concepts are most readily defined for stationary stochastic processes, it is possible to extend these concepts to non-stationary processes which are composed, roughly speaking, of a stationary part and a transient part (see Parzen [1961 c]).

Ergodic theorems. In order that the theory of stochastic processes be useful as a method of describing physical systems, it is necessary that one be able to measure from observations of a stochastic process $\{X(t), t \geq 0\}$ probabilistic quantities such as the mean value function

$$m(t) = E[X(t)],$$

the covariance kernel

$$K(s,t) = \text{Cov}[X(s), X(t)],$$

and the one-dimensional distribution function

$$F_{X(t)}(x) = P[X(t) \leq x].$$

Consequently, the question naturally arises: if one has observed a single finite record $\{X(t), t = 1, 2, \cdots, T\}$ of a discrete parameter stochastic process $\{X(t), t = 1, 2, \cdots\}$ or a finite record $\{X(t), 0 \leq t \leq T\}$ of a continuous parameter stochastic process $\{X(t), t \geq 0\}$, under what circumstances (if any) is it possible to use this record to estimate ensemble averages, such as those above, by means of estimates which become increasingly accurate as the length T of record available becomes increasingly large?

The problem of determining conditions under which averages computed from a sample of a stochastic process can be ultimately identified with corresponding ensemble averages first arose in statistical mechanics. Physical systems possessing properties of this kind were called *ergodic* (see ter Haar [1954], p. 356, for the origin of the word "ergodic"). The first satisfactory mathematical proofs of an ergodic theorem (a theorem stating conditions under which sample averages may be identified with corresponding ensemble averages) were given by Birkhoff (1931) and von Neumann (1932). Birkhoff showed that if $\{X(t), t = 1, 2, \cdots\}$ is a strictly stationary process, then for any function $g(\cdot)$, such that the ensemble average

$$E[g(X(t))]$$

exists, the sample averages

$$\frac{1}{T} \sum_{t=1}^{T} g[X(t)]$$

converge (as $T \to \infty$) with probability one. In order that the limit of the sample averages be the ensemble average $E[g(X(t))]$, which it should be noted does not depend on t, one must impose certain supplementary conditions (called "metric transitivity"). An elegant proof of the Birkhoff ergodic theorem is given by Khintchine (1949); see also Doob (1953) and Loève (1960).

A discussion of the ergodic theorem for strictly stationary stochastic processes is beyond the scope of this book. However, a stochastic process need not be strictly stationary in order for it to obey an ergodic theorem, as we now show.

Given any discrete parameter stochastic process $\{X(t), t = 1, 2, \cdots\}$ one can consider the sequence of sample means

$$M_T = \frac{1}{T} \sum_{t=1}^{T} X(t) \tag{2.6}$$

formed from increasingly larger samples. The sequence of sample means $\{M_T, T = 1, 2, \cdots\}$ may be said to be *ergodic* if

$$\lim_{T \to \infty} \text{Var}[M_T] = 0. \tag{2.7}$$

In words, Eq. 2.7 may be interpreted as follows: the successive sample means, formed from a sample function of a stochastic process, have variances which tend to zero as the sample size T increases to ∞. One may then write that, for large enough sample sizes T, M_T is approximately equal to its ensemble mean

$$M_T \approx E[M_T] = \frac{1}{T} \sum_{t=1}^{T} m(t) \tag{2.8}$$

for almost all possible sample functions that could have been observed.

In general, a stochastic process is said to be ergodic if it has the property that sample (or time) averages formed from an observed record of the process may be used as an approximation to the corresponding ensemble (or population) averages. The general nature of the conditions which a stochastic process has to satisfy in order to be ergodic is indicated by the following theorem (see also Parzen [1958]).

THEOREM 2A

Necessary and sufficient conditions that the sample means of a stochastic process be ergodic. Let $\{X(t), t = 1, 2, \cdots\}$ be a stochastic process whose covariance kernel

$$K(s,t) = \text{Cov}[X(s), X(t)]$$

is a bounded function; that is, there is a constant K_0 such that

$$\text{Var}[X(t)] = K(t,t) \leq K_0 \quad \text{for all } t = 1, 2, \cdots. \tag{2.9}$$

For $t = 1, 2, \cdots$ let $C(t)$ be the covariance between the tth sample mean M_t and the tth observation $X(t)$:

$$C(t) = \text{Cov}[X(t), M_t] = \frac{1}{t} \sum_{s=1}^{t} K(s,t). \tag{2.10}$$

In order for Eq. 2.7 to hold, it is necessary and sufficient that

$$\lim_{t \to \infty} C(t) = 0. \tag{2.11}$$

In words, the sample means are ergodic if and only if as the sample size t is increased there is less and less correlation (covariance) between the sample mean M_t and the last observation $X(t)$.

Remark. For a covariance stationary process with covariance function $R(v)$,

$$C(t) = \frac{1}{t} \sum_{s=1}^{t} R(t-s) = \frac{1}{t} \sum_{v=0}^{t-1} R(v). \tag{2.12}$$

Now, the sequence $\{R(v), v = 0, 1, \cdots\}$ is said to converge to 0, as v tends to ∞, if

$$\lim_{v \to \infty} R(v) = 0 \tag{2.13}$$

and is said to converge to 0 in Cesàro mean as v tends to ∞ if

$$\lim_{t \to \infty} \frac{1}{t} \sum_{v=0}^{t-1} R(v) = 0. \tag{2.14}$$

It may be shown that Eq. 2.13 implies Eq. 2.14. We consequently obtain from Theorem 2A the following assertions.

The sample means of a covariance stationary stochastic process $\{X(t), t = 1, 2, \cdots\}$ are ergodic if and only if its covariance function $R(v)$ tends to 0 in Cesàro mean, as v tends to ∞; a sufficient condition for the sample means to be ergodic is that $R(v)$ tends to 0 as v tends to ∞.

Proof of Theorem 2A. That Eq. 2.7 implies Eq. 2.11 follows from the fact that by Schwarz's inequality

$$|C(t)|^2 \leq \text{Var}[X(t)] \, \text{Var}[M_t] \leq K_0 \, \text{Var}[M_t].$$

To prove that Eq. 2.11 implies Eq. 2.7, we first prove the following formula:

$$\text{Var}[M_n] = \frac{2}{n^2} \sum_{k=1}^{n} kC_k - \frac{1}{n^2} \sum_{k=1}^{n} \text{Var}[X(k)], \tag{2.15}$$

where $C_k = C(k)$. To prove Eq. 2.15 we write

$$\begin{aligned}
n^2 \, \text{Var}[M_n] &= \sum_{k=1}^{n} \text{Var}[X(k)] + 2 \sum_{k=1}^{n} \sum_{j=1}^{k-1} \text{Cov}[X(j), X(k)] \\
&= 2 \sum_{k=1}^{n} \sum_{j=1}^{k} \text{Cov}[X(j), X(k)] - \sum_{k=1}^{n} \text{Var}[X(k)] \\
&= 2 \sum_{k=1}^{n} kC_k - \sum_{k=1}^{n} \text{Var}[X(k)].
\end{aligned}$$

In view of Eq. 2.15, to show that Eq. 2.11 implies Eq. 2.7, it suffices to show that Eq. 2.11 implies that

$$\lim_{n \to \infty} \frac{1}{n^2} \sum_{k=1}^{n} kC_k = 0.$$

Now, for any $n > N > 0$,

$$\frac{1}{n^2} \sum_{k=1}^{n} kC_k \leq \frac{1}{n^2} \sum_{k=1}^{N} |kC_k| + \sup_{N < k} |C_k|,$$

which tends to 0 as first n tends to ∞ and then N tends to ∞. The proof of Theorem 2A is now complete.

The notion of ergodicity of sample means can be extended to continuous parameter time series. Given a sample $\{X(t), 0 \leq t \leq T\}$, the sample mean is defined by

$$M_T = \frac{1}{T} \int_0^T X(t)\, dt.$$

In order to discuss the behavior of M_T, one must first discuss the meaning and properties of integrals whose integrands are stochastic processes. This question is treated in the next section.

EXERCISES

For each of the stochastic processes $\{X(t), t \geq 0\}$ defined in Exercises 2.1 to 2.12, compute

(i) The mean value function $m(t) = E[X(t)]$,

(ii) The covariance kernel $K(s,t) = \text{Cov}[X(s), X(t)]$,

(iii) The covariance function $R(v)$ if the process is covariance stationary.

2.1 $X(t) = A + Bt$, in which A and B are independent random variables, each uniformly distributed on the unit interval.

2.2 $X(t) = A + Bt + Ct^2$, in which A, B, C are independent random variables, each with mean 1 and variance 1.

2.3 $X(t) = A \cos \omega t + B \sin \omega t$, in which ω is a positive constant, and A and B are uncorrelated random variables with means 0 and variances σ^2.

2.4 $X(t) = \cos(\omega t + \theta)$, in which ω is a positive constant, and θ is uniformly distributed on the interval 0 to 2π.

In Exercises 2.5 to 2.7, let $\{W(t), t \geq 0\}$ be the Wiener process with parameter σ^2.

2.5 $X(t) = W(t + L) - W(t)$, in which L is a positive constant.

2.6 $X(t) = At + W(t)$, in which A is a positive constant.

2.7 $X(t) = At + W(t)$, in which A is a random variable independent of the process $\{W(t), t \geq 0\}$, normally distributed with mean m and variance σ_1^2.

2.8 Let $X(t) = f(t + \theta)$, where $f(t)$ is a periodic function with period L, and θ is uniformly distributed on 0 to L.

2.9 *Real sine wave with random frequency and phase.* Let $X(t) = \cos(At + \theta)$, where A and θ are independent random variables, θ is uniformly distributed on 0 to 2π, and A has probability density function $f_A(x) = \{\pi(1 + x^2)\}^{-1}$.

2.10 *Sine wave whose phase is a stochastic process.* Define $X(t) = \cos(\omega t + \varphi(t))$, where ω is a positive constant and $\{\varphi(t), t \geq 0\}$ is a stochastic process defined by

$$\varphi(t) = W(t + 1) - W(t),$$

where $\{W(t), t \geq 0\}$ is the Wiener process with parameter σ^2.

2.11 Let $X(t) = \sum_{j=1}^{q} (A_j \cos \omega_j t + B_j \sin \omega_j t)$, where q is an integer, $\omega_1, \cdots, \omega_q$ are positive constants, and $A_1, B_1, \cdots, A_q, B_q$ are uncorrelated random variables with means 0 and respective variances $\sigma_j^2 = E[A_j^2] = E[B_j^2]$.

2.12 $\{X(t), t \geq 0\}$ is the random telegraph signal (defined in Section 1–4).

2.13 Let $\{Y(t), t \geq 0\}$ be a stochastic process whose values are A and B successively, the times at which the values change being distributed according to a Poisson process at mean rate ν. The values A and B may be any real numbers. Show that the process $Y(t)$ may be represented by Eq. 4.1 of Chapter 1 in which $X(t)$ is the random telegraph signal. Find the mean value function, covariance kernel, and two-dimensional characteristic function of $Y(t)$.

2.14 Let $\{N(t), t \geq 0\}$ be a stochastic process with stationary independent increments satisfying $N(0) = 0$ and

$$\varphi_{N(t)}(u) = \exp[\nu t\{\tfrac{1}{3}(e^{iu} + 2e^{iu}) - 1\}].$$

Find the mean value function, covariance kernel, and two-dimensional characteristic function of the one-minus-one process $\{X(t), t \geq 0\}$ defined by Eq. 4.4 of Chapter 1.

2.15 *Poisson square wave with random amplitudes.* Let $X(t)$ be a stochastic process whose graph consists of horizontal segments and jumps. The times at which $X(t)$ changes value occur in accord with a Poisson process at mean rate ν. The successive values which the process takes each time it changes value are independent random variables identically distributed as a random variable A with finite second moments.

2.16 Let $X(t) = \sin \omega t$, where ω is uniformly distributed on 0 to 2π. (i) Show that $\{X(t), t = 1, 2, \cdots\}$ is covariance stationary but is not strictly stationary. (ii) Show that $\{X(t), t \geq 0\}$ is neither covariance stationary nor strictly stationary.

2.17 Show that any sequence $\{X(t), t = 1, 2, \cdots\}$ of identically distributed random variables is strictly stationary of order 1.

2.18 Let X and Y be independent normally distributed random variables, with means 0 and variances 1. Let the sequence $\{X(t), t = 1, 2, \cdots\}$ be defined as follows: $X(t) = X$ if t is not a multiple of 4, and $X(t) = Y$ if t is a multiple of 4. Show that $\{X(t), t = 1, 2, \cdots\}$ is strictly stationary of order 1, but is not strictly stationary of order 2.

2.19 Show that if $\{X(t), t \, \epsilon \, T\}$ is strictly stationary of order k, then $\{X(t), t \, \epsilon \, T\}$ is strictly stationary of order k', for any integer k' less than k.

2.20 Show that the random telegraph signal is a strictly stationary process.

2.21 Prove that the process $\{X(t), t \geq 0\}$ defined in Complement 2A of Chapter 1 is (i) covariance stationary, (ii) strictly stationary.

3-3 INTEGRATION AND DIFFERENTIATION OF STOCHASTIC PROCESSES

If one observes a stochastic process $\{X(t), t \geq 0\}$ continuously over an interval $0 \leq t \leq T$ it is of interest to form the sample mean

$$M_T = \frac{1}{T} \int_0^T X(t) \, dt. \tag{3.1}$$

Before considering the problem of finding the mean and variance of M_T, we must first discuss the definition of the integral in Eq. 3.1 and determine conditions for the integral to exist.

A natural way of defining the integral $\int_a^b X(t) \, dt$ of a stochastic process over an interval $a < t \leq b$ is as a limit of approximating sums:

$$\int_a^b X(t) \, dt = \lim \sum_{k=1}^n X(t_k)\{t_k - t_{k-1}\}, \tag{3.2}$$

the limit being taken over subdivisions of the interval $(a,b]$ into sub-intervals, by points $a = t_0 < t_1 < \cdots < t_n = b$, as the maximum length $\max_{k=1,\ldots,n} (t_k - t_{k-1})$ of a sub-interval tends to 0.

In order to define the limiting operation employed in Eq. 3.2, we must consider the various modes of convergence in which a sequence of random variables can converge. The most important modes of convergence are convergence with probability one, convergence in probability, and convergence in mean square.

Given random variables $Z, Z_1, Z_2, \cdots$ we say that (i) the sequence $\{Z_n\}$ converges to Z *with probability one* if

$$P[\lim_{n \to \infty} Z_n = Z] = 1, \tag{3.3}$$

(ii) the sequence $\{Z_n\}$ converges to Z *in probability* if for every $\epsilon > 0$

$$\lim_{n \to \infty} P[\mid Z_n - Z \mid > \epsilon] = 0, \tag{3.4}$$

(iii) the sequence $\{Z_n\}$ converges to Z in *mean square*, if each random variable Z_n has a finite mean square and if

$$\lim_{n \to \infty} E[\mid Z_n - Z \mid^2] = 0. \tag{3.5}$$

For a discussion of the relation between these modes of convergence, the reader is referred to *Mod Prob*, p. 415. In this book we shall mainly be concerned with convergence in mean square.

It may be shown (see Loève [1960], p. 472) that in order for the family of approximating sums on the right-hand side of Eq. 3.2 to have a limit, in the sense of convergence in mean square, it is necessary and sufficient that the product moment $E[X(s) X(t)]$, regarded as a function of (s,t), be Riemann integrable over the set $\{(s,t)\colon a \le s \le b, a \le t \le b\}$. In this book we will make extensive use of the following theorem, which we state without proof.

THEOREM 3A

Let $\{X(t), t \ge 0\}$ be a continuous parameter stochastic process with finite second moments, whose mean value function and covariance kernel,

$$m(t) = E[X(t)], \tag{3.6}$$
$$K(s,t) = \mathrm{Cov}[X(s), X(t)], \tag{3.7}$$

are continuous functions of s and t. Then, for any $b > a \ge 0$, the integral $\int_a^b X(t)\,dt$ is well defined as a limit in mean square of the usual approximating sums of the form of Eq. 3.2. The mean, mean square, and variance of the integral are given by

$$E\left[\int_a^b X(t)\,dt\right] = \int_a^b E[X(t)]\,dt = \int_a^b m(t)\,dt, \tag{3.8}$$

$$E\left[\,\left|\int_a^b X(t)\,dt\right|^2\right] = E\left[\int_a^b \int_a^b X(s)X(t)\,ds\,dt\right] = \int_a^b \int_a^b E[X(s)X(t)]\,ds\,dt, \tag{3.9}$$

$$\mathrm{Var}\left[\int_a^b X(t)\,dt\right] = \int_a^b \int_a^b \mathrm{Cov}[X(s), X(t)]\,ds\,dt = \int_a^b \int_a^b K(s,t)\,ds\,dt. \tag{3.10}$$

Further, for any non-negative real numbers a, b, c, and d

$$E\left[\int_a^b X(s)\,ds \int_c^d X(t)\,dt\right] = \int_a^b ds \int_c^d dt\, E[X(s)X(t)], \tag{3.11}$$

$$\mathrm{Cov}\left[\int_a^b X(s)\,ds, \int_c^d X(t)\,dt\right] = \int_a^b ds \int_c^d dt\, K(s,t). \tag{3.12}$$

One may state Eqs. 3.8, 3.9, and 3.11 intuitively: the linear operations of integration and forming expectations commute. From Eqs. 3.8 and 3.9 one may deduce Eq. 3.10:

$$\begin{aligned}
\mathrm{Var}\left[\int_a^b X(t)\,dt\right] &= E\left[\,\left|\int_a^b X(t)\,dt\right|^2\right] - E^2\left[\int_a^b X(t)\,dt\right] \\
&= \int_a^b \int_a^b E[X(s)X(t)]\,ds\,dt - \int_a^b \int_a^b m(s)m(t)\,ds\,dt \\
&= \int_a^b \int_a^b \{E[X(s)X(t)] - m(s)m(t)\}\,ds\,dt \\
&= \int_a^b \int_a^b K(s,t)\,ds\,dt.
\end{aligned}$$

Similarly, from Eqs. 3.8 and 3.11 one may deduce 3.12.

Because $K(s,t)$ is a symmetric function [that is, $K(t,s) = K(s,t)$] one may deduce from Eq. 3.10 the following useful expression:

$$\text{Var}\left[\int_a^b X(t)\, dt\right] = 2 \int_a^b dt \int_a^t ds\, K(s,t). \tag{3.13}$$

EXAMPLE 3A

The displacement of a particle in free Brownian motion. Consider a particle moving on a straight line in response to its collisions with other particles. Suppose that the number $N(t)$ of collisions that the particle undergoes up to time t is a Poisson process with mean rate ν. Suppose that every time the particle suffers a collision it reverses its velocity. Consequently the particle's velocity is either v or $-v$, where v is a given constant. If $V(t)$ denotes the particle's velocity at time t, and if the particle's initial velocity $V(0)$ is equally likely to be v or $-v$, then

$$V(t) = V(0)(-1)^{N(t)}$$

is, up to a multiplicative constant v, the random telegraph signal. Consequently $\{V(t), t \geq 0\}$ has mean value function and covariance kernel given by

$$E[V(t)] = 0, \qquad E[V(s)V(t)] = v^2 e^{-\beta|s-t|},$$

where $\beta = 2\nu$. If $X(t)$ denotes the displacement of the particle at time t from its position at time 0, then

$$X(t) = \int_0^t V(t')\, dt'.$$

Formally,

$$E\left[\left|\int_0^t V(t')\, dt'\right|^2\right] = E\left[\int_0^t \int_0^t V(t_1)V(t_2)\, dt_1\, dt_2\right]$$
$$= \int_0^t \int_0^t E[V(t_1)V(t_2)]\, dt_1\, dt_2. \tag{3.14}$$

Since $E[V(s)\,V(t)]$ is a continuous function of (s,t), it follows from Theorem 3A that the integral $\int_0^t V(t')\, dt'$ exists (as a limit in mean square of the usual approximating sums to Riemann integrals) and has a second moment satisfying Eq. 3.14. Consequently, the mean square displacement is given by

$$E[\,|X(t)|^2\,] = 2 \int_0^t dt_1 \int_0^{t_1} dt_2 \{ v^2 e^{-\beta(t_1-t_2)} \}$$
$$= \frac{2v^2}{\beta^2}(e^{-\beta t} - 1 + \beta t).$$

For a very long or a very short time interval the mean square displacement is given by

$$E[\ |X(t)\ |^2\ ] = \frac{2v^2}{\beta} t \qquad (t \to \infty)$$

$$= v^2 t^2 \qquad (t \to 0),$$

in the sense that

$$\lim_{t \to \infty} \frac{1}{t} E[\ |X(t)\ |^2\ ] = \frac{2v^2}{\beta},$$

$$\lim_{t \to 0} \frac{1}{t^2} E[\ |X(t)\ |^2\ ] = v^2.$$

EXAMPLE 3B

   The integrated Wiener process. Let $\{X(t), t \geq 0\}$ be the Wiener process with parameter σ^2. Define a new stochastic process $\{Z(t), t \geq 0\}$ by

$$Z(t) = \int_0^t X(s)\ ds.$$

We call $\{Z(t), t \geq 0\}$ the integrated Wiener process.

   As a first step to finding the probability law of the process $\{Z(t), t \geq 0\}$ one finds its mean value function and covariance kernel. Since $E[X(s)] = 0$ for all s, and $K_X(s,t) = E[X(s)\ X(t)] = \sigma^2 s$ for $t > s \geq 0$, it follows by Theorem 3A that

$$E[Z(t)] = \int_0^t E[X(s)]\ ds = 0 \qquad \text{for all} \quad s. \tag{3.15}$$

Before determining the covariance kernel of $Z(t)$ let us determine its variance:

$$\mathrm{Var}[Z(t)] = \int_0^t dv \int_0^t du\ K_X(u,v) = 2 \int_0^t dv \int_0^v du\ K_X(u,v)$$

$$= \int_0^t dv \int_0^v du\ 2\sigma^2 u = \sigma^2 \int_0^t dv\ v^2.$$

Consequently,

$$\mathrm{Var}[Z(t)] = \sigma^2 \frac{t^3}{3}. \tag{3.16}$$

To determine the covariance kernel of $Z(t)$ we write, for $t > s \geq 0$

$$Z(t) = Z(s) + \int_s^t \{X(v) - X(s)\}\ dv + (t-s)X(s).$$

Because $\{X(t), t \geq 0\}$ has independent increments it follows that

$$E[Z(s)Z(t)] = E[Z^2(s)] + (t-s)E[Z(s)X(s)]. \tag{3.17}$$

Now,

$$E[X(s)Z(s)] = E[X(s) \int_0^s X(u) \, du] = \int_0^s E[X(u)X(s)] \, du$$
$$= \sigma^2 \int_0^s u \, du = \tfrac{1}{2} \sigma^2 s^2.$$
(3.18)

From Eqs. 3.16, 3.17, and 3.18 it follows that, for $t > s > 0$

$$E[Z(s)Z(t)] = \tfrac{1}{3} \sigma^2 s^3 + (t - s) \tfrac{1}{2} \sigma^2 s^2$$
$$= \frac{\sigma^2}{6} s^2 (3t - s).$$
(3.19)

EXAMPLE 3C

The mean and variance of a sample mean. Consider a stochastic process $\{X(t), t \geq 0\}$ which is being observed continuously over an interval $0 \leq t \leq T$. From Theorem 3A, it follows that the sample mean M_T, defined by Eq. 3.1, has mean and variance given by

$$E[M_T] = \frac{1}{T} \int_0^T m(t) \, dt,$$
(3.20)

$$\text{Var}[M_T] = \frac{2}{T^2} \int_0^T dt \int_0^t ds \, K(s,t),$$
(3.21)

where $m(t)$ and $K(s,t)$ are respectively the mean value function and the covariance kernel of the process.

Suppose, as an example, that $X(t)$ represents the number of servers in an infinite server queue who are busy at time t. It may be shown (see Section 4–5) that the mean value function and covariance kernel of $X(t)$ are given by

$$m(t) = \nu\mu \quad \text{for all} \quad t,$$
$$K(s,t) = \nu\mu e^{-(|t-s|/\mu)} \quad \text{for all} \quad s, t \geq 0,$$
(3.22)

if it is assumed that (i) customer arrivals are events of Poisson type with intensity ν, (ii) the service times of customers are independent random variables, each exponentially distributed with mean μ, and (iii) the queue has been in operation for a long time. From Eqs. 3.20, 3.21, and 3.22 it follows that the sample mean M_T has mean and variance given by

$$E[M_T] = \nu\mu,$$
$$\text{Var}[M_T] = \frac{2\nu\mu}{T^2} \int_0^T dt \int_0^t ds \, e^{-(t-s)/\mu}$$
$$= \frac{2\nu\mu^2}{T^2} \int_0^T dt(1 - e^{-t/\mu})$$
$$= \frac{2\nu\mu^2}{T} - \frac{2\nu\mu^3}{T^2} (1 - e^{-T/\mu}).$$
(3.23)

Using Eq. 3.23 and some prior information about the parameters ν and μ one can determine how long an observation interval to take in order to estimate these parameters with preassigned accuracy.

Derivatives of stochastic processes. Let $\{X(t), t \geq 0\}$ be a stochastic process with finite second moments. The derivative $X'(t)$ is defined by

$$X'(t) = \lim_{h \to 0} \frac{X(t+h) - X(t)}{h}, \tag{3.24}$$

where the limit is to be taken in the sense of convergence in mean square. It may be shown that the right-hand side in Eq. 3.24 exists as a limit in mean square if and only if the limits

$$\lim_{h \to 0} \frac{E[X(t+h) - X(t)]}{h}, \tag{3.25}$$

$$\lim_{h \to 0,\, h' \to 0} \mathrm{Cov}\left[\frac{X(t+h) - X(t)}{h}, \frac{X(t+h') - X(t)}{h'} \right]$$

exist. A sufficient condition for Eq. 3.25 to hold is that (i) the mean value function $m(t)$ be differentiable and (ii) the mixed second derivative

$$\frac{\partial^2}{\partial s\, \partial t} K(s,t) \tag{3.26}$$

exist and be continuous. Under these conditions it may be shown that

$$E[X'(t)] = E\left[\frac{d}{dt} X(t) \right] = \frac{d}{dt} E[X(t)] = m'(t), \tag{3.27}$$

$$\mathrm{Cov}[X'(s), X'(t)] = \mathrm{Cov}\left[\frac{d}{ds} X(s), \frac{d}{dt} X(t) \right] = \frac{d}{ds} \frac{d}{dt} \mathrm{Cov}[X(s), X(t)] \tag{3.28}$$

$$= \frac{\partial^2}{\partial s\, \partial t} K(s,t),$$

$$\mathrm{Cov}[X'(s), X(t)] = \mathrm{Cov}\left[\frac{d}{ds} X(s), X(t) \right] = \frac{d}{ds} \mathrm{Cov}[X(s), X(t)] \tag{3.29}$$

$$= \frac{\partial}{\partial s} K(s,t).$$

Intuitively, Eqs. 3.27, 3.28, and 3.29 state that the operations of differentiation and expectation may be interchanged, in the same way that Eqs. 3.8 and 3.9 state that the operations of integration and expectation may be interchanged.

A stochastic process $\{X(t), t \geq 0\}$ is said to be *differentiable in mean square* if for all $t \geq 0$ the derivative $X'(t)$ exists as a limit in mean square.

Now, let $\{X(t), t \geq 0\}$ be a covariance stationary process whose derivative $X'(t)$ exists for all $t \geq 0$. The question arises: is the derivative process $\{X'(t), t \geq 0\}$ covariance stationary? Let

$$K_X(s,t) = R_X(s-t) \tag{3.30}$$

be the covariance kernel of $\{X(t), t \geq 0\}$. Then the covariance kernel of $\{X'(t), t \geq 0\}$ is given by

$$K_{X'}(s,t) = -R_X''(s-t), \tag{3.31}$$

where

$$R_X''(v) = \frac{d^2}{dv^2} R_X(v) \tag{3.32}$$

is the second derivative of the covariance function $R_X(v)$. Consequently, the derivative process $\{X'(t), t \geq 0\}$ is covariance stationary with covariance function $R_{X'}(v)$ given by

$$R_{X'}(v) = -R_X''(v). \tag{3.33}$$

Note that a covariance stationary stochastic process with differentiable mean value function is differentiable in mean square if and only if its covariance function is twice differentiable. Thus a covariance stationary process with covariance function

$$R(v) = e^{-\alpha|v|}$$

is not differentiable in mean square since $R(v)$ is not differentiable at $v = 0$.

Stochastic differential equations. In many physical systems, the output $X(t)$ (which can be displacement, deflection, velocity, current) produced by an input $I(t)$ (force, voltage, current) is related to the input by a differential equation. If the input $I(t)$ is a stochastic process, then the output $X(t)$ is a stochastic process. One is thus led to consider stochastic processes $X(t)$ which arise as the solutions of linear differential equations

$$a_0(t)X^{(n)}(t) + a_1(t)X^{(n-1)}(t) + \cdots + a_n(t)X(t) = I(t) \tag{3.34}$$

in which the forcing function $I(t)$ is a stochastic process, and the coefficients $a_k(t)$ are non-random functions (which may be either constants or functions of t). The stochastic process $X(t)$ satisfying Eq. 3.34 is said to arise as the solution of a stochastic differential equation. Such stochastic processes play an important role in time series analysis and statistical communication theory (see Example 6B).

Chebyshev inequality for stochastic processes. The mean value function of a stochastic process may be regarded as a kind of average function such that the various realizations of the process are grouped around it and oscillate in its neighborhood. Under certain circumstances, one can determine neighborhoods of the mean value function in which with high probability a realization of the process will lie. For example, if $\{X(t), t \geq 0\}$ is a stochastic process differentiable in mean square one may prove the following theorem, which can be regarded as a Chebyshev inequality for stochastic processes (compare Whittle [1958]).

THEOREM 3B

Let $\{X(t), a \leq t \leq b\}$ be a stochastic process which is differentiable in mean square. Let

$$C(t) = \{E[\,|\,X(t)\,|^2\,]\}^{1/2},$$
$$C_1(t) = \{E[\,|\,X'(t)\,|^2\,]\}^{1/2}. \tag{3.35}$$

Then

$$E[\sup_{a \leq t \leq b} X^2(t)] \leq \tfrac{1}{2}\{C^2(a) + C^2(b)\} + \int_a^b C(t)C_1(t)\,dt. \tag{3.36}$$

Remark. For any $\epsilon > 0$

$$P[\sup_{a \leq t \leq b} |\,X(t)\,| > \epsilon] \leq \frac{1}{\epsilon^2}E[\sup_{a \leq t \leq b} X^2(t)].$$

Therefore Eq. 3.36 provides an upper bound to the probability of the event

$$|\,X(t)\,| > \epsilon \qquad \text{for some } t \text{ in } a \leq t \leq b.$$

If one wants an upper bound for the probability

$$P[\,|\,X(t) - m(t)\,| \leq \epsilon \qquad \text{for all } t \text{ in } a \leq t \leq b]$$

that the stochastic process be within a preassigned region about its mean value function over the interval $a \leq t \leq b$, it can be obtained by using the inequality

$$P[\,|\,X(t) - m(t)\,| \leq \epsilon \text{ for all } t \text{ in } a \leq t \leq b]$$
$$\geq 1 - \left\{\frac{\sigma^2[X(a)] + \sigma^2[X(b)]}{2\epsilon^2} + \frac{1}{\epsilon^2}\int_a^b \sigma[X(t)]\sigma[X'(t)]\,dt\right\}.$$

To illustrate these results, let us consider a covariance stationary stochastic process $\{X(t), t \geq 0\}$ with zero means and covariance function

$$R(v) = e^{-\alpha|v|}(1 + \alpha\,|\,v\,|\,),$$

where α is a positive constant. Since

$$R'(v) = -\alpha^2 v e^{-\alpha|v|},$$
$$R''(v) = \alpha^2 e^{-\alpha|v|}(\alpha|v| - 1),$$

it follows that $\{X(t), t \geq 0\}$ is differentiable and that

$$E[|X'(t)|^2] = -R''(0) = \alpha^2.$$

Since $\sigma^2[X(t)] = R(0) = 1$ for all t, it follows that for any interval a to b

$$E[\sup_{a \leq t \leq b} X^2(t)] \leq 1 + (b - a)\alpha$$

Proof of Theorem 3B. First note that

$$X^2(t) = X^2(a) + 2 \int_a^t X'(u)X(u)\,du = X^2(b) - 2\int_t^b X'(u)X(u)\,du.$$

Therefore, for any t in the interval a to b

$$2X^2(t) = X^2(a) + X^2(b) + 2\int_a^t X'(u)X(u)\,du - 2\int_t^b X'(u)X(u)\,du$$
$$\leq X^2(a) + X^2(b) + 2\int_a^b |X(u)X'(u)|\,du.$$

Consequently, we may write

$$\sup_{a \leq t \leq b} X^2(t) \leq \tfrac{1}{2}\{X^2(a) + X^2(b)\} + \int_a^b |X'(u)X(u)|\,du.$$

Taking expectations

$$E[\sup_{a \leq t \leq b} X^2(t)] \leq \tfrac{1}{2}\{E[X^2(a)] + E[X^2(b)]\} + \int_a^b E[|X'(u)X(u)|]\,du$$
$$\leq \tfrac{1}{2}\{E[X^2(a)] + E[X^2(b)]\} + \int_a^b \{E[X^2(u)]E[X'^2(u)]\}^{1/2}\,du.$$

EXERCISES

In Exercises 3.1 to 3.4, find the mean and variance of the sample mean

$$M_T = \frac{1}{T}\int_0^T X(t)\,dt,$$

where $\{X(t), t \geq 0\}$ is the stochastic process described in the exercise.

3.1 $\{X(t), t \geq 0\}$ is the Poisson process with intensity ν;

3.2 $\{X(t), t \geq 0\}$ is the random telegraph signal (defined in Section 1–4);

3.3 $\{X(t), t \geq 0\}$ is the increment process of a Poisson process (defined in Example 1C);

3.4 $\{X(t), t \geq 0\}$ is the integrated Wiener process (defined in Example 3B).

3.5 Show that the Wiener process is not differentiable in mean square.

In Exercises 3.6 to 3.11, consider a covariance stationary stochastic process $\{X(t), t \geq 0\}$ with a differentiable mean value function and the covariance function $R(v)$ given: (i) Determine whether or not the stochastic derivative $X'(t)$ exists (as a limit in mean square); (ii) if it does exist find for any real numbers t and v

(a) $E[X(t)X'(t)]$,
(b) $E[X(t)X'(t + v)]$, $E[X'(t)X(t + v)]$,
(c) $E[X'(t)X'(t + v)]$.

Note. α and β are positive constants.

3.6 $R(v) = e^{-\alpha|v|} \left\{ \cos \beta v + \dfrac{\alpha}{\beta} \sin \beta \, | \, v \, | \right\}$;

3.7 $R(v) = e^{-\alpha|v|} \{1 + \alpha \, | \, v \, |\}$;

3.8 $R(v) = \dfrac{1}{\beta} e^{-\beta|v|} - \dfrac{1}{\alpha} e^{-\alpha|v|}$, where $\alpha \geq \beta$;

3.9 $R(v) = e^{-\alpha|v|}$;

3.10 $R(v) = \dfrac{1}{\alpha^2 + v^2}$;

3.11 $R(v) = \dfrac{\sin \alpha v}{v}$.

In Exercises 3.12 to 3.14 find the covariance kernel of the stochastic process $\{Y(t), t \geq 0\}$ defined by

$$Y(t) = \frac{1}{L} \int_{t}^{t+L} X(s) \, ds,$$

where L is a positive constant and $\{X(t), t \geq 0\}$ has the covariance kernel given.

3.12 $\mathrm{Cov}[X(s), X(t)] = \sigma^2 \min (s,t)$.

3.13 $\{X(t), t \geq 0\}$ is covariance stationary with covariance function $R(v) = e^{-\alpha|v|}$.

3.14 $\{X(t), t \geq 0\}$ is covariance stationary with covariance function

$$\begin{aligned} R(v) &= 1 - | \, v \, | & &\text{if} \quad | \, v \, | \leq 1, \\ &= 0, & &\text{otherwise.} \end{aligned}$$

3-4   NORMAL PROCESSES

The basic role which the normal (or Gaussian) distribution plays in probability theory arises from the fact that

(i) many random variables which arise in applications of probability theory may be considered to be approximately normally distributed;

(ii) the normal probability law is particularly tractable and convenient to deal with.

Similarly, in the theory of stochastic processes normal processes play a central role since

(i) many important stochastic processes can be approximated by normal processes;

(ii) many questions can be answered for normal processes more easily than for other processes.

Jointly normally distributed random variables. One says that the n random variables $X_1, \cdots, X_n$ are jointly normally distributed if their joint characteristic function is given, for any real numbers $u_1, \cdots, u_n$, by

$$\varphi_{X_1,\ldots,X_n}(u_1, \cdots, u_n) = \exp\left\{ i \sum_{j=1}^{n} u_j m_j - \tfrac{1}{2} \sum_{j,k=1}^{n} u_j K_{jk} u_k \right\} \quad (4.1)$$

in which, for $j, k = 1, 2, \cdots, n$

$$m_j = E[X_j], \quad K_{jk} = \mathrm{Cov}[X_j, X_k]. \quad (4.2)$$

If the matrix of covariances

$$K = \begin{bmatrix} K_{11} & K_{12} & \cdots & K_{1n} \\ K_{21} & K_{22} & \cdots & K_{2n} \\ \cdots\cdots\cdots\cdots\cdots\cdots \\ K_{n1} & K_{n2} & \cdots & K_{nn} \end{bmatrix} \quad (4.3)$$

possesses an inverse matrix

$$K^{-1} = \begin{bmatrix} K^{11} & \cdots & K^{1n} \\ \vdots & & \vdots \\ K^{n1} & \cdots & K^{nn} \end{bmatrix},$$

then it can be shown that $X_1, \cdots, X_n$ possess a joint probability density given, for any real numbers $x_1, \cdots, x_n$, by

$$f_{X_1,\ldots,X_n}(x_1, x_2, \cdots, x_n)$$
$$= \frac{1}{(2\pi)^{n/2}} \frac{1}{|K|^{1/2}} \exp\left\{ -\tfrac{1}{2} \sum_{j,k=1}^{n} (x_j - m_j) K^{jk} (x_k - m_k) \right\}, \quad (4.4)$$

where $| K |$ is the determinant of the matrix K. To prove Eq. 4.4 one uses the following inversion formula for the joint probability density of n random variables $X_1, \cdots, X_n$:

$$f_{X_1,\ldots,X_n}(x_1, x_2, \cdots, x_n)$$
$$= \frac{1}{(2\pi)^n} \int_{-\infty}^{\infty} du_1 \cdots \int_{-\infty}^{\infty} du_n \, e^{-i(x_1u_1+\ldots+x_nu_n)} \varphi_{X_1,\ldots,X_n}(u_1, \cdots, u_n) \quad (4.5)$$

in the case that the integrand in Eq. 4.5 is absolutely integrable. Consequently, to prove Eq. 4.4 it suffices to prove that for any real numbers $y_1, \cdots, y_n$ and positive definite matrix $\{K_{jk}\}$ with inverse matrix $\{K^{jk}\}$

$$\int_{-\infty}^{\infty} \cdots \int_{-\infty}^{\infty} \exp\left\{ i \sum_{j=1}^{n} u_j y_j - \tfrac{1}{2} \sum_{j,k=1}^{n} u_j K_{jk} u_k \right\} du_1 \cdots du_n$$
$$= \frac{(2\pi)^{n/2}}{\sqrt{| K |}} \exp\left\{ -\tfrac{1}{2} \sum_{j,k=1}^{n} y_j K^{jk} y_k \right\} . \quad (4.6)$$

For a derivation of Eq. 4.6 the reader may consult Cramér (1946), p. 118, or Friedman (1956), p. 105.

Definition of a normal process. A stochastic process $\{X(t), t \,\epsilon\, T\}$ is said to be a *normal* process if for any integer n and any subset $\{t_1, t_2, \cdots, t_n\}$ of T the n random variables $X(t_1), \cdots X(t_n)$ are jointly normally distributed in the sense that their joint characteristic function is given, for any real numbers $u_1, u_2, \cdots, u_n$, by

$$\varphi_{X(t_1),X(t_2),\ldots,X(t_n)}(u_1, u_2, \cdots, u_n)$$
$$= E[\exp i\{u_1X(t_1) + \cdots + u_nX(t_n)\}]$$
$$= \exp\left\{ i \sum_{j=1}^{n} u_j E[X(t_j)] - \tfrac{1}{2} \sum_{j,k=1}^{n} u_j u_k \, \text{Cov}[X(t_j), X(t_k)] \right\} . \quad (4.7)$$

From Eq. 4.7 one sees that for a normal process one obtains a knowledge of the complete probability law of the process from a knowledge of the mean value function $E[X(t)]$ and the covariance kernel $\text{Cov}[X(s), X(t)]$.

The Wiener process is an example of a normal process. In view of Eq. 4.7, to prove that a Wiener process $\{X(t), t \geq 0\}$ is a normal process it suffices to show that for any n time points $t_1, \cdots, t_n$, the joint characteristic function of $X(t_1), \cdots, X(t_n)$ is of the form of Eq. 4.7:

$$\varphi_{X(t_1),\ldots,X(t_n)}(u_1, \cdots, u_n) = \exp\left\{ -\tfrac{1}{2}\sigma^2 \sum_{j,k=1}^{n} u_j u_k \min (t_j,t_k) \right\} .$$

In practice the fact that a stochastic process is a normal process is not usually established by using the definition, Eq. 4.7, but by using Theorems 4A and 4B.

Linear operations on normal processes. The usefulness of normal processes follows from the fact that any stochastic process, such as $\int_0^t X(s)\,ds$, $X'(t)$, or $X(t+1) - X(t)$, derived by means of linear operations from a normal process, is itself a normal process. The validity of this assertion is an immediate consequence of the following theorems.

THEOREM 4A

A fundamental property of normally distributed random variables. Let $X_1, \cdots, X_n$ be jointly normally distributed random variables. Let $Y_1, \cdots, Y_m$ be linear functions of $X_1, \cdots, X_n$:

$$
\begin{aligned}
Y_1 &= c_{11}X_1 + c_{12}X_2 + \cdots + c_{1n}X_n \\
&\;\;\vdots \qquad\qquad\qquad\qquad\;\; \vdots \\
Y_m &= c_{m1}X_1 + c_{m2}X_2 + \cdots + c_{mn}X_n.
\end{aligned}
\tag{4.8}
$$

Then $Y_1, \cdots, Y_m$ are jointly normally distributed with joint characteristic function

$$
\varphi_{Y_1,\ldots,Y_m}(u_1, \cdots, u_m) = \exp\left\{ i\sum_{j=1}^m u_j E[Y_j] - \tfrac{1}{2} \sum_{j,k=1}^m u_j u_k \,\mathrm{Cov}[Y_j, Y_k] \right\},
\tag{4.9}
$$

in which for $j, k = 1, 2, \cdots, m$

$$
E[Y_j] = c_{j1}E[X_1] + \cdots + c_{jn}E[X_n],
\tag{4.10}
$$

$$
\mathrm{Cov}[Y_j, Y_k] = \sum_{s,\,t=1}^n c_{js}c_{kt}\,\mathrm{Cov}[X_s, X_t].
\tag{4.11}
$$

Proof. First verify that

$$
\sum_{j=1}^m u_j Y_j = \sum_{j=1}^m u_j \sum_{t=1}^n c_{jt}X_t = \sum_{t=1}^n \left\{ \sum_{j=1}^m u_j c_{jt} \right\} X_t.
\tag{4.12}
$$

Let $v_t = \sum_{j=1}^m u_j c_{jt}$ for $t = 1, \cdots, n$. From Eq. 4.12 it follows that

$$
\varphi_{Y_1,\cdots,Y_m}(u_1, \cdots, u_m) = \varphi_{X_1,\cdots,X_n}(v_1, \cdots, v_n).
\tag{4.13}
$$

Evaluating the right-hand side of Eq. 4.13 by Eq. 4.1 one obtains the desired conclusion.

EXAMPLE 4A

The fact that the Wiener process $\{X(t), t \geq 0\}$ is a normal process follows from Theorem 4A by the following argument. Given n time points $t_1 < t_2 < \cdots < t_n$, let $V_1 = X(t_1)$, $V_2 = X(t_2) - X(t_1)$, $\cdots$, $V_n = X(t_n) - X(t_{n-1})$. By hypothesis, $V_1, V_2, \cdots, V_n$ are independent and normally distributed. Therefore, they are jointly normally distributed. One may write the random variables $X(t_1), \cdots, X(t_n)$ as a linear combination of $V_1, \cdots, V_n$. Therefore, $X(t_1), \cdots, X(t_n)$ are jointly normally distributed, and $\{X(t), t \geq 0\}$ is a normal process.

THEOREM 4B

Let $\{Z_n\}$ be a sequence of normally distributed random variables which converge in mean square to a random variable Z. Then Z is normally distributed.

Proof. To prove that Z is normal it suffices to prove that the characteristic function of Z may be expressed in terms of its mean and variance by

$$\varphi_Z(u) = \exp\{iuE[Z] - \tfrac{1}{2}u^2 \operatorname{Var}[Z]\}.$$

Consequently, to prove that Z is normal it suffices to prove that

$$\lim_{n \to \infty} \varphi_{Z_n}(u) = \varphi_Z(u), \tag{4.14}$$

$$\lim_{n \to \infty} E[Z_n] = E[Z], \quad \lim_{n \to \infty} \operatorname{Var}[Z_n] = \operatorname{Var}[Z]. \tag{4.15}$$

From the fact that

$$\lim_{n \to \infty} E[\,|\,Z_n - Z\,|^2\,] = 0$$

it follows that, as n tends to ∞,

$$|\,E[Z_n] - E[Z]\,| \leq E[\,|\,Z_n - Z\,|\,] \leq E^{1/2}[\,|\,Z_n - Z\,|^2\,] \to 0,$$
$$|\,\varphi_{Z_n}(u) - \varphi_Z(u)\,| = |\,E[e^{iuZ_n} - e^{iuZ}]\,| \leq |\,u\,|\,E[\,|\,Z_n - Z\,|\,] \to 0,$$
$$|\,\sigma[Z_n] - \sigma[Z]\,|^2 \leq \sigma^2[Z_n - Z] = E[\,|\,Z_n - Z\,|^2\,] + E^2[Z_n - Z] \to 0.$$

The proof of Theorem 4B is complete.

EXAMPLE 4B

Let $\{X(t), -\infty < t < \infty\}$ be a covariance stationary normal process with zero means and covariance function $R(v)$. Assume that $R(v)$ is four times differentiable. Then $X(t)$ is twice differentiable in mean square and, for any t, the random variables $X(t)$, $X'(t)$, and $X''(t)$ are jointly normally distributed with means 0 and covariance matrix

$$\begin{bmatrix} E[X^2(t)] & E[X(t)X'(t)] & E[X(t)X''(t)] \\ E[X'(t)X(t)] & E[\{X'(t)\}^2] & E[X'(t)X''(t)] \\ E[X''(t)X(t)] & E[X''(t)X'(t)] & E[\{X''(t)\}^2] \end{bmatrix}$$
$$= \begin{bmatrix} R(0) & 0 & -R^{(2)}(0) \\ 0 & -R^{(2)}(0) & 0 \\ -R^{(2)}(0) & 0 & R^{(4)}(0) \end{bmatrix} \quad (4.16)$$

where, for $k = 1, 2, \cdots$, $R^{(k)}(v)$ denotes the kth derivative of $R(v)$.

Non-linear operations on normal processes. Stochastic processes $\{Y(t), t > 0\}$ which arise from normal processes by means of non-linear operations are of great interest in applications where one is concerned with the effect of passing a stochastic process through a device which performs a non-linear operation on its input. Here we can only discuss the simplest problem associated with non-linear operations on normal processes, namely the problem of finding the mean value function and covariance kernel of a stochastic process $\{Y(t), t \geq 0\}$ obtained from a normal process by means of quadratic operations.

THEOREM 4C

Let $\{X(t), t \geq 0\}$ be a normal process with zero means and co-variance kernel $K(s,t)$. Then for any non-negative numbers s, t, and h

$$E[X^2(t)] = K(t,t), \quad (4.17)$$

$$\text{Var}[X^2(t)] = 2K^2(t,t), \quad (4.18)$$

$$\text{Cov}[X^2(s), X^2(t)] = 2K^2(s,t), \quad (4.19)$$

$$E[X(t)X(t+h)] = K(t,t+h), \quad (4.17')$$

$$\text{Var}[X(t)X(t+h)] = K(t,t)K(t+h, t+h) + K^2(t,t+h), \quad (4.18')$$

$$\text{Cov}[X(s)X(s+h), X(t)X(t+h)]$$
$$= K(s,t)K(s+h, t+h) + K(s,t+h)K(s+h,t). \quad (4.19')$$

Proof. That Eqs. 4.17 and 4.17' hold is immediate. To prove Eqs. 4.18, 4.19, 4.18', and 4.19' it suffices to prove 4.19'. Now,

$$\text{Cov}[X(s)X(s+h), X(t)X(t+h)] = E[X(s)X(s+h)X(t)X(t+h)]$$
$$- E[X(s)X(s+h)]E[X(t)X(t+h)]. \quad (4.20)$$

To evaluate the first term on the right-hand side of Eq. 4.20, we use the following *general formula for the fourth product moment of normal random variables with zero means*:

$$E[X_1X_2X_3X_4] = E[X_1X_2]E[X_3X_4]$$
$$+ E[X_1X_3]E[X_2X_4]$$
$$+ E[X_1X_4]E[X_2X_3]. \qquad (4.21)$$

To prove Eq. 4.21 one uses the fact that

$$E[X_1X_2X_3X_4] = \frac{\partial^4}{\partial u_1\,\partial u_2\,\partial u_3\,\partial u_4}\,\varphi_{X_1,X_2,X_3,X_4}(0, 0, 0, 0). \qquad (4.22)$$

Now,

$$\varphi_{X_1,X_2,X_3,X_4}(u_1, u_2, u_3, u_4) = \exp\left\{-\tfrac{1}{2}\sum_{i,j=1}^{4} u_i u_j E[X_i X_j]\right\}. \qquad (4.23)$$

For ease of writing, we let φ denote the right-hand side of Eq. 4.23 and define

$$L_i = \sum_{j=1}^{4} u_j E[X_i X_j].$$

Then

$$\frac{\partial\varphi}{\partial u_1} = -\varphi L_1,$$

$$\frac{\partial^2\varphi}{\partial u_2\,\partial u_1} = \varphi\{L_1 L_2 - E[X_1 X_2]\},$$

$$\frac{\partial^3\varphi}{\partial u_3\,\partial u_2\,\partial u_1} = \varphi\{-L_1 L_2 L_3 + L_3 E[X_1 X_2] + L_2 E[X_1 X_3] + L_1 E[X_2 X_3]\},$$

$$\frac{\partial^4\varphi}{\partial u_4\,\partial u_3\,\partial u_2\,\partial u_1} = \varphi\{L_1 L_2 L_3 L_4 - L_1 L_2 E[X_3 X_4] - L_1 L_3 E[X_2 X_4]$$
$$- L_1 L_4 E[X_2 X_3] - L_2 L_3 E[X_1 X_4]$$
$$- L_2 L_4 E[X_1 X_3] - L_3 L_4 E[X_1 X_2]$$
$$+ E[X_1 X_2]E[X_3 X_4] + E[X_1 X_3]E[X_2 X_4]$$
$$+ E[X_1 X_4]E[X_2 X_3]\}. \qquad (4.24)$$

Combining Eqs. 4.22 and 4.24, one obtains Eq. 4.21.

It should be noted that Eq. 4.21 holds even if some of the random variables X_1, X_2, X_3, X_4 coincide. An extension of Eq. 4.21 is given in Complement 4E.

From Eq. 4.21 it follows that the right-hand side of Eq. 4.20 is equal to

$$E[X(s)X(t)]E[X(s+h)X(t+h)] + E[X(s)X(t+h)]E[X(s+h)X(t)].$$

The proof of Eq. 4.19' is now complete.

EXAMPLE 4C

Let $\{X(t), -\infty < t < \infty\}$ be a covariance stationary normal process with zero means and covariance function $R(v)$. Then

$$\{X^2(t), -\infty < t < \infty\}$$

is a covariance stationary process with mean value function

$$m_{X^2}(t) = E[X^2(t)] = R(0) \tag{4.25}$$

and covariance function

$$R_{X^2}(v) = \text{Cov}[X^2(t), X^2(t+v)] = 2R^2(v). \tag{4.26}$$

A variety of questions can be more simply treated for normal processes than for other kinds of stochastic processes. Examples of such questions are the problem of determining the distribution of the number of zeros of a stochastic process $\{X(t), t \geq 0\}$ in an interval a to b, or the problem of determining the distribution of the supremum

$$\sup_{a \leq t \leq b} |X(t)|$$

of a stochastic process over an interval a to b. These questions arise naturally in radio propagation, vibration, fatigue, and reliability studies. Even for a normal process the treatment of such questions is far from completely resolved. An exposition of these topics would require a level of mathematical sophistication beyond the scope of this book. In Example 5B it is shown how such questions also have applications in statistics.

COMPLEMENTS

4A For many purposes one desires to express the joint probability density function of jointly normally distributed random variables $X_1, X_2, \cdots, X_n$ in terms of their correlations ρ_{jk} and variances σ_j^2:

$$\rho_{jk} = \frac{K_{jk}}{\sigma_j \sigma_k}, \quad \sigma_j^2 = \text{Var}[X_j] = K_{jj}.$$

Show that in terms of the inverse

$$\rho^{-1} = \begin{bmatrix} \rho^{11} & \cdots & \rho^{1n} \\ \vdots & & \vdots \\ \rho^{n1} & \cdots & \rho^{nn} \end{bmatrix}$$

of the matrix ρ of correlations, one has the following expression for the joint probability density function of n jointly normal random variables:

$$f_{X_1,X_2,\ldots,X_n}(x_1, x_2, \cdots, x_n) = \frac{1}{(2\pi)^{n/2}\sigma_1\sigma_2\cdots\sigma_n \mid \rho \mid^{1/2}}$$

$$\times \exp\left\{-\frac{1}{2}\sum_{j,k=1}^{n}\left(\frac{x_j - m_j}{\sigma_j}\right)\rho^{jk}\left(\frac{x_k - m_k}{\sigma_k}\right)\right\},$$

where $\mid \rho \mid$ is the determinant of the matrix ρ.

4B Show that for jointly normal random variables X_1 and X_2, with zero means, variances $\sigma_1{}^2$ and $\sigma_2{}^2$ respectively, and correlation coefficient ρ,

$$E[e^{i(u_1X_1{}^2+u_2X_2{}^2)}] = (1 - 2i(u_1\sigma_1{}^2 + u_2\sigma_2{}^2) - 4u_1u_2(1 - \rho^2)\sigma_1{}^2\sigma_2{}^2)^{-1/2}. \tag{4.27}$$

Hint. Show and use the fact that

$$\frac{1}{2\pi}\int_{-\infty}^{\infty}\int_{-\infty}^{\infty}e^{-(1/2)(ax^2+2bxy+cy^2)}\,dx\,dy = (ac - b^2)^{-1/2}.$$

4C Show that for jointly normal random variables X_1 and X_2 with zero means

$$\mathrm{Cov}[X_1{}^2,X_2{}^2] = 2\{E[X_1X_2]\}^2.$$

4D *The fourth product moment of squares of normal random variables.* Show that if X_1, X_2, X_3, X_4 are jointly normally distributed with zero means, and if $U_j = X_j{}^2 - E[X_j{}^2]$, $j = 1, \cdots, 4$, then

$$\begin{aligned}
E[U_1U_2U_3U_4] = {}&4E^2[X_1X_2]E^2[X_3X_4]+4E^2[X_1X_3]E^2[X_2X_4]\\
&+ 4E^2[X_1X_4]E^2[X_2X_3]\\
&+ 16E[X_1X_2]E[X_1X_4]E[X_2X_3]E[X_3X_4]\\
&+ 16E[X_1X_2]E[X_1X_3]E[X_2X_4]E[X_3X_4]\\
&+ 16E[X_1X_3]E[X_1X_4]E[X_2X_3]E[X_2X_4].
\end{aligned}$$

4E *The higher moments of a normal process.* Let $\{X(t), t \in T\}$ be a normal process with zero mean value function, $E[X(t)] = 0$. Show that all odd order moments of $X(t)$ vanish, and that the even order moments may be expressed in terms of the second order moments by the following formula. Let n be an even integer, and let $t_1, \cdots, t_n$ be points in T, some of which may coincide. Then

$$E[X(t_1) \cdots X(t_n)] = \Sigma E[X(t_{i_1})X(t_{i_2})] \cdots E[X(t_{i_{n-1}})X(t_{i_n})], \tag{4.28}$$

in which the sum is taken over all possible ways of dividing the n points into $n/2$ combinations of pairs. The number of terms in the summation is equal to $1 \cdot 3 \cdot 5 \cdots (n - 3)(n - 1)$.

Hint. Differentiate the characteristic function. For references to the fundamental role which Eq. 4.28 plays in the theory of Brownian motion, see Wang and Uhlenbeck (1945), p. 332, Eq. 42.

4F *A class of normal processes related to the Wiener process.* Let $\{X(t), t \geq 0\}$ be a normal process, with zero means, $E[X(t)] = 0$, and covariance kernel of the form

$$K(s,t) = E[X(s)X(t)] = u(s)v(t) \quad \text{for} \quad s \leq t$$

for some continuous functions $u(\cdot)$ and $v(\cdot)$. If the ratio

$$\frac{u(t)}{v(t)} = a(t)$$

is continuous and monotone increasing with inverse function $a_1(t)$, then the process

$$Y(t) = \frac{X(a_1(t))}{v(a_1(t))}$$

is a Wiener process (with $\sigma = 1$). As an example, suppose $K(s,t) = s(1-t)$ for $s < t$. Then

$$u(s) = s, \ v(t) = 1 - t, \ a_1(t) = t/(t+1), \ Y(t) = (t+1)X(t/\{1+t\}).$$

(These results are due to Doob [1949 b].)

EXERCISES

For each of the stochastic processes $\{X(t), t \geq 0\}$ defined in Exercises 4.1 to 4.5,

   (i) compute $m(t) = E[X(t)]$,
   (ii) compute $K(s,t) = \text{Cov}[X(s), X(t)]$,
   (iii) state $R(v)$ if the process is covariance stationary,
   (iv) state whether or not the process is a normal process.

In Exercises 4.1 to 4.3, let $\{W(t), t \geq 0\}$ be the Wiener process with parameter σ^2.

4.1 $\begin{aligned} X(t) &= (1-t)W(t/\{1-t\}) &&\text{for} \ \ 0 < t < 1, \\ &= 0 &&\text{for} \ \ t \geq 1. \end{aligned}$

4.2 $X(t) = e^{-\beta t}W(e^{2\beta t})$, in which β is a positive constant.

4.3 $X(t) = W^2(t)$.

4.4 *A process with perfect correlation.* $X(t) = X$, where X is $N(m,\sigma^2)$.

4.5 $X(t) = Y(t+1) - Y(t)$, where $\{Y(t), t \geq 0\}$ is a normal process with $E[Y(t)] = \alpha + \beta t$, $\text{Cov}[Y(t), Y(t+v)] = e^{-\lambda|v|}$.

4.6 Give an example to show that two random variables X_1 and X_2 may be individually normally distributed without being jointly normally distributed.

   Hint. Let Y_1 and Y_2 be independent $N(0,1)$. Define

$$\begin{aligned} (X_1, X_2) &= (Y_1, \ |\ Y_2\ |) &&\text{if} \ \ Y_1 \geq 0, \\ &= (Y_1, -\ |\ Y_2\ |) &&\text{if} \ \ Y_1 < 0. \end{aligned}$$

4.7 *The Ornstein-Uhlenbeck process.* Models for Brownian motion which are somewhat more realistic than the Wiener process can be constructed.

One such model is the Ornstein-Uhlenbeck process (after Uhlenbeck and Ornstein [1930]; see also Doob [1942]). A stochastic process $\{X(t), t \geq 0\}$ is said to be an Ornstein-Uhlenbeck process with parameters $\alpha > 0$ and $\beta > 0$ if it is a normal process satisfying

$$E[X(t)] = 0, \quad \text{Cov}[X(s), X(t)] = \alpha e^{-\beta|s-t|}.$$

Using Complement 4F, express the Ornstein-Uhlenbeck process in terms of the Wiener process.

3-5   NORMAL PROCESSES AS LIMITS OF STOCHASTIC PROCESSES

It is well known that a random variable X which is Poisson distributed with mean λ is approximately normally distributed if λ is large. More precisely, let

$$X^* = \frac{X - E[X]}{\sigma[X]} \tag{5.1}$$

be the standardization of X (that is, X^* is that linear function of X which has mean 0 and variance 1). Then X^* is approximately normally distributed with mean 0 and variance 1 in the sense that, for every real number u,

$$\lim_{\lambda \to \infty} \log \varphi_{X^*}(u) = -\tfrac{1}{2}u^2. \tag{5.2}$$

To prove Eq. 5.2 we use the following observation. Let X be a random variable whose log-characteristic function possesses the following expansion for any real number u,

$$\log \varphi_X(u) = imu - \tfrac{1}{2}u^2\sigma^2 + K\theta \mid u \mid^3, \tag{5.3}$$

where $m = E[X]$, $\sigma^2 = \text{Var}[X]$, K is some positive constant independent of u and θ denotes a quantity depending on u of modulus less than 1. Then the standardized random variable $X^* = (X - m)/\sigma$ has log-characteristic function

$$\log \varphi_{X^*}(u) = -\tfrac{1}{2}u^2 + \frac{K}{\sigma^3}\theta \mid u \mid^3. \tag{5.4}$$

That Eq. 5.3 implies Eq. 5.4 follows immediately from the fact that

$$\varphi_{X^*}(u) = \varphi_X\left(\frac{u}{\sigma}\right)e^{-i(m/\sigma)u}.$$

To prove that a Poisson random variable X whose mean λ is large is approximately normal, one may argue as follows. The log-characteristic function of X is

$$\log \varphi_X(u) = \lambda(e^{iu} - 1) = iu\lambda - \tfrac{1}{2}u^2\lambda + \lambda\theta \mid u \mid^3. \tag{5.5}$$

Therefore, Eq. 5.3 holds with $\sigma^2 = \lambda$ and $K = \lambda$. By Eq. 5.4

$$\log \varphi_{X^*}(u) = -\tfrac{1}{2}u^2 + \frac{1}{\sqrt{\lambda}} \theta \mid u \mid^3. \tag{5.6}$$

From Eq. 5.6 one sees that X^* is approximately $N(0,1)$ for large values of λ.

Approximately normal stochastic processes. A stochastic process $\{X(t), t \in T\}$ is said to be approximately normal if for every finite set of indices $\{t_1, t_2, \cdots, t_n\}$ the random variables $X(t_1), X(t_2), \cdots, X(t_n)$ are approximately jointly normally distributed. More precisely, a stochastic process $\{X(t), t \in T\}$ with finite second moments, whose probability law depends on some parameter, is said to be asymptotically normal as that parameter tends to a given limit, if the following condition holds for every finite set of indices $t_1, t_2, \cdots, t_n$ in the index set T: the joint characteristic function of the standardized random variables $X^*(t_1), \cdots, X^*(t_n)$,

$$X^*(t) = \frac{X(t) - E[X(t)]}{\sigma[X(t)]}, \tag{5.7}$$

satisfies (for every $u_1, \cdots, u_n$)

$$\log \varphi_{X^*(t_1), X^*(t_2), \ldots, X^*(t_n)}(u_1, u_2, \cdots, u_n) \to -\tfrac{1}{2}\sum_{j,k=1}^{n} u_j\rho(X(t_j), X(t_k))u_k, \tag{5.8}$$

as the parameter tends to the given limit, where

$$\rho(X(t_j), X(t_k)) = \frac{\text{Cov}[X(t_j), X(t_k)]}{\sigma[X(t_j)]\sigma[X(t_k)]} \tag{5.9}$$

is the correlation between $X(t_j)$ and $X(t_k)$.

EXAMPLE 5A

The Wiener process as a model for Brownian motion. We consider the following model for the motion of a particle on a line arising from its numerous random impacts with other particles. At time 0 let the particle be at 0. Assume that the times between successive impacts are independent random variables, exponentially distributed with mean $1/\nu$. It may be shown (see Section 5-2) that the number $N(t)$ of impacts in the time interval 0 to t is a Poisson process, with intensity ν (that is, the mean number of impacts per unit time is ν). Assume that the effect of an impact on the particle is to change its position by either a or $-a$, each possibility having probability 1/2. The position $X(t)$ of the particle at time t may be represented by

$$X(t) = \sum_{n=1}^{N(t)} Y_n,$$ (5.10)

where Y_n is the particle's change of position as a result of the nth impact. The random variables $\{Y_n\}$ are assumed to be independent random variables identically distributed as a random variable Y such that

$$P[Y = a] = P[Y = -a] = \tfrac{1}{2},$$

where a is a given positive number. It may be shown (see the theory of compound Poisson processes in Section 4–2) that $\{X(t), t \geq 0\}$ is a stochastic process with stationary independent increments, and one-dimensional characteristic function

$$\log \varphi_{X(t)}(u) = \nu t\, E[e^{iuY} - 1].$$ (5.11)

We now use the expansion

$$E[e^{iuY} - 1] = iuE[Y] - \tfrac{1}{2}u^2 E[Y^2] + \theta \mid u \mid^3 E[\mid Y \mid^3]$$
$$= -\tfrac{1}{2}u^2 a^2 + \theta \mid u \mid^3 a^3,$$

where θ denotes some quantity such that $\mid \theta \mid \leq 1$. If we define

$$\sigma^2 = a^2\nu,$$

then the log-characteristic function of $X(t)$ is given by

$$\log \varphi_{X(t)}(u) = -\tfrac{1}{2}u^2\sigma^2 t + a\theta \mid u \mid^3 \sigma^2 t.$$ (5.12)

Now let the density ν of impacts tend to infinity, and the magnitude a of each impact tend to 0, in such a way that the product νa^2 (representing the total mean square displacement of the particle per unit time) is constant. Under these conditions it follows that

$$\log \varphi_{X(t)}(u) \quad \text{tends to} \quad -\tfrac{1}{2}u^2\sigma^2 t.$$

Therefore $\{X(t), t \geq 0\}$ is approximately the Wiener process, since it is approximately normally distributed and has stationary independent increments.

EXAMPLE 5B

The use of normal processes to find the asymptotic distribution of certain goodness of fit tests for distribution functions. Given n independent observations $X_1, \cdots, X_n$ with a common (unknown) continuous distribution function $F(x)$, the hypothesis that a given $F(x)$ is the true distribution function can be tested by comparing the sample distribution function $F_n(x)$ with $F(x)$. We define $F_n(x)$ to be the fraction of observations less than or equal to x. Two possible measures of the deviation between $F_n(x)$ and $F(x)$ are the Kolmogorov-Smirnov statistic

$$D_n = \sqrt{n} \sup_{-\infty < x < \infty} | F_n(x) - F(x) |$$

and the Cramér-von Mises statistic

$$W_n^2 = n \int_{-\infty}^{\infty} [F_n(x) - F(x)]^2 \, dF(x).$$

If the observations are arranged in increasing order of magnitude, $x_1 \leq x_2 \leq \cdots \leq x_n$, it may be shown that

$$\sqrt{n} \, D_n = \max_{j=1,\ldots,n} [nF(x_j) - (j-1), j - nF(x_j)],$$

$$W_n^2 = \frac{1}{12n} + \sum_{j=1}^{n} \left[F(x_j) - \frac{2j-1}{2n} \right]^2.$$

    The distributions of the random variables D_n and W_n^2, under the (null) hypothesis that the observations $X_1, \cdots, X_n$ have $F(x)$ as their distribution function, does not depend on $F(x)$, if it is assumed that $F(x)$ is continuous, since then $F(X_1), \cdots, F(X_n)$ are uniformly distributed on the interval 0 to 1 (see *Mod Prob*, p. 314). To derive the distribution theory of D_n and W_n^2, we may therefore assume that $F(x)$ is the uniform distribution on the interval 0 to 1 (so that $F(x) = x$ for $0 \leq x \leq 1$), and that the independent random variables $X_1, \cdots, X_n$ are uniformly distributed on the interval 0 to 1. Let us now define for $0 \leq t \leq 1$

$$Y_n(t) = \sqrt{n} \, [F_n(t) - t] \tag{5.13}$$

where $F_n(t)$ is the fraction of observations (among $X_1, \cdots, X_n$) which are less than or equal to t. To give a symbolic representation of $F_n(t)$, define

$$h_t(x) = 1 \quad \text{if} \quad x \leq t$$
$$= 0 \quad \text{if} \quad x > t.$$

Then

$$Y_n(t) = \frac{1}{\sqrt{n}} \sum_{j=1}^{n} \{ h_t(X_j) - t \}. \tag{5.14}$$

It is easily verified that for $0 \leq s, t \leq 1$ and $j, k = 1, \cdots, n$

$$E[h_t(X_j)] = t, \qquad (5.15)$$

$$
\begin{aligned}
E[h_s(X_j)h_t(X_k)] &= \min\,(s,t) & \text{if } \quad j = k \\
&= st & \text{if } \quad j \neq k, \\
\mathrm{Cov}[h_s(X_j),\, h_t(X_k)] &= \min\,(s,t) - st & \text{if } \quad j = k \\
&= 0 & \text{if } \quad j \neq k.
\end{aligned}
$$

Therefore, $E[Y_n(t)] = 0$, and

$$
\begin{aligned}
\mathrm{Cov}[Y_n(s),\, Y_n(t)] &= \frac{1}{n}\sum_{j,k=1}^{n}\mathrm{Cov}[h_s(X_j),\, h_t(X_k)] \qquad (5.16) \\
&= \min\,(s,t) - st \\
&= s(1-t) \quad \text{if } \ s \le t.
\end{aligned}
$$

Let $\{W(t),\, t \ge 0\}$ be a Wiener process. We define a new stochastic process $\{Y(t),\, 0 \le t \le 1\}$ by

$$Y(t) = (1-t)W\!\left(\frac{t}{1-t}\right). \qquad (5.17)$$

It is clear that $Y(t)$ is normal, $Y(0) = 0$, $E[Y(t)] = 0$, and

$$
\begin{aligned}
\mathrm{Cov}[Y(s),\, Y(t)] &= (1-s)(1-t)\,\mathrm{Cov}\!\left[W\!\left(\frac{s}{1-s}\right),\, W\!\left(\frac{t}{1-t}\right)\right] \qquad (5.18) \\
&= (1-s)(1-t)\min\left(\frac{s}{1-s},\,\frac{t}{1-t}\right) \\
&= s(1-t) \quad \text{if } \ s \le t.
\end{aligned}
$$

Comparing Eqs. 5.16 and 5.18 one sees that, for each n, the stochastic process $Y_n(t)$ has the same mean value function and covariance kernel as a normal process $Y(t)$ which is essentially a Wiener process except for a change of time scale. The only difference between $Y_n(t)$ and $Y(t)$ is that $Y_n(t)$ is *not* a normal process. However, since $Y_n(t)$ is for each t a sum of independent identically distributed random variables with finite variances, it follows by the central limit theorem that $Y_n(t)$ is *asymptotically a normal process* as n tends to ∞, in the sense that for any integer k, any positive numbers $t_1, \cdots, t_k$, and real numbers $u_1, \cdots, u_k$

$$
\begin{aligned}
\lim_{n\to\infty}\varphi_{Y_n(t_1),\ldots,Y_n(t_k)}(u_1,\cdots,u_k) &= \exp\left\{-\tfrac{1}{2}\sum_{i,j=1}^{k}u_i u_j\,[\min\,(t_i,t_j) - t_i t_j]\right\} \\
&= \varphi_{Y(t_1),\ldots,Y(t_k)}(u_1,\cdots,u_k). \qquad (5.19)
\end{aligned}
$$

It seems reasonable that consequently the probability distribution of the Kolmogorov-Smirnov statistic

$$D_n = \max_{0 \leq t \leq 1} |Y_n(t)|$$

tends as $n \to \infty$ to the probability distribution of

$$D = \max_{0 \leq t \leq 1} |Y(t)|,$$

and that the probability distribution of the Cramér-von Mises statistic

$$W_n^2 = \int_0^1 Y_n^2(t)\, dt$$

tends as $n \to \infty$ to the probability distribution of

$$W^2 = \int_0^1 Y^2(t)\, dt.$$

It may be shown that these assertions are correct (see Donsker [1950]). Consequently, one way to find the asymptotic distribution of the goodness of fit statistics D_n and W_n is by finding the probability distributions of D and W which are functionals defined on a normal process $Y(t)$ intimately related to the Wiener process. The important problem of determining the distribution of functionals defined on stochastic processes is beyond the scope of this book (for a discussion of this problem and references to the literature see Kac [1951] and Anderson and Darling [1952]).

For discussions of the Kolmogorov-Smirnov and Cramér-von Mises tests, the reader is referred respectively to Birnbaum (1952) and Anderson and Darling (1952). For a numerical illustration of the use of these tests (on data of Poisson type) see Barnard (1953) and Maguire, Pearson, and Wynn (1953).

EXERCISES

5.1 Let S_n be a random variable with characteristic function

$$\varphi_{S_n}(u) = (pe^{iu} + q)^n,$$

where $0 < p < 1$, $q = 1 - p$ and n is an integer. Let

$$(S_n)^* = \frac{S_n - E[S_n]}{\sigma[S_n]}.$$

(a) Find the mean, variance, third central moment, and fourth central moment of S_n and $(S_n)^*$.

(b) Show that $\lim_{n \to \infty} \log \varphi_{(S_n)^*}(u) = -(\tfrac{1}{2})u^2$.

5.2 Let S_n and T_n be random variables with joint characteristic function

$$\varphi_{S_n, T_n}(u,v) = \{q_1 q_2 + p_1 e^{iu} + p_2 e^{iv} + p_1 p_2 e^{i(u+v)}\}^n,$$

where $0 < p_1 < 1, 0 < p_2 < 1, q_1 = 1 - p_1, q_2 = 1 - p_2, n$ is an integer.

Let $(S_n)^* = \dfrac{S_n - E[S_n]}{\sigma[S_n]}, (T_n)^* = \dfrac{T_n - E[T_n]}{\sigma[T_n]}$.

Evaluate $\lim\limits_{n\to\infty} \log \varphi_{(S_n)^*,(T_n)^*}(u,v)$.

5.3 Show that the Poisson process $\{N(t), t \geq 0\}$ with intensity ν is asymptotically normal as ν tends to ∞.

5.4 Let $\{X(t), t \geq 0\}$ be a stochastic process with stationary independent increments. Assume that $X(0) = 0$ and that

$$\log \varphi_{X(t)}(u) = \nu t \int_{-\infty}^{\infty} (e^{iux} - 1)f(x)\, dx,$$

where $\nu > 0$ and $f(x)$ is a probability density function. Let

$$\beta = \int_{-\infty}^{\infty} |x|^3 f(x)\, dx \Big/ \left\{ \int_{-\infty}^{\infty} x^2 f(x)\, dx \right\}^{3/2}.$$

Show that in order for $\{X(t), t \geq 0\}$ to be asymptotically normal any one of the following conditions is sufficient:

(i) $\nu \to \infty$;
(ii) $\beta \to 0$;
(iii) $\beta^2/\nu \to 0$.

5.5 Let $\{X_n\}$ be a sequence of independent identically distributed random variables with means 0 and variances 1.
Let $S_n = X_1 + X_2 + \cdots + X_n$ be the sequence of their consecutive sums. For $n = 1, 2, \cdots$ define a stochastic process $\{Y_n(t), 0 \leq t \leq 1\}$ as follows: $Y_n(0) = 0$ and for $0 < t \leq 1$

$$Y_n(t) = \frac{1}{\sqrt{n}} S_k \quad \text{if} \quad \frac{k-1}{n} < t \leq \frac{k}{n}.$$

Alternately one may write, for $t > 0$,

$$Y_n(t) = \frac{1}{\sqrt{n}} S_{[tn]},$$

where $[tn]$ denotes the largest integer less than or equal to tn. Show that, as n tends to ∞, $\{Y_n(t), 0 \leq t \leq 1\}$ is asymptotically normal, and indeed is approximately a Wiener process.

3-6 HARMONIC ANALYSIS OF STOCHASTIC PROCESSES

In the theory of time series analysis and statistical communications theory, a central role is played by the notion of the spectrum of a sto-- chastic process. For the time series analyst, the spectrum represents a basic tool for determining the mechanism generating an observed time series. For the communication theorist, the spectrum provides the major concept in terms of which one may study the behavior of stochastic processes (representing either signal or noise) passing through linear (and, to some extent, non-linear) devices. In order to understand the notion of spectrum one must introduce the notion of *filter* (or, more precisely, of a time-invariant linear filter). This notion, which grew to maturity in the hand of electronic engineers, is now coming to play an important role in many scientific fields.

Consider a "black box" which in response to an input represented by a function $\{x(t), -\infty < t < \infty\}$ yields an output represented by a function $\{y(t), -\infty < t < \infty\}$.

The "black box" is said to be *time invariant* if it has the property that translating the input function by an amount τ translates the output function by an amount τ; in symbols, if $\{y(t), -\infty < t < \infty\}$ is the response to $\{x(t), -\infty < t < \infty\}$ then $\{y(t+\tau), -\infty < t < \infty\}$ is the response to $\{x(t+\tau), -\infty < t < \infty\}$.

The "black box" is said to be a *linear filter* if an input function

$$x(t) = a_1 x_1(t) + a_2 x_2(t), \tag{6.1}$$

which is a linear combination of two functions $x_1(t)$ and $x_2(t)$ with responses $y_1(t)$ and $y_2(t)$, respectively, has as its response function

$$y(t) = a_1 y_1(t) + a_2 y_2(t). \tag{6.2}$$

In summary, any procedure or process (be it computational, physical, or purely conceptual) is called a time-invariant linear filter if it can be regarded as converting an input to an output and satisfies two conditions:

(i) The output corresponding to the superposition of two inputs is the superposition of the corresponding outputs;

(ii) The only effect of delaying an input by a fixed time is a delay in the output by the same time.

Some examples of time invariant linear filters [where $y(\cdot)$ represents the output function to an input function $x(\cdot)$] are

(i) moving linear combinations, such as

$$y(t) = a_0 x(t) + a_1 x(t-1) + \cdots + a_h x(t-h),$$
$$y(t) = x(t-h), \cdot$$
$$y(t) = x(t) - x(t-1);$$

(ii) combinations of derivatives, such as

$$y(t) = a_0 x(t) + a_1 x'(t) + \cdots + a_n x^{(n)}(t),$$
$$y(t) = x'(t);$$

(iii) integral operators, such as

$$y(t) = \int_{-\infty}^{t} e^{-\beta(t-s)} x(s)\, ds, \tag{6.3}$$

$$y(t) = \int_{-\infty}^{t} w(t-s) x(s)\, ds = \int_{0}^{\infty} w(s) x(t-s)\, ds. \tag{6.4}$$

The function $w(u)$ is called the impulse response function of the filter; in each case in which one considers a filter of integral operator type one must state the mathematical conditions which $w(u)$ must possess in order that the mathematical manipulations performed be meaningful.

There are various ways of describing a time-invariant linear filter. One way is to give an explicit formula for the output function corresponding to any input function; examples of such descriptions are (i)–(iii). A somewhat more intuitively meaningful way of describing time-invariant linear filters is in terms of their *frequency response*. A function of the form

$$x(t) = A e^{i\omega t} \tag{6.5}$$

for some real number ω is called a complex harmonic with amplitude A and frequency ω. The functions

$$x(t) = A \cos \omega t \quad \text{and} \quad x(t) = A \sin \omega t \tag{6.6}$$

are called real harmonics with amplitude A and frequency ω. Because $e^{i\omega t} = \cos \omega t + i \sin \omega t$, there is an obvious close relation between the complex harmonic in Eq. 6.5 and the real harmonics in Eq. 6.6.

THEOREM 6A

If a complex harmonic of unit amplitude and frequency ω is applied to a time-invariant linear filter, then the corresponding output (assuming it exists) is a complex harmonic of amplitude A and frequency ω. The amplitude A depends on ω and consequently may be written as $A(\omega)$.

Proof. Let $y(t)$ be the output corresponding to $x(t) = e^{i\omega t}$. Then the translated input $x(t + \tau)$ has as its output the function $y(t + \tau)$ which is $y(t)$ translated by τ. Now, the translated input

$$x(t + \tau) = e^{i\omega(t+\tau)} = e^{i\omega\tau} x(t)$$

is a constant multiple of the original input. Therefore, by linearity

$$y(t + \tau) = e^{i\omega\tau}y(t). \tag{6.7}$$

Since Eq. 6.7 holds for all τ and t it follows that (setting $t = 0$)

$$y(\tau) = e^{i\omega\tau}y(0). \tag{6.8}$$

Defining $A(\omega) = y(0)$, it follows that the output function $y(t)$ is a complex harmonic of amplitude $A(\omega)$ and frequency ω. The proof of Theorem 6A is now complete.

Given a time-invariant linear filter which, for each frequency ω, possesses an output $y(t)$ in response to a complex harmonic of unit amplitude and frequency ω applied for an infinitely long time to the filter, we define the *frequency response function* of the filter to be the function $A(\omega)$ having as its value at the frequency ω the quantity $A(\omega)$ satisfying the relation

$$y(t) = A(\omega)e^{i\omega t}. \tag{6.9}$$

EXAMPLE 6A

The frequency response function $A(\omega)$ of the time-invariant linear filter described in the time domain by $y(t) = x'(t)$ is immediately seen to be given by

$$A(\omega) = i\omega, \tag{6.10}$$

since to an input $x(t) = e^{i\omega t}$ the filter yields the output

$$y(t) = x'(t) = i\omega e^{i\omega t} = A(\omega)x(t). \tag{6.11}$$

The frequency response function $A(\omega)$ of the time-invariant linear filter defined by Eq. 6.3 is given by

$$A(\omega) = \int_0^\infty e^{-i\omega u}e^{-\beta u}\,du = \frac{1}{\beta + i\omega}, \tag{6.12}$$

since to an input $x(t) = e^{i\omega t}$ the filter yields the output

$$y(t) = \int_{-\infty}^t e^{-\beta(t-s)}e^{i\omega s}\,ds = e^{i\omega t}\int_0^\infty e^{-\beta u}e^{-i\omega u}\,du.$$

For the filter defined by Eq. 6.4, the Fourier transform

$$A(\omega) = \int_0^\infty e^{-i\omega s}w(s)\,ds \tag{6.13}$$

is the frequency response function of the filter, since to a harmonic input $x(t) = e^{i\omega t}$ one clearly obtains a harmonic output $y(t) = e^{i\omega t}A(\omega)$.

The question naturally arises: why is the frequency response function the most intuitively meaningful way of describing a filter? To answer this question, let us consider a filter of integral operator type, Eq. 6.4. For the sake of ease of manipulation, we define $w(s)$ for $s < 0$ by

$$w(s) = 0 \tag{6.14}$$

and we may then write

$$y(t) = \int_{-\infty}^{\infty} w(s)x(t-s) \, ds. \tag{6.15}$$

Let us now define the *Fourier transforms of the input and output functions*:

$$A_x(\omega) = \int_{-\infty}^{\infty} e^{-i\omega t} x(t) \, dt, \tag{6.16}$$

$$A_y(\omega) = \int_{-\infty}^{\infty} e^{-i\omega t} y(t) \, dt. \tag{6.17}$$

For the moment we are ignoring questions of rigor and are assuming that the integrals in Eqs. 6.16 and 6.17 exist and are finite. Taking the Fourier transform of both sides of Eq. 6.15 we find that

$$\begin{aligned}
A_y(\omega) &= \int_{-\infty}^{\infty} dt \, e^{-i\omega t} \int_{-\infty}^{\infty} ds \, w(s)x(t-s) \\
&= \int_{-\infty}^{\infty} ds \, w(s)e^{-i\omega s} \int_{-\infty}^{\infty} dt \, e^{-i\omega(t-s)} x(t-s) \\
&= \int_{-\infty}^{\infty} ds \, w(s)e^{-i\omega s} \int_{-\infty}^{\infty} dt' \, e^{-i\omega t'} x(t') \\
&= A(\omega)A_x(\omega). \tag{6.18}
\end{aligned}$$

In words, we may express Eq. 6.18 as follows:

$$\begin{Bmatrix} \text{Fourier transform} \\ \text{of output} \end{Bmatrix} = \begin{Bmatrix} \text{Frequency response} \\ \text{function of filter} \end{Bmatrix} \begin{Bmatrix} \text{Fourier transform} \\ \text{of input} \end{Bmatrix}. \tag{6.19}$$

We see, therefore, that for those inputs which possess Fourier transforms and which give rise to outputs possessing Fourier transforms, the effect of the filtering operation can be very simply expressed by Eq. 6.19. To understand the meaning of Eq. 6.19 consider an input $x(t)$ which may be regarded as the sum of two time functions,

$$x(t) = s(t) + n(t), \tag{6.20}$$

where $s(t)$ represents "signal" and $n(t)$ represents "noise" [that is, any unwanted part of the record $x(t)$]. One desires a filter which will pass as much as possible of the "signal" time function $s(t)$ and pass as little as possible of the "noise" time function $n(t)$. If the Fourier transform $A_s(\omega)$ of the "signal" is appreciably non-zero only at frequencies at which the Fourier transform of the "noise" is small, then one can design such a filter by choosing its frequency response function $A(\omega)$ to be as close as possible to one at frequencies ω where $A_s(\omega)$ is not near zero and as close as possible to zero at other frequencies.

Finally, we come to the relevance of the foregoing considerations to the theory of stochastic processes. For many electrical networks and mechanical systems, the input must be regarded as a stochastic process. For example, if one is to design a successful ballistic missile, the environment of the missile during flight must first be described as a stochastic process and the behavior of the missile may then be treated as the response of a linear mechanical system to random inputs. The question now arises: can one define for a stochastic process $\{X(t), -\infty < t < \infty\}$ a Fourier transform

$$A_X(\omega) = \int_{-\infty}^{\infty} e^{-it\omega} X(t) \, dt. \tag{6.21}$$

Unfortunately, no meaning can be attached to the integral in Eq. 6.21 for many stochastic processes — such as covariance stationary processes — since their sample functions are non-periodic undamped functions and therefore do not belong to the class of functions dealt with in the usual theories of Fourier series and Fourier integrals. Nevertheless, it is possible to define a notion of *harmonic analysis of stochastic processes* (that is, a method of assigning to each frequency ω a measure of its contribution to the "content" of the process) as was first shown by Wiener (1930) and Khintchine (1934). The following quotation from Wiener (1930, p. 126) indicates the spirit of this extension of the notion of harmonic analysis.

> The two theories of harmonic analysis embodied in the classical Fourier series development and the theory of Plancherel [Fourier integral development] do not exhaust the possibilities of harmonic analysis. The Fourier series is restricted to the very special class of periodic functions, while the Plancherel theory [Fourier integral] is restricted to functions which are quadratically summable, and hence tend on the average to zero as their argument tends to infinity. Neither is adequate for the treatment of a ray of white light which is supposed to endure for an indefinite time. Nevertheless, the physicists who first were faced with the problem of analyzing white light into its components had to employ one or the other of these tools. Gouy accordingly represented white light by a Fourier series, the period of which he allowed to grow without limit; and by focussing his attention on the average values of the energies concerned, he was able to arrive at results in agreement with the experiments. Lord Rayleigh, on the other hand, achieved much the same purpose by using the Fourier integral, and what we

now should call Plancherel's theorem. In both cases, one is astonished by the skill with which the authors use clumsy and unsuitable tools to obtain the right results, and one is led to admire the unfailing heuristic insight of the true physicist.

An exposition of the theory of harmonic analysis of stochastic processes is beyond the scope of this book, since it properly belongs to a course on time series analysis or on statistical communication theory. However, let us state without proof the essentials of this theory. Since many people find integrals easier to manipulate than sums, we consider the case of continuous parameter time series $\{X(t), t \geq 0\}$.

Given a finite sample $\{X(t), 0 \leq t \leq T\}$ of the process, one defines two functions: the *sample covariance function*,

$$R_T(v) = \frac{1}{T} \int_0^{T-v} X(t)\, X(t+v)\, dt, \quad 0 \leq v < T \qquad (6.22)$$
$$= 0, \qquad\qquad\qquad\qquad\qquad v > T$$
$$= R_T(-v), \qquad\qquad\qquad\qquad v < 0$$

and the *sample spectral density function*,

$$f_T(\omega) = \frac{1}{2\pi T} \left| \int_0^T e^{-it\omega} X(t)\, dt \right|^2, \quad -\infty < \omega < \infty. \qquad (6.23)$$

It may be shown (using the theory of Fourier integrals) that $f_T(\omega)$ and $R_T(v)$ are Fourier transforms of each other:

$$R_T(v) = \int_{-\infty}^{\infty} e^{iv\omega} f_T(\omega)\, d\omega, \quad -\infty < v < \infty, \qquad (6.24)$$

$$f_T(\omega) = \frac{1}{2\pi} \int_{-T}^{T} e^{-iv\omega} R_T(v)\, dv = \frac{1}{\pi} \int_0^T \cos v\omega\, R_T(v)\, dv. \qquad (6.25)$$

From Eq. 6.24 it follows that the mean of the sample covariance function is the Fourier transform of the mean of the sample spectral density function:

$$E[R_T(v)] = \int_{-\infty}^{\infty} e^{iv\omega} E[f_T(\omega)]\, d\omega, \quad -\infty < v < \infty. \qquad (6.26)$$

Now, suppose that there exists a continuous function $R(v)$ such that

$$\lim_{T \to \infty} E[R_T(v)] = R(v). \qquad (6.27)$$

In the case of a covariance stationary stochastic process with zero means, Eq. 6.27 holds with $R(v)$ equal to the covariance function of the process since (for $0 < v < T$)

$$E[R_T(v)] = \frac{1}{T} \int_0^{T-v} R(v) \, dt = \left(1 - \frac{v}{T}\right) R(v).$$

If Eq. 6.27 holds, it may be shown (using the continuity theorem of probability theory, *Mod Prob*, p. 425) that there exists a bounded nondecreasing function of a real variable ω, denoted $F(\omega)$, and called the *spectral distribution function*, such that

$$R(v) = \int_{-\infty}^{\infty} e^{iv\omega} \, dF(\omega), \quad -\infty < v < \infty. \tag{6.28}$$

For any frequencies $\omega_1 < \omega_2$, $F(\omega_2) - F(\omega_1)$ may be interpreted as a measure of the contribution of the frequency band ω_1 to ω_2 to the "content" of the stochastic process.

Now, suppose that the spectral distribution function $F(\omega)$ is everywhere differentiable. Its derivative is denoted by $f(\omega)$ and is called the *spectral density function* of the stochastic process. Instead of Eq. 6.28, one has then

$$R(v) = \int_{-\infty}^{\infty} e^{iv\omega} f(\omega) \, d\omega. \tag{6.29}$$

A sufficient condition for $F(\omega)$ to be differentiable, and for Eq. 6.29 to hold, is that

$$\int_{-\infty}^{\infty} |R(v)| \, dv < \infty. \tag{6.30}$$

An explicit formula for the spectral density function is then given by

$$f(\omega) = \frac{1}{2\pi} \int_{-\infty}^{\infty} e^{-iv\omega} R(v) \, dv. \tag{6.31}$$

Eqs. 6.29 and 6.31 are basic relations in the theory of harmonic analysis of stochastic processes and together constitute what are usually called the *Wiener-Khintchine relations* in commemoration of the pioneering work of Wiener (1930) and Khintchine (1934).

The spectral density function derives its importance from the fact that in terms of it one can easily express the response of linear systems to random excitations. The following relation may be proved:

$$\left\{ \begin{matrix} \text{spectral density} \\ \text{function of output} \\ \text{stochastic process} \end{matrix} \right\} = \left| \left\{ \begin{matrix} \text{frequency response} \\ \text{function of filter} \end{matrix} \right\} \right|^2 \times \left\{ \begin{matrix} \text{spectral density} \\ \text{function of input} \\ \text{stochastic process} \end{matrix} \right\} \tag{6.32}$$

From this relation, one can determine the filtering effects of various networks through which a stochastic process is passed. Further, if a process

$X(\cdot)$ may be regarded as the sum, $X(t) = S(t) + N(t)$, of two processes $S(\cdot)$ and $N(\cdot)$ representing signal and noise, respectively, then from a knowledge of the respective spectral density functions of the signal and noise, one may develop schemes for filtering out the signal from the noise.

Many aspects of a covariance stationary time series (stochastic process) are best understood in terms of its spectral density function. The spectral density function enables one to investigate the physical mechanisms generating a time series, and possibly to simulate a time series. Other uses of the spectrum are as operational means (i) of transmitting or detecting signals, (ii) of classifying records of phenomena such as brain waves, (iii) of studying radio propagation phenomena, and (iv) of determining characteristics of control systems. The theory of statistical spectral analysis is concerned with the problem of estimating, from a finite length of record, the spectral density function of a time series which possesses one. For surveys of this theory, see Jenkins (1961) and Parzen (1961 a).

EXAMPLE 6B

The motion of a pendulum in a turbulent fluid. The equation of motion of a pendulum is given by

$$X''(t) + 2\alpha X'(t) + (\omega_0{}^2 + \alpha^2)X(t) = I(t) \qquad (6.33)$$

in which $X(t)$ is the displacement of the pendulum from its rest position, α is a damping factor, $2\pi/\omega_0$ is the damped period of the pendulum, and $I(t)$ is the exciting force per unit mass of pendulum. If the motion has been continuing for a long time, it may be shown that $X(t)$ may be expressed as the output of the filter (of integral operator type) described by

$$X(t) = \int_{-\infty}^{t} e^{-\alpha(t-s)} \frac{\sin \omega_0(t-s)}{\omega_0} I(s) \, ds. \qquad (6.34)$$

From Eq. 6.34 it follows, using Theorem 3A, that

$$E[X(t)] = \int_{-\infty}^{t} e^{-\alpha(t-s)} \frac{\sin \omega_0(t-s)}{\omega_0} E[I(s)] \, ds,$$

$$\text{Cov}[X(u), X(v)] = \int_{-\infty}^{u} ds \int_{-\infty}^{v} dt \, e^{-\alpha(u-s)} e^{-\alpha(v-t)}$$

$$\times \frac{\sin \omega_0(u-s)}{\omega_0} \frac{\sin \omega_0(v-t)}{\omega_0} \text{Cov}[I(s), I(t)]. \qquad (6.35)$$

Suppose now that the input $\{I(t), -\infty < t < \infty\}$ is covariance stationary, with covariance function $R_I(\cdot)$, and possesses a spectral density function $f_I(\cdot)$; equivalently, suppose that the covariance kernel of $I(\cdot)$ may be written as a Fourier integral

$$\text{Cov}[I(s), I(t)] = R_I(s - t) = \int_{-\infty}^{\infty} e^{i\omega(s-t)} f_I(\omega)\, d\omega. \qquad (6.36)$$

Substituting Eq. 6.36 in Eq. 6.35, one may write

$$\text{Cov}[X(u), X(v)] = \int_{-\infty}^{\infty} d\omega\, f_I(\omega) \int_{-\infty}^{u} ds\, e^{is\omega} e^{-\alpha(u-s)} \frac{\sin \omega_0(u - s)}{\omega_0}$$

$$\times \int_{-\infty}^{v} dt\, e^{-it\omega} e^{-\alpha(v-t)} \frac{\sin \omega_0(v - t)}{\omega_0}$$

$$= \int_{-\infty}^{\infty} d\omega\, f_I(\omega) e^{i\omega(u-v)} \left| \int_{0}^{\infty} dx\, e^{ix\omega} e^{-\alpha x} \frac{\sin \omega_0 x}{\omega_0} \right|^2 .$$

We thus see that the covariance kernel of $X(t)$ may be written as a Fourier integral

$$\text{Cov}[X(u), X(v)] = \int_{-\infty}^{\infty} e^{i\omega(u-v)} f_X(\omega)\, d\omega,$$

where

$$f_X(\omega) = f_I(\omega)\, |\, A(\omega)\, |^2, \qquad (6.37)$$

$$A(\omega) = \int_{0}^{\infty} dx\, e^{ix\omega} e^{-\alpha x} \frac{\sin \omega_0 x}{\omega_0}. \qquad (6.38)$$

Now, $A(\omega)$ is the frequency response function of the time invariant linear filter whose output $X(t)$ is related to its input $I(t)$ by the differential equation 6.33. Consequently, Eq. 6.37 yields an example of the validity of Eq. 6.32.

To evaluate the integral in Eq. 6.38, one proceeds as follows. Corresponding to an input

$$I(t) = e^{i\omega t}$$

the linear system described by the differential equation 6.33 has output

$$X(t) = A(\omega)e^{i\omega t}, \quad A(\omega) = \int_{0}^{\infty} e^{-i\omega x} e^{-\alpha x} \frac{\sin \omega_0 x}{\omega_0}\, dx,$$

with derivatives

$$X'(t) = (i\omega)X(t), \quad X''(t) = (i\omega)^2 X(t).$$

Substituting these expressions into Eq. 6.33, we obtain the following equation for the frequency response function $A(\omega)$:

$$\{(i\omega)^2 + 2\alpha(i\omega) + (\omega_0^2 + \alpha^2)\} e^{i\omega t} A(\omega) = e^{i\omega t},$$

so that

$$A(\omega) = \frac{1}{\{(i\omega)^2 + 2\alpha(i\omega) + (\omega_0^2 + \alpha^2)\}}.$$

The square modulus of the frequency response function is given by

$$| A(\omega) |^2 = \frac{1}{(\omega_0^2 + \alpha^2 - \omega^2)^2 + 4\alpha^2\omega^2}. \tag{6.39}$$

Consequently, if the input $\{I(t), -\infty < t < \infty\}$ to the stochastic differential equation 6.33 is covariance stationary with spectral density function $f_I(\omega)$, then the output stochastic process $\{X(t), -\infty < t < \infty\}$ is covariance stationary with spectral density function $f_X(\omega)$ given by Eq. 6.37, where $| A(\omega) |^2$ is given by Eq. 6.39.

White noise. By analogy with the continuous energy distribution in white light from an incandescent body, a covariance stationary stochastic process which has equal power in all frequency intervals over a wide frequency range is called *white noise*. Mathematically, $\{X(t), t \geq 0\}$ is said to be a *white noise process* if it possesses a constant spectral density function: for some $C > 0$,

$$f_X(\omega) = C \text{ for all } \omega. \tag{6.40}$$

Of course, a constant spectral density function is non-integrable and is therefore a mathematical fiction. There are several possible ways of giving a precise mathematical definition of white noise. For most purposes it is sufficient to regard white noise as a normal process with covariance function

$$R(t) = \rho e^{-\rho|t|} \tag{6.41}$$

and spectral density function

$$f(\omega) = \frac{1}{\pi} \frac{1}{1 + (\omega/\rho)^2} \tag{6.42}$$

in which ρ is so large as to be considered infinite. The spectral density function in Eq. 6.42 is then effectively a constant (and the covariance function in Eq. 6.41 is then the Dirac delta function).

EXAMPLE 6C
Prediction theory and the representation of a stationary process as the output of a filter with white noise input. Let $X(t)$ be a stationary process with zero means and covariance function $R(v)$. Does there always exist a filter, with impulse response function $w(s)$, such that

$$X(t) = \int_{-\infty}^{t} w(t - s)I(s)\, ds, \tag{6.43}$$

where the input $I(s)$ is white noise? A necessary condition for Eq. 6.43 to hold is that the covariance function of $X(t)$ satisfies, for some positive constant C and square integrable function $w(s)$,

$$E[X(u)X(v)] = C \int_{-\infty}^{\min (u,v)} w(u-s)w(v-s) \, ds, \qquad (6.44)$$

or, equivalently,

$$R(v) = C \int_0^\infty w(y)w(y+v) \, dy. \qquad (6.45)$$

It may be shown that Eqs. 6.44 and 6.45 are also sufficient conditions for Eq. 6.43 to hold. Thus Eq. 6.45 provides a criterion in terms of the covariance function for Eq. 6.43 to hold. It is also possible to give a criterion in terms of the spectral density function $f(\omega)$. It may be shown that a necessary and sufficient condition that Eq. 6.43 holds is that

$$\int_{-\infty}^\infty \frac{|\log f(\omega)|}{1+\omega^2} \, d\omega < \infty. \qquad (6.46)$$

If one can represent a process $X(t)$ in the form of Eq. 6.43, then one can give an explicit answer to the following problem of prediction: given the values of a stationary normal time series $X(t)$ for $-\infty < t \le t_0$, use these values to form a predictor of $X(t_1)$, for a given time $t_1 > t_0$, which has minimum mean square error among all possible predictors. Write

$$X(t_1) = \int_{-\infty}^{t_0} w(t_1-s)I(s) \, ds + \int_{t_0}^{t_1} w(t_1-s)I(s) \, ds. \qquad (6.47)$$

It may be shown that the best predictor of $X(t_1)$, given the values of $X(t)$ for $-\infty < t \le t_0$, is given by the first integral in Eq. 6.47. The reader who studies problems of prediction and regression analysis of time series will find that an important tool used is the representation of a stochastic process as the output of a filter with white noise input.

EXERCISES

6.1 Suppose that $\{X(t), t \ge 0\}$ has constant mean $E[X(t)] = m$ and is covariance stationary with spectral density function $f_X(\omega)$. Let h be a fixed positive constant and let $Y(t) = X(t+h) - X(t)$. Find the spectral density function of $\{Y(t), t \ge 0\}$. Differencing is a method often employed to reduce a time series with non-zero, but constant mean, to a time series with zero mean. However, the operation of differencing affects the spectral density function and account should be taken of this fact.

6.2 Let $\{X(t), -\infty < t < \infty\}$ be a covariance stationary process with zero means, covariance function $R(v)$, and spectral density function $f(\omega)$ satisfying

$$\int_{-\infty}^{\infty} \omega^2 f(\omega)\, d\omega < \infty.$$

Express in terms of $f(\omega)$ the spectral density function of

(i) $Y(t) = X'(t)$,

(ii) $Y(t) = \int_{-\infty}^{t} e^{-\beta(t-s)} X'(s)\, ds$,

(iii) $Y(t) = \int_{-\infty}^{t} e^{-\alpha(t-s)} \dfrac{\sin \omega(t-s)}{\omega} X'(s)\, ds$.

6.3 *Real sine wave of random amplitude.* Let $X(t) = A(t) \cos \omega t$, where $\{A(t), t \ge 0\}$ is a covariance stationary process with covariance function $R_A(\,\cdot\,)$, and ω is a positive constant. Express the spectral density function of $\{X(t), t \ge 0\}$ in terms of the spectral density function of $\{A(t), t \ge 0\}$. Note that $\{X(t), t \ge 0\}$ is not covariance stationary. Nevertheless, its spectral density function is defined by Eq. 6.29, where $R(v)$ is defined by Eq. 6.27.

6.4 *A physical example of white noise.* Let $\{X(t), -\infty < t < \infty\}$ be the shot noise current whose covariance function is given by Eq. 5.36 of Chapter 4. (i) Show that it is covariance stationary with spectral density function

$$f(\omega) = \frac{eI}{\pi} \frac{2}{(\omega T)^4} \{(\omega T)^2 + 2(1 - \cos \omega T - \omega T \sin \omega T)\}.$$

(ii) Show that for frequencies ω such that $(\omega T)^2$ is very small, the spectral density function is approximately equal to a constant:

$$f(\omega) = \frac{eI}{2\pi}.$$

Thus, if the transit time T of an electron is of the order of 10^{-9} seconds, then the spectral density function can be regarded as constant for frequencies up to 100 megacycles/second. From the point of view of its behavior when passed through a filter whose frequency response function is zero for frequencies beyond 100 megacycles/second, the shot noise current $\{X(t), -\infty < t < \infty\}$ can be considered to be white noise.

6.5 *The covariance function (or spectral density function) does not uniquely characterize a covariance stationary process.* Show that

$$R(t) = Ce^{-\beta|t|}$$

is for suitable values of C and β the covariance function of (i) the random telegraph signal, (ii) the process defined in Exercise 4.7, and (iii) the process of shot effect type (defined in Section 4–5)

$$X(t) = \sum_{-\infty < \tau_n \le t} e^{-\alpha(t-\tau_n)}.$$

6.6 *The sample spectral density function is not a consistent estimate of the spectral density function.* Prove the following theorem. If $\{X(t), t \geq 0\}$ is a covariance stationary normal process with zero means and with continuous spectral density function $f(\omega)$, then for any frequency ω

$$\lim_{T \to \infty} \varphi_{f_T(\omega)}(u) = (1 - iuf(\omega))^{-1}.$$

In words, the sample spectral density function $f_T(\omega)$ of a normal process is for large sample sizes T exponentially distributed with mean $f(\omega)$. Consequently, $f_T(\omega)$ does not converge to $f(\omega)$ as T tends to ∞. *Hint.* Using the formula given by Eq. 4.27 for the joint characteristic function of the squares of two jointly normal random variables with zero means, show that the sample spectral density function has characteristic function

$$\varphi_{f_T(\omega)}(u)$$

$$= \left\{ 1 - i\left(\frac{uT}{\pi}\right)[\sigma_1{}^2(T) + \sigma_2{}^2(T)] - \left(\frac{uT}{\pi}\right)^2 [1 - \rho^2(T)]\sigma_1{}^2(T)\sigma_2{}^2(T) \right\}^{-1/2},$$

where

$$\sigma_1{}^2(T) = \mathrm{Var}\left[\frac{1}{T} \int_0^T \cos \omega t \, X(t) dt\right], \quad \sigma_2{}^2(T) = \mathrm{Var}\left[\frac{1}{T} \int_0^T \sin \omega t \, X(t) dt\right],$$

$$\rho(T) = \rho\left[\frac{1}{T} \int_0^T \cos \omega t \, X(t) dt, \frac{1}{T} \int_0^T \sin \omega t \, X(t) dt\right].$$

Consequently,

$$\lim_{T \to \infty} \varphi_{f_T(\omega)}(u) = (1 - 2iuf(\omega) - u^2 f^2(\omega))^{-1/2}.$$

Counting processes
and Poisson processes

By a counting process we mean an integer-valued process $\{N(t),\, t \geq 0\}$ which counts the number of points occurring in an interval, these points having been distributed by some stochastic mechanism. In a typical case, the points represent the times $\tau_1, \tau_2, \cdots$ at which events of a specified character have occurred, where $0 < \tau_1 < \tau_2 < \cdots$. The random variables

$$T_1 = \tau_1,\, T_2 = \tau_2 - \tau_1,\, \cdots,\, T_n = \tau_n - \tau_{n-1},\, \cdots$$

are called the successive inter-arrival times. If for $t \geq 0$, $N(t)$ represents the number of points lying in the interval $(0,t]$, then $\{N(t),\, t \geq 0\}$ is called the counting process of the series of points.

To define a counting process two main methods are available:

(i) one can directly define $\{N(t),\, t \geq 0\}$ as either a process with independent increments or a Markov process (see Chapter 7), or

(ii) one can derive the properties of the counting process from assumptions made concerning the inter-arrival times.

Among counting processes the Poisson process plays a prominent part. The Poisson process not only arises frequently as a model for the counts of random events, but also provides a basic building block with which other useful stochastic processes can be constructed.

In Chapters 4 and 5 our aim is to study both the general properties of counting processes and the special properties of Poisson processes.

4-1   AXIOMATIC DERIVATIONS OF THE POISSON PROCESS

There are several ways one can give systems of axioms that a stochastic process needs to satisfy in order to be a Poisson process. The Poisson process can be characterized as

(i) the pure birth process with constant birthrate (see Section 7–2),

(ii) the renewal counting process with exponentially distributed inter-arrival times (see Section 5–2),

(iii) an integer-valued process with stationary independent increments which has unit jumps (the model of independent increments).

In this section we discuss model (iii) and state conditions under which an integer-valued stochastic process is a Poisson process.

Consider events occurring in time on the interval 0 to ∞. For $t > 0$, let $N(t)$ be the number of events that have occurred in the interval $(0,t]$, open at 0 and closed at t. Consequently, $N(t)$ and, for any $h > 0$, $N(t + h) - N(t)$ assume only non-negative integer values.

We now make the following **assumptions.**

Axiom 0.   Since we begin counting events at time 0, we define $N(0) = 0$.

Axiom 1.   The process $\{N(t), t \geq 0\}$ has independent increments.

Axiom 2.   For any $t > 0$, $0 < P[N(t) > 0] < 1$. In words, in any interval (no matter how small) there is a positive probability that an event will occur, but it is not certain that an event will occur.

Axiom 3.   For any $t \geq 0$

$$\lim_{h \to 0} \frac{P[N(t + h) - N(t) \geq 2]}{P[N(t + h) - N(t) = 1]} = 0.$$

In words, in sufficiently small intervals, at most one event can occur; that is, it is not possible for events to happen simultaneously.

Axiom 4.   The counting process $N(t)$ has stationary increments; that is, for any 2 points $t > s \geq 0$ and any $h > 0$, the random variables $N(t) - N(s)$ and $N(t + h) - N(s + h)$ are identically distributed.

It should be noted that various modifications of the foregoing scheme are of interest.

If Axiom 3 is dropped, we are led to the generalized Poisson process, defined in Section 4–2.

If Axiom 4 is dropped we are led to the non-homogeneous Poisson process, defined in Section 4–2.

Counting processes $\{N(t), -\infty < t < \infty\}$, defined for all real values of t, are also of interest. These arise when one assumes that the events being counted started up a very long time ago. The Poisson process $\{N(t), -\infty < t < \infty\}$ defined on the infinite line is characterized by Axioms 1 to 4, with t and s assumed to satisfy $-\infty < s \leq t < \infty$. In this case, $N(t) - N(0)$, rather than $N(t)$, represents the number of events that have occurred in the interval $(0,t]$.

THEOREM 1A

A counting process $\{N(t), t \geq 0\}$ satisfying Axioms 0 to 4 is the Poisson process (as defined in Section 1-3).

Proof. To prove the theorem it suffices to show that there is a positive constant ν such that, for any $t > 0$, $N(t)$ is Poisson distributed with mean νt. We prove this by showing that there is a positive constant ν such that the probability generating function of $N(t)$,

$$\psi(z,t) = \sum_{n=0}^{\infty} z^n P[N(t) = n], \tag{1.1}$$

defined for $|z| \leq 1$, satisfies

$$\psi(z,t) = e^{\nu t(z-1)}, \quad |z| < 1, \tag{1.2}$$

which is the probability generating function of a random variable that is Poisson distributed with mean νt.

We first show that Eq. 1.2 holds under the assumption that there exists a positive constant ν satisfying

$$\lim_{h \to 0} \frac{1 - P[N(h) = 0]}{h} = \nu, \tag{1.3}$$

$$\lim_{h \to 0} \frac{P[N(h) = 1]}{h} = \nu, \tag{1.4}$$

$$\lim_{h \to 0} \frac{P[N(h) \geq 2]}{h} = 0. \tag{1.5}$$

Because $\{N(t), t \geq 0\}$ has independent increments it follows that for any non-negative t and h

$$E[z^{N(t+h)}] = E[z^{N(t+h)-N(t)}]E[z^{N(t)}]. \tag{1.6}$$

Because $\{N(t), t \geq 0\}$ has stationary increments, it follows from Eq. 1.6 that

$$\psi(z, t + h) = \psi(z,t)\psi(z,h), \tag{1.7}$$

$$\frac{1}{h}\{\psi(z, t + h) - \psi(z,t)\} = \psi(z,t)\frac{1}{h}\{\psi(z,h) - 1\}. \tag{1.8}$$

We now show that from Eqs. 1.3–1.5 it follows that

$$\lim_{h \to 0} \frac{1}{h} \{\psi(z,h) - 1\} = \nu(z - 1). \tag{1.9}$$

One may write

$$\frac{1}{h} \{\psi(z,h) - 1\} = \frac{1}{h} \{P[N(h) = 0] - 1\} + z \frac{1}{h} P[N(h) = 1] \tag{1.10}$$

$$+ \frac{1}{h} \sum_{n=2}^{\infty} z^n P[N(h) = n].$$

For $|z| < 1$,

$$\sum_{n=2}^{\infty} z^n P[N(h) = n] \le P[N(h) \ge 2].$$

From Eqs. 1.10, 1.3, 1.4, and 1.5 one immediately obtains Eq. 1.9.

Next let h tend to 0 in Eq. 1.8. By Eq. 1.9 the limit of the right-hand side of Eq. 1.8 exists. Consequently, the limit of the left-hand side exists. We have thus shown that the probability generating function of $N(t)$ satisfies the differential equation, for $|z| < 1$, and all $t \ge 0$,

$$\frac{\partial}{\partial t} \psi(z,t) = \nu(z - 1)\psi(z,t). \tag{1.11}$$

From Eq. 1.11, and the initial condition $\psi(z,0) = 1$, one obtains Eq. 1.2.

To complete the proof of the theorem we show that there exists a positive constant ν satisfying Eqs. 1.3–1.5. Define

$$P_0(t) = P[N(t) = 0]. \tag{1.12}$$

Since

$$P[N(t_1 + t_2) = 0] = P[N(t_1 + t_2) - N(t_1) = 0 \quad \text{and} \quad N(t_1) = 0], \tag{1.13}$$

it follows from the assumption of stationary independent increments that

$$P_0(t_1 + t_2) = P_0(t_1)P_0(t_2). \tag{1.14}$$

Since $P_0(t)$ is a bounded function, it follows by Theorem 1B below that either for some constant ν

$$P_0(t) = e^{-\nu t} \tag{1.15}$$

or $P_0(t)$ vanishes identically. By Axiom 2, $0 < P_0(t) < 1$ for all t; consequently, Eq. 1.15 holds with $\nu > 0$. From Eq. 1.15 one obtains Eq. 1.3. To obtain Eq. 1.4 define $P_1(t) = P[N(t) = 1]$ and $Q(t) = P[N(t) \ge 2]$. We may write

$$\frac{P_1(h)}{h}\left\{1 + \frac{Q(h)}{P_1(h)}\right\} = \frac{1 - P_0(h)}{h}.\qquad(1.16)$$

By Axiom 3,

$$\lim_{h \to 0}\frac{Q(h)}{P_1(h)} = 0.\qquad(1.17)$$

From Eqs. 1.16 and 1.17 one obtains Eq. 1.4. From Eqs. 1.4 and 1.17 one obtains Eq. 1.5. The proof of Theorem 1A is now complete.

The solution of certain functional equations. In studying the properties of stochastic processes with stationary independent increments, one frequently encounters functions satisfying the following functional equations:

$$g(t_1 + t_2) = g(t_1) + g(t_2) \quad \text{for} \quad t_1, t_2 > 0,\qquad(1.18)$$
$$f(t_1 + t_2) = f(t_1)f(t_2) \quad \text{for} \quad t_1, t_2 > 0.\qquad(1.19)$$

The solution of Eq. 1.18 is given in Complement 1D. We discuss the solution of Eq. 1.19.

THEOREM 1B

Let $\{f(t), t > 0\}$ be a real valued function satisfying Eq. 1.19 and which is bounded in every finite interval. Then either $f(t)$ vanishes identically or there exists a constant ν such that

$$f(t) = e^{-\nu t}, t > 0.\qquad(1.20)$$

Proof. We first give a proof of the theorem under the stronger assumption that the function $f(t)$ is continuous, or is non-decreasing. From Eq. 1.19 it follows that for any integers m and n and $t > 0$

$$f(mt) = [f(t)]^m, \quad f\left(\frac{m}{n}\right) = \left[f\left(\frac{1}{n}\right)\right]^m, \quad f\left(\frac{1}{n}\right) = [f(1)]^{1/n}.$$

Consequently, for any rational number t (that is, a number t of the form $t = m/n$ for some integers m and n),

$$f(t) = \{f(1)\}^t.\qquad(1.21)$$

We next show that Eq. 1.21 holds for any real number t. Let $\{t_n\}$ be a sequence of rational numbers, each less than t, such that the sequence $\{t_n\}$ converges to t. It then follows that

$$\lim_{n \to \infty} \{f(1)\}^{t_n} = \{f(1)\}^t.$$

If $f(t)$ is continuous, then $\lim_{n \to \infty} f(t_n) = f(t)$ which proves that Eq. 1.21 holds. If $f(t)$ is non-decreasing, then

$$t_n \leq t \quad \text{implies} \quad f(t_n) \leq f(t) \quad \text{implies} \quad \{f(1)\}^t \leq f(t).$$

To show that $f(t) \leq \{f(1)\}^t$, and consequently $f(t) = \{f(1)\}^t$, choose a sequence $\{t_n^*\}$ of rational numbers, each greater than t, such that the sequence $\{t_n^*\}$ converges to t. Then

$$t_n^* \geq t \quad \text{implies} \quad f(t_n^*) \geq f(t) \quad \text{implies} \quad \{f(1)\}^t \geq f(t).$$

Since $f(1) = [f(1/n)]^n \geq 0$ for n even, in the case that $f(t)$ does not vanish identically, there is a real number ν such that $f(1) = e^{-\nu}$, and from Eq. 1.21 one may write $f(t)$ in the form of Eq. 1.20.

We next prove Theorem 1B under the assumption that $f(t)$ is bounded in every finite interval.

If $f(t)$ does not vanish identically, then there exists a point $t_0 > 0$ such that $f(t_0) > 0$, since $f(2t) = f^2(t)$. Let $F(t) = [f(t_0)]^{-t} f(t_0 t)$. To prove that Eq. 1.20 holds for some ν it suffices to prove that

$$F(t) = 1 \quad \text{for all} \quad t > 0, \tag{1.22}$$

since Eq. 1.22 implies that

$$f(t_0 t) = [f(t_0)]^t \quad \text{for all} \quad t > 0$$

and therefore (letting $u = t_0 t$),

$$f(u) = [\{f(t_0)\}^{1/t_0}]^u = e^{-\nu u} \text{ for all} \quad u > 0,$$

where $\nu = -(1/t_0) \log f(t_0)$.

To prove that Eq. 1.22 holds we argue as follows. Clearly, $F(1) = 1$, and $F(t)$ satisfies the functional equation Eq. 1.19. Therefore, $F(m/n) = 1$ for any integers m and n, so that $F(t) = 1$ for all rational numbers t. Now for any number $t > 0$ there exists a rational number r such that $t' = t - r$ satisfies $0 < t' \leq 1$, from which it follows that $F(t) = F(t')F(r) = F(t')$. Consequently, any value which $F(\cdot)$ assumes for some value of t it assumes for a value of t in the interval $0 < t \leq 1$. Now, $F(\cdot)$ is bounded in the unit interval $0 < t \leq 1$ since $f(\cdot)$ is assumed to be bounded in every finite interval. Therefore, $F(\cdot)$ is a bounded function; that is, there is a constant K such that

$$|F(t)| \leq K \quad \text{for all} \quad t. \tag{1.23}$$

Suppose that $F(\tau) = c \neq 1$ for some $\tau > 0$. It may be assumed that $0 < \tau < 1$ and $c > 1$, since $F(1 - \tau) = c^{-1}$. Since $F(N\tau) = c^N$, and c^N

can be made arbitrarily large by choosing N sufficiently large, it follows that if Eq. 1.22 did not hold then there does not exist any constant K such that Eq. 1.23 holds. The proof of the theorem is now complete.

COMPLEMENTS

1A *An alternative derivation of the Poisson process.* Write out a proof of Theorem 1A along the following lines. Define $P_n(t) = P[N(t) = n]$. Show that for any t and s

$$P_n(t + s) = \sum_{j=0}^{n} P_j(t)P_{n-j}(s).$$

Assuming that Eqs. 1.3–1.5 hold, show that $P_n(t)$ satisfies the system of differential equations for $n = 1, 2, \cdots$

$$\frac{dP_n(t)}{dt} = -\nu P_n(t) + \nu P_{n-1}(t),$$

with initial conditions $P_n(0) = 0$. Consequently, show that

$$P_n(t) = e^{-\nu t} \frac{(\nu t)^n}{n!}.$$

1B *The exponential distribution is the unique distribution without memory.* A non-negative random variable T is said to be without memory if and only if for any positive numbers x and y

$$P[T > x + y \mid T > x] = P[T > y]. \tag{1.24}$$

Show that Eq. 1.24 holds if and only if T is exponentially distributed.

1C *Characterization of the geometric distribution.* Show that if T is positive and integer-valued then Eq. 1.24 holds for non-negative integers x and y if and only if for some constant p

$$P[T = x] = p(1 - p)^{x-1}, \quad x = 1, 2, \cdots.$$

1D Prove the following theorem (compare *Mod Prob*, p. 263). Let $g(\,\cdot\,)$ be a real-valued function, defined for $t \geq 0$, which satisfies the functional equation Eq. 1.18. The function $g(\,\cdot\,)$ is a linear function,

$$g(t) = ct, \quad t \geq 0,$$

for some constant c if any one of the following conditions is satisfied:

(i) $g(t)$ is continuous for $t \geq 0$;
(ii) $g(t) \geq 0$ for all $t \geq 0$;
(iii) $g(t)$ is bounded in the interval 0 to 1;
(iv) $g(t)$ is bounded in a finite interval.

1E The characteristic function of a stochastic process with stationary independent increments is of a definite form. If one considers the characteristic function of a sum S_n of independent identically distributed random variables $X_1, \cdots, X_n$, one sees that

$$\varphi_{S_n}(u) = \varphi_{X_1}(u) \cdots \varphi_{X_n}(u) = [\varphi_{X_1}(u)]^n,$$

so that $\varphi_{S_n}(u)$ is the nth power of a characteristic function. Prove the following analogous assertion. If (i) $\{X(t), t \geq 0\}$ is a stochastic process with stationary independent increments, (ii) $X(0) = 0$, and (iii) for every real number u, $\varphi_{X(t)}(u)$ is a continuous function of t, then

$$\varphi_{X(t)}(u) = [\varphi_{X(1)}(u)]^t.$$

Consequently, show that, for any fixed t, $\varphi_{X(t)}(u)$ has the property that for every ʀeal number $r \geq 0$

$$[\varphi_{X(t)}(u)]^r \text{ is a characteristic function.}$$

1F *Infinitely divisible laws.* · A characteristic function $\varphi(u)$ which has the property that, for every $r \geq 0$, $[\varphi(u)]^r$ is a characteristic function, is said to be *infinitely divisible.* Show that the following characteristic functions are infinitely divisible:

(i) $\varphi(u) = \exp[ium - \frac{1}{2} u^2 \sigma^2]$,
(ii) $\varphi(u) = \exp[\lambda(e^{iu} - 1)]$, where $\lambda > 0$,
(iii) $\varphi(u) = \exp\left[\lambda \int_{-\infty}^{\infty} (e^{iux} - 1) f(x) \, dx\right]$,
where $f(x)$ is a probability density function and $\lambda > 0$,
(iv) $\varphi(u) = (1 - q)(1 - q \, e^{iu})^{-1}$, where $0 < q < 1$.

1G Show that an infinitely divisible characteristic function never vanishes.

4-2 NON-HOMOGENEOUS, GENERALIZED, AND COMPOUND POISSON PROCESSES

An integer-valued process $\{N(t), t \geq 0\}$ which has independent increments and unit jumps can be shown to have characteristic function, for $t \geq 0$,

$$\varphi_{N(t)}(u) = \exp[m(t)\{e^{iu} - 1\}] \tag{2.1}$$

for some non-decreasing function $m(\cdot)$. In words, $N(t)$ is Poisson distributed with mean $m(t)$.

For a Poisson process $\{N(t), t \geq 0\}$ with stationary increments,

$$m(t) = E[N(t)] = \nu t, \tag{2.2}$$

so that $m(t)$ is directly proportional to t, with proportionality factor ν, the mean rate at which counts are being made. In the case that Eq. 2.2 does *not* hold, we call $\{N(t), t \geq 0\}$ a *non-homogeneous Poisson process* (or a

Poisson process with non-stationary increments). A Poisson process with stationary increments is called a homogeneous Poisson process.

The mean value function $m(t)$ is always assumed to be continuous. Usually $m(t)$ is also assumed to be differentiable, with derivative denoted by

$$\nu(t) = \frac{d}{dt} m(t). \tag{2.3}$$

We call $\nu(t)$ the *intensity* function. For a discussion of non-continuous mean value functions, and a general axiomatic derivation of Eq. 2.1, the reader is referred to Khinchin (1956). We shall here derive Eq. 2.1 under the assumption of the existence of an intensity function.

THEOREM 2A

Axiomatic derivation of the non-homogeneous Poisson process. Let $\{N(t), t \geq 0\}$ be an integer-valued process satisfying axioms 0, 1, 2, and 3 of Section 4–1. Suppose further that for some function $\nu(t)$,

Axiom 4. $\displaystyle\lim_{h \to 0} \frac{1 - P[N(t+h) - N(t) = 0]}{h} = \nu(t).$

Then $N(t)$ has probability generating function

$$\psi(z,t) = \exp\{m(t)(z - 1)\} \tag{2.4}$$

where

$$m(t) = \int_0^t \nu(t') \, dt'. \tag{2.5}$$

Remark. We call $\nu(t)$ the intensity function of the Poisson process. Note that $\nu(t)h$ represents the approximate probability that an event will occur in the time interval $(t, t+h)$. A homogeneous Poisson process has constant intensity function:

$$\nu(t) = \nu \qquad \text{for all } t.$$

For many events one expects a decreasing intensity of occurrence. (For example, accidents in coal mines should occur less frequently as more safety precautions are introduced.) A form of intensity function frequently chosen is then

$$\nu(t) = \alpha e^{-\beta t}$$

for some positive constants α, β to be determined empirically.

Proof. Because $\{N(t), t \geq 0\}$ has independent increments, it follows that

$$\psi(z, t + h) = \psi(z,t) E[z^{N(t+h)-N(t)}]. \tag{2.6}$$

Now,

$$\frac{1}{h} \{E[z^{N(t+h)-N(t)}] - 1\} = \frac{1}{h} \{P[N(t+h) - N(t) = 0] - 1\}$$

$$+ z \frac{1}{h} P[N(t+h) - N(t) = 1]$$

$$+ \frac{1}{h} \sum_{n=2}^{\infty} z^n P[N(t+h) - N(t) = n]. \quad (2.7)$$

From Axioms 3 and 4 one may show that

$$\lim_{h \to 0} \frac{1}{h} \{E[z^{N(t+h)-N(t)}] - 1\} = \nu(t)\{z - 1\}. \quad (2.8)$$

From Eqs. 2.6 and 2.8 one may infer that $\psi(z,t)$ satisfies the differential equation, for $|z| < 1$ and all $t \geq 0$

$$\frac{\partial}{\partial t} \psi(z,t) = \psi(z,t)\nu(t)\{z - 1\}. \quad (2.9)$$

Under the initial condition $\psi(z,0) = 1$, the solution of Eq. 2.9 is given by Eq. 2.4.

The non-homogeneous Poisson process can be transformed into a homogeneous Poisson process. Since the mean value function $m(t)$ is continuous and non-decreasing, one can define its inverse function $m^{-1}(u)$ as follows: for $u > 0$, $m^{-1}(u)$ is the smallest value of t satisfying the condition $m(t) \geq u$. The stochastic process $\{M(u), u \geq 0\}$ defined by

$$M(u) = N(m^{-1}(u)), \qquad u \geq 0 \quad (2.10)$$

is a Poisson process with mean value function

$$E[M(u)] = E[N(m^{-1}(u))] = m(m^{-1}(u)) = u. \quad (2.11)$$

Consequently, $\{M(u), u \geq 0\}$ is a homogeneous Poisson process, with intensity $\nu = 1$. The transformation Eq. 2.10 can often be profitably used in studying the properties of non-homogeneous Poisson processes (see Complement 4B).

Generalized Poisson process. An integer-valued stochastic process $\{N(t), t \geq 0\}$ with stationary independent increments is called a *generalized Poisson process*. It may be shown (see Feller [1957], p. 271) that a generalized Poisson process necessarily has a characteristic function of the form

$$\varphi_{N(t)}(u) = e^{\nu t [\varphi(u)-1]} \tag{2.12}$$

for some positive constant ν, and some characteristic function

$$\varphi(u) = \sum_{k=1}^{\infty} p_k e^{iku}, \tag{2.13}$$

which is the characteristic function of a non-negative integer valued random variable with probability distribution $\{p_k\}$. The Poisson process corresponds to the case $\varphi(u) = e^{iu}$. Various other important cases of $\varphi(u)$ which have been considered are

$$\varphi(u) = e^{\lambda(e^{iu}-1)}, \qquad \text{Poisson distribution,}$$

$$\varphi(u) = \left(\frac{p}{1 - qe^{iu}}\right)^r, \qquad \text{negative binomial distribution.}$$

We discuss below a variety of random phenomena for which the generalized Poisson process arises naturally as a model. First we give an axiomatic derivation of the generalized Poisson process.

THEOREM 2B

An axiomatic derivation of the generalized Poisson process. Let $\{N(t), t \geq 0\}$ be a counting process satisfying Axioms 0, 1, 2, and 4. In addition, assume that there exists a sequence $\{p_k\}$ such that, for $k = 1, 2, \cdots$ and $t \geq 0$,

$$\lim_{h \to 0} P[N(t+h) - N(t) = k \mid N(t+h) - N(t) \geq 1] = p_k. \tag{2.14}$$

Then $\{N(t), t \geq 0\}$ is a process with stationary independent increments satisfying Eq. 2.12 for some constant ν.

Remark. Intuitively p_k represents the conditional probability that exactly k events occur simultaneously at a given time, given that at least one has occurred.

Proof. As in the proof of Theorem 1A it follows that the probability generating function

$$\psi(z,t) = \sum_{n=0}^{\infty} z^n P[N(t) = n]$$

satisfies the differential equation

$$\frac{\partial}{\partial t} \psi(z,t) = \psi(z,t) \lim_{h \to 0} \frac{1}{h} \{\psi(z,h) - 1\}. \tag{2.15}$$

Now

$$\frac{1}{h}\{\psi(z,h) - 1\} = \frac{1}{h}\{P[N(h) = 0] - 1\}$$

$$+ \frac{P[N(h) \geq 1]}{h} \sum_{n=1}^{\infty} z^n P[N(h) = n \mid N(h) \geq 1].   (2.16)$$

As in the proof of Theorem 1A one may prove that there exists a constant $\nu > 0$ such that

$$\lim_{h \to 0} \frac{1}{h}\{1 - P[N(h) = 0]\} = \lim_{h \to 0} \frac{1}{h} P[N(h) \geq 1] = \nu.   (2.17)$$

From Eq. 2.14 it follows that

$$\lim_{h \to 0} \sum_{n=1}^{\infty} z^n P[N(h) = n \mid N(h) \geq 1] = \sum_{n=1}^{\infty} z^n p_n;   (2.18)$$

that the limit and summation in Eq. 2.18 may be interchanged follows from the dominated convergence theorem (see Section 6–10). From Eqs. 2.15–2.18 it follows that

$$\frac{\partial}{\partial t} \psi(z,t) = \psi(z,t)\nu\{\psi(z) - 1\},   (2.19)$$

where

$$\psi(z) = \sum_{n=1}^{\infty} z^n p_n.   (2.20)$$

Consequently,

$$\psi(z, t) = e^{\nu t\{\psi(z)-1\}},   (2.21)$$

which is equivalent to Eq. 2.12.

Compound Poisson processes. A stochastic process $\{X(t), t \geq 0\}$ is said to be a *compound Poisson process* if it can be represented, for $t \geq 0$, by

$$X(t) = \sum_{n=1}^{N(t)} Y_n,   (2.22)$$

in which $\{N(t), t \geq 0\}$ is a Poisson process, and $\{Y_n, n = 1, 2, \cdots\}$ is a family of independent random variables identically distributed as a random variable Y. The process $\{N(t), t \geq 0\}$ and the sequence $\{Y_n\}$ are assumed to be independent.

It should be noted that the right-hand side of Eq. 2.22 is the sum of a random number of independent identically distributed random variables.

Before deriving the properties of compound Poisson processes, let us consider several examples of how such processes arise.

EXAMPLE 2A

The total claims on an insurance company. Suppose that policy holders of a certain life insurance company die at times $\tau_1, \tau_2, \cdots$, where $0 < \tau_1 < \tau_2 < \cdots$. Deaths are assumed to be events of Poisson type with intensity ν. The policy holder dying at time τ_n carries a policy for an amount Y_n, which is paid to his beneficiary at the time of his death. The insurance company is interested in knowing $X(t)$, the total amount of claims it will have to pay in the time period 0 to t, in order to determine how large a reserve to have on hand to meet the claims it will be called upon to pay. One may represent $X(t)$ in the form of Eq. 2.22. Therefore $\{X(t), t \geq 0\}$ is a compound Poisson process.

EXAMPLE 2B

Drifting of stones on river beds. Suppose that a stone on a river bed suffers displacements at times $\tau_1 < \tau_2 < \cdots$, where displacements are assumed to be events of Poisson type with intensity ν. The displacement the stone undergoes at time τ_n is a random variable Y_n. Let $X(t)$ be the total distance the stone has drifted at time t. One may represent $X(t)$ in the form of Eq. 2.22. Consequently, $\{X(t), t \geq 0\}$ is a compound Poisson process.

EXAMPLE 2C

A model for Brownian motion. A model for the motion of a particle, arising from its numerous random impacts with other particles, can be constructed as follows. Let the times at which it suffers impacts be $0 < \tau_1 < \tau_2 < \cdots$, where impacts are assumed to be events of Poisson type with intensity ν. Assume that as a result of the impact at time τ_n the particle changes its position by a random amount Y_n. The position $X(t)$ of the particle at time t is then given by Eq. 2.22, assuming the particle to be at 0 at time 0. It may be shown that the stochastic process $\{X(t), t \geq 0\}$ is asymptotically the Wiener process as the number of collisions per unit time is increased indefinitely (see Section 3-5).

EXAMPLE 2D

Distribution of galaxies. Suppose (i) that "centers" (such as the center of a cluster of galaxies in the stellar system or the center of an animal population) are distributed at random (in accord with a Poisson process with density ν on a plane or in three-dimensional space), (ii) that each center gives rise to a number of galaxies or offspring (the actual number being a random variable Y with probability law $P[Y = k] = p_k$) and (iii) that the galaxies or offspring from any one center are distributed in the space around that center independently of one another. The total number $X(R)$ of galaxies or offspring occurring in a region R (of area or volume t) may be written

$$X(R) = \sum_{\tau_n \in R} Y_n, \qquad (2.23)$$

where Y_n is the number of galaxies or offspring arising from the center at τ_n. To find the probability law of the random variables defined by Eq. 2.23 one may use the same methods as are used to treat random variables of the form of Eq. 2.22. Random variables of the form of Eq. 2.23 play an important role in the theory of spatial distribution of galaxies recently developed by J. Neyman and E. L. Scott (for a popular account of this theory, see Neyman and Scott [1957]; for a technical account, see Neyman and Scott [1958]).

THEOREM 2C

A compound Poisson process $\{X(t), t \geq 0\}$ has stationary independent increments, and characteristic function, for any $t \geq 0$,

$$\varphi_{X(t)}(u) = e^{\nu t \{\varphi_Y(u) - 1\}}, \qquad (2.24)$$

where $\varphi_Y(u)$ is the common characteristic function of the independent identically distributed random variables $\{Y_n\}$ and ν is the mean rate of occurrence of events. If $E[Y^2] < \infty$, then $X(t)$ has finite second moments given by

$$E[X(t)] = \nu t \, E[Y], \qquad (2.25)$$
$$\mathrm{Var}[X(t)] = \nu t \, E[Y^2], \qquad (2.26)$$
$$\mathrm{Cov}[X(s), X(t)] = \nu E[Y^2] \min (s,t). \qquad (2.27)$$

Remark. From Eq. 2.24 it follows that if Y is integer-valued, then the compound Poisson process $\{X(t), t \geq 0\}$ is a generalized Poisson process. Conversely, any generalized Poisson process can be represented as a compound Poisson process. However, a generalized Poisson process can arise in other ways as well; for example it can be represented as an infinite linear combination of Poisson processes (see Complement 2A).

Proof. Since the Poisson process $\{N(t), t \geq 0\}$ has independent increments, and since $\{Y_n\}$ is a sequence of independent identically distributed random variables, it is (intuitively) clear that $\{X(t), t \geq 0\}$ has independent increments (we do not give a formal proof). To prove that $\{X(t), t \geq 0\}$ has stationary increments, and that Eq. 2.24 holds, it suffices to prove that for any $t > s \geq 0$

$$\varphi_{X(t)-X(s)}(u) = \exp[\nu(t - s)\{\varphi_Y(u) - 1\}]. \qquad (2.28)$$

Now, for $n = 0, 1, 2, \cdots$

$$E[e^{iu\{X(t)-X(s)\}} \mid N(t) - N(s) = n] = \{\varphi_Y(u)\}^n,$$

since, given that n events have occurred in the interval $(s,t]$, $X(t) - X(s)$ is the sum of n independent random variables identically distributed as Y. Now,

$$\varphi_{X(t)-X(s)}(u) = \sum_{n=0}^{\infty} E[e^{iu\{X(t)-X(s)\}} \mid N(t) - N(s) = n]P[N(t) - N(s) = n]$$

$$= \sum_{n=0}^{\infty} \{\varphi_Y(u)\}^n e^{-\nu(t-s)} \frac{\{\nu(t-s)\}^n}{n!}$$

$$= e^{-\nu(t-s)} \exp[\nu(t-s)\varphi_Y(u)].$$

The proof of Eq. 2.28 is complete. To prove Eqs. 2.25 and 2.26 one may either differentiate Eq. 2.24 or use the formulas (given in Section 2–2) for the mean and variance of the sum of a random number of independent identically distributed random variables:

$$E[X(t)] = E[N(t)]E[Y] = \nu t\, E[Y],$$
$$\mathrm{Var}[X(t)] = E[N(t)]\,\mathrm{Var}[Y] + \mathrm{Var}[N(t)]E^2[Y] = \nu t\, E[Y^2].$$

It is left to the reader to prove Eq. 2.27.

COMPLEMENTS

2A *Representation of a generalized Poisson process as an infinite linear combination of Poisson processes.* Let $\{N_k(t), t \geq 0\}$, $k = 1, 2, \cdots$, be a sequence of independent Poisson processes with respective mean rates λ_k. Assume that

$$\nu = \sum_{k=1}^{\infty} \lambda_k < \infty. \text{ Define } p_k = \lambda_k/\nu. \text{ Define}$$

$$N(t) = N_1(t) + 2N_2(t) + \cdots + kN_k(t) + \cdots.$$

Show that $\{N(t), t \geq 0\}$ is a generalized Poisson process, with characteristic function given by Eq. 2.12.

EXERCISES

For each of the stochastic processes $\{X(t), t \geq 0\}$ described in Exercises 2.1 to 2.7 find, for $t > 0$, (i) the mean $E[X(t)]$, (ii) the variance $\mathrm{Var}[X(t)]$, (iii) the probability $P[X(t) = 0]$, (iv) the characteristic function $\varphi_{X(t)}(u)$.

2.1 Suppose that pulses arrive at a Geiger counter in accord with a Poisson process at a rate of 6 arrivals per minute. Each particle arriving at the counter has probability 2/3 of being recorded. Let $X(t)$ be the number of pulses recorded in time t.

2.2 A housewife sells magazine subscriptions by mail. Her customers respond, in accord with a Poisson process, at a mean rate of 6 a day. They may subscribe for 1, 2, or 3 years, which they do independently of one another with probabilities 1/2, 1/3, and 1/6 respectively. For each subscription sold, she is paid $1 a subscription year at the time the subscription is ordered. Let $X(t)$ be her total commission from subscriptions sold in the period 0 to t.

2.3 (Modification of Exercise 2.2) Do exercise 2.2, assuming now that the housewife is absent-minded, and does not transmit all subscriptions ordered through her. Rather only 2/3 of the subscriptions ordered (chosen at random) are actually transmitted to the magazine and paid for by the customers.

2.4 Let $\{N(t), t \geq 0\}$ be a Poisson process, with mean rate ν per unit time, representing the number of cosmic ray showers arriving in the time 0 to t at a certain observatory. Suppose that the number of particles in each cosmic ray shower are independent identically distributed random variables, each obeying geometric distribution with mean λ. Let $X(t)$ be the total number of particles recorded in the time 0 to t.

2.5 Let $X(t)$ be the total displacement in time t of the stone described in Example 2B. Assume that displacements obey a gamma distribution with parameters r and λ.

2.6 Let $X(t)$ be the total claims on an insurance company as described in Example 2A. Assume that the amount of a policy is a random variable (i) uniformly distributed between $1,000 and $10,000, (ii) exponentially distributed with mean $5,000.

2.7 Let $X(t)$ be the total quantity of a certain item ordered from a certain manufacturer in t days. Assume that orders are events of Poisson type, with intensity ν per day, while the amount ordered is exponentially distributed with mean μ.

**4-3   INTER-ARRIVAL TIMES
          AND WAITING TIMES**

In experimentally observing the emission of particles from a radio-active substance, it is customary to clock the length of time taken to register a fixed number of particles, rather than to count the number of particles emitted during a fixed interval of time. One is thus led to consider the random variable W_n, called the *waiting time* to the nth event, which represents the time it takes to register n events if one is observing a series of events occurring in time.

Given events occurring in the interval 0 to ∞ the *successive inter-arrival* times $T_1, T_2, \cdots$ are defined as follows: T_1 is the time from 0 to the first event and for $j > 1$, T_j is the time from the $(j-1)$st to the jth event. In terms of the waiting times $\{W_n\}$, the inter-arrival times $\{T_n\}$ are given by

$$T_1 = W_1, \; T_2 = W_2 - W_1, \cdots, T_n = W_n - W_{n-1}, \cdots. \quad (3.1)$$

In terms of the inter-arrival times $\{T_n\}$, the waiting times $\{W_n\}$ are given by

$$W_n = T_1 + T_2 + \cdots + T_n \qquad \text{for } n \geq 1. \quad (3.2)$$

There exists a basic relation between a counting process $\{N(t), t \geq 0\}$ and the corresponding sequence of waiting times $\{W_n\}$: for any $t > 0$ and $n = 1, 2, \cdots$

$$N(t) \leq n \quad \text{if and only if} \quad W_{n+1} > t. \quad (3.3)$$

In words, Eq. 3.3 states that the number of events occurring in the interval $(0,t]$ is less than or equal to n if and only if the waiting time to the $(n+1)$st event is greater than t. From Eq. 3.3 it follows that

$$N(t) = n \quad \text{if and only if} \quad W_n \leq t \quad \text{and} \quad W_{n+1} > t. \quad (3.4)$$

In words, Eq. 3.4 states that in the interval $(0,t]$ exactly n events occur if and only if the waiting time to the nth event is less than or equal to t, and the waiting time to the $(n+1)$st event is greater than t.

From Eqs. 3.3 and 3.4 two important probability relations follow:

$$P[N(t) \leq n] = P[W_{n+1} > t], \; n = 0, 1, \cdots; \quad (3.5)$$
$$P[N(t) = n] = P[W_n \leq t] - P[W_{n+1} \leq t], \; n = 1, 2, \cdots,$$
$$P[N(t) = 0] = 1 - P[W_1 \leq t]. \quad (3.6)$$

To prove Eq. 3.6 note that $W_n \leq t$ and $W_{n+1} > t$ if and only if $W_n \leq t$ and it is not so that $W_{n+1} \leq t$. One may state Eqs. 3.5 and 3.6 in terms of the distribution functions of the waiting times:

$$F_{N(t)}(n) = 1 - F_{W_{n+1}}(t), \; n = 0, 1, \cdots; \quad (3.7)$$

$$p_{N(t)}(n) = F_{W_n}(t) - F_{W_{n+1}}(t), \; n = 1, 2, \cdots,$$

$$p_{N(t)}(0) = 1 - F_{W_1}(t). \quad (3.8)$$

The Poisson process, gamma distributed waiting times, and independent exponentially distributed inter-arrival times. Let W_n be the waiting time to the nth event in a series of events occurring on the interval

0 to ∞ in accord with a Poisson process at mean rate ν. Then W_n obeys the gamma probability law with parameters n and ν. More precisely,

$$f_{W_n}(t) = \nu e^{-\nu t} \frac{(\nu t)^{n-1}}{(n-1)!}, \quad t > 0 \tag{3.9}$$

$$= 0, \qquad\qquad t < 0;$$

$$F_{W_n}(t) = 1 - e^{-\nu t}\left(1 + \nu t + \cdots + \frac{(\nu t)^{n-1}}{(n-1)!}\right), \quad t > 0; \tag{3.10}$$

$$\varphi_{W_n}(u) = \left(1 - i\frac{u}{\nu}\right)^{-n}; \tag{3.11}$$

$$E[W_n] = \frac{n}{\nu}; \tag{3.12}$$

$$\mathrm{Var}[W_n] = \frac{n}{\nu^2}. \tag{3.13}$$

To prove these assertions, it suffices to prove Eq. 3.10, which follows from the fact that

$$1 - F_{W_n}(t) = P[W_n > t] = P[N(t) < n] = \sum_{k=0}^{n-1} e^{-\nu t}\frac{(\nu t)^k}{k!}.$$

EXAMPLE 3A

Length of life of machine parts and structures. Consider a structure, consisting of m components, which is subjected to sudden shocks of Poisson type with intensity ν. In the course of time, one component after another fails. If W_n is the life length of the structure until n components have failed, then W_n is gamma distributed with parameters n and ν. On the basis of this argument the gamma distribution is often adopted as a model for the length of life of systems subject to failure. Thus suppose one is studying the length of life of pipes used in a municipal water system. More precisely, let T be the number of years a foot of pipe will serve before corroding. Suppose that municipal records indicate that T has mean 36 years and standard deviation 18 years. If one hypothesizes that T obeys a gamma distribution with parameters n and ν, then one may estimate n and ν by the method of moments; that is, estimate n and ν to be the values satisfying

$$E[T] = \frac{n}{\nu} = 36, \quad \mathrm{Var}[T] = \frac{n}{\nu^2} = (18)^2. \tag{3.14}$$

Therefore

$$\nu = \frac{E[T]}{\mathrm{Var}[T]} = \frac{1}{9}, \quad n = \frac{E^2[T]}{\mathrm{Var}[T]} = 4. \tag{3.15}$$

The method of moments is not in general the most efficient method for estimating parameters [see Cramér (1946), p. 498]. For references to other methods of estimating the parameters of the gamma distribution, see Gupta and Groll (1961).

We next show that the *successive inter-arrival times* T_1, T_2, $\cdots$ of events of Poisson type with intensity ν are *independent identically distributed random variables, each obeying an exponential probability law with mean* $1/\nu$. This fact provides a method of testing the hypothesis that a sequence of events occurring in time are events of Poisson type. The observed inter-arrival times T_1, T_2, $\cdots$ are assumed to be independent observations of a random variable T. Using various goodness of fit tests (such as the χ^2 tests or the tests described in Section 3–5) one may test the hypothesis that T is exponentially distributed. In this way, one tests the hypothesis that the observed events are of Poisson type since the *Poisson process* $\{N(t), t \geq 0\}$ *is characterized by the fact that the inter-arrival times* $\{T_n\}$ *are independent exponentially distributed random variables.*

THEOREM 3A

The successive inter-arrival times T_1, T_2, $\cdots$ of events of Poisson type of intensity ν are independent identically distributed random variables, obeying an exponential probability law with mean $1/\nu$.

Proof. To prove the theorem it suffices to prove that for any integer n, and real numbers t_1, t_2, $\cdots$, t_n

$$P[T_1 > t_1, T_2 > t_2, \cdots, T_n > t_n] = e^{-\nu t_1} e^{-\nu t_2} \cdots e^{-\nu t_n}. \quad (3.16)$$

It is clear that Eq. 3.16 holds for $n = 1$, since

$$P[T_1 > t_1] = P[N(t_1) = 0] = e^{-\nu t_1}. \quad (3.17)$$

To prove Eq. 3.16 in general, one way is to show that for any $n > 1$, and real numbers y, x_1, $\cdots$, x_{n-1}

$$P[T_n > y \mid T_1 = x_1, \cdots, T_{n-1} = x_{n-1}] = e^{-\nu y}. \quad (3.18)$$

From Eq. 3.18 and the fact that

$$P[T_1 > t_1, \cdots, T_n > t_n]$$
$$= \int_{t_1}^{\infty} \cdots \int_{t_{n-1}}^{\infty} P[T_n > t_n \mid T_1 = x_1, \cdots, T_{n-1} = x_{n-1}]$$
$$f_{T_1, \ldots, T_{n-1}}(x_1, \cdots, x_{n-1}) \, dx_1 \cdots dx_{n-1}$$

one may infer Eq. 3.16 by mathematical induction. Unfortunately, a complete proof of Eq. 3.18 is beyond the scope of this book. We indicate the argument by considering the case in which $n = 2$. Now,

$$P[T_2 \geq y \mid T_1 = x] = P[N(y + T_1) - N(T_1) = 0 \mid T_1 = x]. \quad (3.19)$$

Since the Poisson process $N(\cdot)$ has stationary independent increments, it seems intuitively plausible that

$$P[N(y + T_1) - N(T_1) = 0 \mid T_1 = x] = P[N(y) - N(0) = 0] = e^{-\nu y} \quad (3.20)$$

in view of the fact that T_1 depends only on the values of $N(t)$ in the interval $0 \le t \le T_1$, which are independent of $N(y + T_1) - N(T_1)$. A rigorous proof of Eq. 3.20 is beyond the scope of this book. (For proofs of related but more general results, see Chung [1960] or Loève [1960].) If the validity of Eq. 3.20 is accepted, and, more generally, if the validity of the assertion that for $n > 1$ and non-negative numbers $y, x_1, \cdots, x_{n-1}$

$$P[T_n > y \mid T_1 = x_1, \cdots, T_{n-1} = x_{n-1}]$$
$$= P[N(y + x_1 + \cdots + x_{n-1}) - N(x_1 + \cdots + x_{n-1}) = 0] = e^{-\nu y}, \quad (3.21)$$

then the proof of Theorem 3A is complete.

An alternate proof of Theorem 3A is given in Section 4–4. A converse of Theorem 3A is given in Section 5–2.

The results of this section play an important role in the theory of statistical inference on Poisson processes. We mention two possible applications.

EXAMPLE 3B

Estimation of the parameter ν of a Poisson process. If one observes a Poisson process for a predetermined observation time t, then the number of counts $N(t)$ can be used to form point estimates and confidence intervals for ν, using the fact that $N(t)$ is Poisson distributed with mean νt. On the other hand, if observation of the Poisson process is continued until a specified number m of events has been counted, then the amount W_m of observation required to obtain the m events can be used to form confidence intervals for ν, using the fact that $2\nu W_m$ is χ^2 distributed with $2m$ degrees of freedom. Let C and D be values such that if Z has a χ^2 distribution with $2m$ degrees of freedom, then $P[Z < C] = P[Z > D] = \alpha/2$. Then

$$1 - \alpha = P[C \le 2\nu W_m \le D] = P\left[\frac{C}{2W_m} \le \nu \le \frac{D}{2W_m}\right].$$

Consequently, $(C/2W_m, D/2W_m)$ is a confidence interval for ν, with confidence coefficient $1 - \alpha$. An extensive discussion of this procedure, and extensions, is given by Girshick, Rubin, and Sitgreaves (1955).

EXAMPLE 3C

Procedures for comparing Poisson processes. Let $N(\cdot)$ and $N'(\cdot)$ be two independent Poisson processes, with respective mean rates ν and ν'.

Theorem 3A provides a means whereby one may test whether $\nu = \nu'$. Let n_1 and n_2 be integers, let W_{n_1} be the waiting time to the n_1th event in the series of events with counting process $N(\,\cdot\,)$, and let W_{n_2}' be the waiting time to the n_2th event in the series of events with counting process $N'(\,\cdot\,)$. Now $2\nu W_{n_1}$ and $2\nu' W_{n_2}'$ are independent χ^2 distributed random variables, with respective degrees of freedom $2n_1$ and $2n_2$. Consequently, under the hypothesis that $\nu = \nu'$ it follows that $n_2 W_{n_1}/n_1 W_{n_2}'$ has an F distribution with $2n_1$ and $2n_2$ degrees of freedom in the numerator and denominator respectively. From this fact one obtains a test of significance for the hypothesis $\nu = \nu'$. A particularly important use of this procedure is to test whether in two different sections of a series of events of Poisson type the mean rate of occurrence is the same. For a numerical illustration of this procedure, see Maguire, Pearson, and Wynn (1952), p. 172. Other methods of estimating and testing hypotheses concerning the ratio ν/ν' are given by Birnbaum (1953, 1954).

COMPLEMENTS

Life testing. The failure of electronic components while in use can often be regarded as events of Poisson type with an intensity $1/\theta$ which is to be determined from experiment. One method of estimating the parameter θ is the following. A convenient sample size n being chosen, n items are put into service simultaneously and kept in operation until exactly r items have failed. For $j = 1, \cdots, r$, let W_j be the time to the jth failure. Let $T_1, \cdots, T_r$ be the successive times between failures (that is, $T_1 = W_1$, $T_j = W_j - W_{j-1}$ for $j \geq 2$). Prove the following assertions. (A full discussion of life testing procedures is given in B. Epstein, *Statistical Techniques in Life Testing*, to be published; see also Epstein [1960].)

3A   For $j = 1, \cdots, r$, T_j is exponentially distributed with mean $\theta/(n - j + 1)$;

3B $T_1, \cdots, T_r$ are independent;

3C   If $V_r = \sum_{j=1}^{r} W_j + (n - r) W_r = \sum_{j=1}^{r} (n - j + 1) T_j$ then $2V_r/\theta$ is χ^2 distributed with $2r$ degrees of freedom;

3D V_r/r is an unbiased estimate of θ in the sense that
$$E\left[\frac{1}{r} V_r\right] = \theta;$$

3E   Show how 3C may be used to form a two-sided confidence interval for θ of significance level $1 - \alpha$.

3F   Suppose that 7 electronic components of a certain type are put into service and run until 4 have failed. The times to the first, second, third, and fourth failures are, respectively, 23.5, 52.8, 72.0, and 158.8 hours. Estimate the mean life θ of the component by (i) a point estimate, (ii) a confidence interval of significance level 90%.

The distribution of inter-arrival times of a non-homogeneous Poisson process. Given events occurring in accord with a non-homogeneous Poisson process with continuous mean value function $m(t)$, one defines the successive inter-arrival times $T_1, T_2, \cdots$, and the successive waiting times $W_1, W_2, \cdots$, as follows: T_1 is the time from 0 to the first event, T_n is the time from the $(n-1)$st to the nth event, W_n is the time from 0 to the nth event. Show that, for $t > 0$,

3G $f_{T_1}(t) = e^{-m(t)} \nu(t)$;

3H $f_{T_2 | T_1}(t \mid s) = e^{-m(t+s) + m(s)} \nu(t+s)$;

3I $f_{T_2}(t) = \int_0^\infty e^{-m(t+s)} \nu(t+s) \nu(s) \, ds$;

3J $f_{W_n}(t) = e^{-m(t)} \dfrac{\{m(t)\}^{n-1}}{(n-1)!} \nu(t)$;

3K $1 - F_{T_k | W_{k-1}}(t \mid s) = e^{-m(s+t) + m(s)}$;

3L $1 - F_{T_k}(t) = \int_0^\infty e^{-m(t+s)} \dfrac{\{m(s)\}^{k-2}}{(k-2)!} \nu(s) \, ds, \, k \geq 2$.

EXERCISES

3.1 Consider a shop at which customer arrivals are events of Poisson type with density 30 per hour. What is the probability that the time interval between successive arrivals will be

(i) longer than 2 minutes?
(ii) shorter than 4 minutes?
(iii) between 1 and 3 minutes?

3.2 Suppose that in a certain α-ray counting experiment conducted on a radioactive source of constant intensity, a total of 18,000 α-rays were counted in 100 hours of continuous observation. The time of arrival of each α-ray is recorded on a tape so that the number of α-rays recorded in each successive 1-minute interval can be determined.

(i) For $k = 0, 1, 2, 3, 4$ find the expected number of 1-minute intervals containing exactly k α-rays, assuming that α-rays are emitted by the source in accord with a Poisson process at a mean rate equal to the observed average number of α-rays per 1-minute interval.

(ii) For $k = 0, 1, 2, 3, 4$ find the expected number of empirically observed inter-arrival times between $10k$ and $10(k+1)$ seconds in length.

3.3 A certain nuclear particle counter records only every second particle actually arriving at the counter. Suppose particles arrive in accord with a Poisson process at a mean rate of 4 per minute. Let T be the inter-arrival time (in minutes) between two successive recorded particles. Find (i) the probability density function of T, (ii) $P[T \geq 1]$, (iii) $E[T]$, $\mathrm{Var}(T)$.

3.4   Do Exercise 3.3 under the assumption that the counter records only every fourth particle arriving.

3.5   Consider a baby who cries in accord with a Poisson process at a mean rate of 12 times per hour. If his parents respond only to every (i) second cry, (ii) third cry, what is the probability that ten or more minutes elapse between successive responses of the parents to the baby?

3.6   A certain nuclear particle counter records only every second particle actually arriving at the counter. Suppose particles arrive in accord with a Poisson process at a mean rate of 2 per minute. Let $N(t)$ denote the number of particles recorded in the first t minutes, assuming that the counter starts counting with the second particle to arrive. Find (i) $P[N(t) = n]$ for $n = 0, 1, \cdots$, (ii) $E[N(t)]$.

3.7   *A geometric distribution.* A certain radioactive source is a mixture of two radioactive substances, which respectively emit α-rays and β-rays. The two substances are assumed to act independently. It is known that the mean rates of emission of α-rays and β-rays are respectively μ and ν. Show that the probability of observing exactly k β-rays in the time interval between two successive α-ray counts is, for $k = 0, 1, \cdots$,

$$\frac{\mu}{\mu + \nu}\left(\frac{\nu}{\mu + \nu}\right)^{k}.$$

3.8   In a series of mine explosions observed in England, it was noted that the time interval between the first and 50th accidents was 7,350 days. Assume that explosions are events of Poisson type. Estimate the rate ν at which explosions are occurring by means of (i) a point estimate, (ii) a confidence interval.

3.9   In a series of 108 mine explosions observed in England (see Maguire, Pearson, and Wynn [1952], p. 214) it was noted that the time interval between the first and 54th accidents was 8,042 days, while the time interval between the 55th and 108th accidents was 16,864 days. If it is assumed that in each of these two periods the mean rate at which explosions occurred were constants ν_1 and ν_2 respectively, perform an F-test for the hypothesis $\nu_1 = \nu_2$. Is the value $F = 16,864/8,042 = 2.10$ significant at the 1% level of significance?

4-4   **THE UNIFORM DISTRIBUTION OF WAITING TIMES OF A POISSON PROCESS**

The Poisson process is often called the random process, or the model of complete randomness, because it represents the distribution at random of an infinite number of points over the infinite interval 0 to ∞. More precisely we have the following theorem.

THEOREM 4A

Let $\{N(t), t \geq 0\}$ be a Poisson process with intensity ν. Under the condition that $N(T) = k$, the k times $\tau_1 < \tau_2 < \cdots < \tau_k$ in the interval 0 to T at which events occur are random variables having the same distribution as if they were the order statistics corresponding to k independent random variables $U_1, \cdots, U_k$ uniformly distributed on the interval 0 to T; we say that $\tau_1, \cdots, \tau_k$ are the order statistics corresponding to $U_1, \cdots, U_k$ if τ_1 is the smallest value among $U_1, \cdots, U_k$, τ_2 is the second smallest value among $U_1, \cdots, U_k$, and so on, so that τ_k is the largest value among $U_1, \cdots, U_k$.

Remark. Intuitively, one usually says that under the condition that k events have occurred in the interval 0 to T the k times $\tau_1, \cdots, \tau_k$ at which events occur, considered as unordered random variables, are distributed independently and uniformly in the interval 0 to T.

It should be noted that the random variable τ_k, representing the time of occurrence of the kth event to occur and the random variable W_k, representing the waiting time to the kth event, are different notations for exactly the same concept.

Proof. We first note that if $U_1, \cdots, U_k$ are independently uniformly distributed on the interval 0 to T, then they have joint probability density function

$$f_{U_1,\ldots,U_k}(u_1, \cdots, u_k) = \frac{1}{T^k}, \quad 0 \leq u_1, \cdots, u_k \leq T \qquad (4.1)$$
$$= 0, \text{ otherwise.}$$

The order statistics $\tau_1, \cdots, \tau_k$ corresponding to $U_1, \cdots, U_k$ have joint probability density function

$$f_{\tau_1,\ldots,\tau_k}(t_1, \cdots, t_k) = \frac{k!}{T^k}, \quad 0 \leq t_1 \leq t_2 \leq \cdots \leq t_k \leq T \qquad (4.2)$$
$$= 0, \text{ otherwise.}$$

Now the conditional probability, given that k events have occurred in $(0,T]$, that in each of k non-overlapping sub-intervals $[t_1, t_1 + h_1], \cdots,$ $[t_k, t_k + h_k]$ of the interval $(0,T]$ exactly 1 event occurs and elsewhere no events occur, is

$$\frac{\lambda h_1 e^{-\lambda h_1} \cdots \lambda h_k e^{-\lambda h_k} e^{-\lambda(T - h_1 - \ldots - h_k)}}{\dfrac{e^{-\lambda T}(\lambda T)^k}{k!}} = \frac{k!}{T^k} \{h_1 h_2 \cdots h_k\}. \qquad (4.3)$$

In symbols, if $\tau_1 < \tau_2 < \cdots < \tau_k$ are the times at which the k events have occurred, then

$$P[t_1 \leq \tau_1 \leq t_1 + h_1, \cdots, t_k \leq \tau_k \leq t_k + h_k \mid N(T) = k] = \frac{k!}{T^k} h_1 \cdots h_k.$$
$$(4.4)$$

Now, by definition of the probability density function, the left-hand side of Eq. 4.4 is approximately equal to

$$f_{\tau_1 \ldots, \tau_k}(t_1, \cdots, t_k) h_1 h_2 \cdots h_k.$$

We have thus proved that the joint probability density of the ordered arrival times $\tau_1, \cdots, \tau_k$ of events is equal to $k!/T^k$, which is what the joint probability density function would be if $\tau_1, \cdots, \tau_k$ were the order statistics corresponding to n independent random variables uniformly distributed on the interval 0 to T. The proof of Theorem 4A is now complete.

EXAMPLE 4A

Testing whether events are of Poisson type. Theorem 4A provides a way of testing whether or not an observed set of events is of Poisson type. Suppose one has observed the process over a period of length T, during which n events have occurred. Suppose that one were to label the n events which have occurred in a random manner (that is, in any manner not based on the order of their occurrence). For $j = 1, \cdots, n$, let U_j denote the time (measured from the start of the period of observation) at which the jth event occurred. If the events have occurred in accord with a Poisson process, the random variables $U_1, \cdots, U_n$ are independent and uniformly distributed over the interval 0 to T. Consequently, one method of testing whether the events are of Poisson type is to test whether the observations $U_1, \cdots, U_n$ are independent and uniformly distributed over 0 to T. To test the latter one may use the Kolmogorov-Smirnov test or the Cramér-von Mises test (see Section 3–5) or one may use the fact that, according to the central limit theorem, for moderately large values of n, the sum

$$S_n = \sum_{i=1}^{n} U_i \qquad (4.5)$$

of n independent random variables, each uniformly distributed on the interval 0 to T, may be considered to be normally distributed with mean

$$E[S_n] = nE[U_1] = n\frac{T}{2} \qquad (4.6)$$

and variance

$$\mathrm{Var}[S_n] = n\,\mathrm{Var}[U_1] = n\frac{T^2}{12}. \qquad (4.7)$$

Thus if in $T = 10$ minutes of observation of a series of events, $n = 12$ events occur, then the sum S_{12} of the times at which the events occur is (approximately) normally distributed with mean 60 and standard deviation 10. Consequently, if S_{12} satisfies the inequalities

$$60 - (1.96)10 \leq S_{12} \leq 60 + (1.96)10, \tag{4.8}$$

then one would accept the statistical hypothesis that the observed events are of Poisson type. This test is said to have a level of significance of 95%; by this is meant that if the events actually are of Poisson type the probability is 95% that one would accept the hypothesis that they are of Poisson type. For a discussion of the properties of this test, see Chapman (1958), p. 667.

Alternate proof of Theorem 3A. Using the extension of Theorem 4A given by Complement 4A, one may give a rigorous proof of Theorem 3A as follows.

Fix $n, t_1, \cdots, t_n$. For $t > 0$, define

$$G(t) = P[T_1 > t_1, \cdots, T_n > t_n \mid W_{n+1} = t] \tag{4.9}$$

Using Eq. 3.1 and Complement 4A, it may be shown, using the argument in *Mod Prob*, pp. 304–306, that for $t > t_1 + \cdots + t_n$

$$G(t) = \frac{n!}{t^n} \int_{t_1}^{t - t_2 - \cdots - t_n} dx_1 \int_{x_1 + t_2}^{t - t_3 - \cdots - t_n} dx_2 \int_{x_{n-1} + t_n}^{t} dx_n$$

$$= \left[1 - \frac{t_1 + \cdots + t_n}{t} \right]^n . \tag{4.10}$$

In words we may express Eq. 4.10 as follows: if a straight line of length t is divided into $(n + 1)$ sub-intervals by n points chosen at random on the line, then the probability $G(t)$, that simultaneously for $j = 1, 2, \cdots, n$, the jth sub-interval has length greater than t_j is given by the last expression in Eq. 4.10. In view of Eqs. 3.9 and 4.10 it follows that

$$P[T_1 > t_1, \cdots, T_n > t_n] = \int_0^\infty G(t) f_{W_{n+1}}(t) \, dt$$

$$= \int_{t_1 + \ldots + t_n}^\infty e^{-\nu t} \frac{(\nu t)^n}{n!} \left(1 - \frac{t_1 + \cdots + t_n}{t} \right)^n \nu \, dt$$

$$= e^{-\nu(t_1 + \ldots + t_n)} .$$

The proof of Eq. 3.16, and therefore of Theorem 3A, is now complete.

COMPLEMENTS

4A *The uniform distribution of waiting times* $W_1, W_2, \cdots, W_{n-1}$, *given that* $W_n = t$. Let W_n be the waiting time to the nth event in a Poisson process. Show that, under the condition that $W_n = t$, the $(n-1)$ waiting times $W_1, \cdots, W_{n-1}$ have the same distribution as if they were the order statistics corresponding to $(n-1)$ independent random variables $U_1, \cdots, U_{n-1}$ uniformly distributed on the interval $(0, t]$.

4B *The distribution of arrival times, given the number of arrivals in an interval, for a non-homogeneous Poisson process.* Let $\{N(t), t \geq 0\}$ be a Poisson process with continuous mean value function $m(t)$. Show that, under the condition that $N(t) = k$, the k times $\tau_1 < \tau_2 < \cdots < \tau_k$ in the interval $(0, t]$ at which events occur are random variables having the same distribution as if they were the order statistics corresponding to k independent random variables $U_1, \cdots, U_k$ with common distribution function

$$F_{U_j}(u) = \frac{m(u)}{m(t)}, \qquad 0 \leq u \leq t.$$

4C *Estimating and testing hypotheses concerning the mean value function of a Poisson process.* Suppose that k events have been observed in the interval $(0, T]$, occurring at times $\tau_1, \cdots, \tau_k$. Show that a test of the hypothesis that the events occur in accord with a Poisson process with true mean value function $m(t)$ may be obtained using either the Kolmogorov-Smirnov test or the Cramér-von Mises test defined in Section 3–5, with

$$n = k, \ F_n(x) = \frac{N(x)}{N(T)}, \ F(x) = \frac{m(x)}{m(T)}.$$

In particular, show that the Cramér-von Mises statistic may be written

$$W_n{}^2 = k \int_0^t \left[\frac{N(u)}{N(t)} - \frac{m(u)}{m(t)} \right]^2 \frac{dm(u)}{m(t)}$$

$$= \frac{1}{12k} + \sum_{j=1}^k \left[\frac{m(\tau_j)}{m(t)} - \frac{2j-1}{2k} \right]^2.$$

EXERCISES

4.1 A newsboy notices that while his customers do not arrive at regular intervals, they do arrive in accordance with a Poisson process, with a mean arrival rate of 1 per minute. One day he had a friend sell his papers for a five-minute period, while he was away. Upon his return he was told that there had been 4 customers in the five-minute period, while he was away. "Aha," he said, "describe them to me, individually, by some characteristic such as the clothes they were wearing, and I will tell you within a minute, the time at which they came." Compute the probability that he will be able to do so, assuming that for each person he states a two-minute interval within which it is claimed that the person arrived.

4.2   Suppose that one observed a series of events for 10 minutes, and noted 12 occurrences of the event, at times 0.20, 0.33, 0.98, 2.02, 3.92, 4.12, 5.74, 6.42, 7.87, 8.49, 9.85, 9.94. Using a statistic of the form of Eq. 4.5, test (at a level of significance of 95%) the hypothesis that the events are of Poisson type.

4.3   Let r and n be integers ($r \leq n$) and t be a real number. Under the condition that n events (of Poisson type) have occurred in time t, show that the waiting time to the rth event has probability density function

$$\frac{n!}{(r-1)!(n-r)!} \left(\frac{x}{t}\right)^{r-1} \left(1 - \frac{x}{t}\right)^{n-r} \frac{1}{t}, \quad 0 < x < t,$$

and has mean

$$\frac{r}{n+1} t.$$

4-5   FILTERED POISSON PROCESSES

Filtered Poisson processes provide models for a wide variety of random phenomena. Intuitively, a stochastic process is called (in this book, at least, since the terminology is not standard) a *filtered Poisson process* if it can be regarded as arising by means of linear operations on a Poisson process. Processes of this kind were first introduced in noise physics, as a model for shot noise. Later, such processes were used in operations research, as a model for the number of busy servers in an infinite-server queue. To motivate our definition of the notion of a filtered Poisson process, we first consider a number of examples from operations research.

EXAMPLE 5A

The number of busy channels in a telephone system.   Consider a telephone exchange with an infinite number of available channels. Each call coming in gives rise to a conversation on one of the free channels. It is assumed that subscribers make calls at instants $\tau_1, \tau_2, \cdots$ where $0 < \tau_1 < \tau_2 < \cdots$. The arrival of calls are assumed to be events of Poisson type with intensity ν. The holding time (duration of conversation) of the subscriber calling at time τ_n is denoted by Y_n. It is assumed that $Y_1, Y_2, \cdots$ are independent identically distributed random variables. We are interested in the number $X(t)$ of channels busy at time t. One way of representing $X(t)$ is as the number of instants τ_n for which $\tau_n \leq t \leq \tau_n + Y_n$. In symbols, one may write this assertion as follows:

$$X(t) = \sum_{n=1}^{N(t)} w_0(t - \tau_n, Y_n), \tag{5.1}$$

where $w_0(s,y)$ is a function defined for $y > 0$ by

$$w_0(s,y) = 1 \quad \text{if} \quad 0 \le s \le y$$
$$= 0 \quad \text{if} \quad s < 0 \quad \text{or} \quad s > y,$$

and $N(t)$ is the number of calls in the interval $(0,t]$.

EXAMPLE 5B

The number of busy servers in an infinite-server queue. Suppose that customers arrive at an infinite-server queue at times $\tau_1 < \tau_2 < \cdots$. The customer arriving at time τ_n requires a random service time Y_n. Let $X(t)$ denote the number of servers busy at time t. The stochastic process $X(t)$ may also be written in the form of Eq. 5.1.

EXAMPLE 5C

The number of claims in force on a workman's compensation insurance policy. Suppose that workmen incur accidents at times $\tau_1 < \tau_2 < \cdots$. The workman injured at time τ_n is disabled for a (random) time Y_n, during which he draws workman's compensation insurance. Let $X(t)$ denote the number of workmen who at time t are drawing compensation insurance. The stochastic process $\{X(t), t \ge 0\}$ may be written in the form of Eq. 5.1.

EXAMPLE 5D

The number of pulses locking a paralyzable counter. Suppose that radioactive particles arrive at a counter at times $\tau_1 < \tau_2 < \cdots$. The particle arriving at time τ_n locks (or paralyzes) the counter for a random time Y_n; that is, a particle arriving during $(\tau_n, \tau_n + Y_n]$ is not counted. Let $X(t)$ denote the number of particles which at time t are causing the counter to be locked (note that the counter is unlocked at time t if and only if $X(t) = 0$). The stochastic process $\{X(t), t \ge 0\}$ satisfies Eq. 5.1.

A stochastic process $\{X(t), t \ge 0\}$ is said to be a *filtered Poisson process* if it can be represented, for $t \ge 0$, by

$$X(t) = \sum_{m=1}^{N(t)} w(t, \tau_m, Y_m) \tag{5.2}$$

where (i) $\{N(t), t \ge 0\}$ is a Poisson process, with intensity ν, (ii) $\{Y_n\}$ is a sequence of independent random variables, identically distributed as a random variable Y, and independent of $\{N(t), t \ge 0\}$, (iii) $w(t,\tau,y)$ is a function of three real variables, called the *response function*. An intuitive interpretation of Eq. 5.2 often adopted is the following: if τ_m represents the time at which an event took place, then Y_m represents the amplitude of a signal associated with the event, $w(t,\tau_m,y)$ represents the value at time t of a signal of magnitude y originating at time τ_m, and $X(t)$

represents the value at time t of the sum of the signals arising from the events occurring in the interval $(0,t]$.

Note that to specify a filtered Poisson process, one must state (i) the intensity ν of the underlying Poisson process, (ii) the common probability distribution of the random variables $\{Y_n\}$, and (iii) the response function.

Typical response functions. It is generally the case that

$$w(t,\tau,y) = w_0(t - \tau,y) \tag{5.3}$$

for some function $w_0(s,y)$. In words, the effect at time t of a signal occurring at time τ depends only on the time difference $(t - \tau)$ between t and τ.

The functions

$$\begin{aligned} w_0(s,y) &= 1 &&\text{for}\quad 0 < s < y \\ &= 0, &&\text{otherwise,} \end{aligned} \tag{5.4}$$

and

$$\begin{aligned} w_0(s,y) &= y - s &&\text{for}\quad 0 < s < y \\ &= 0, &&\text{otherwise} \end{aligned} \tag{5.5}$$

lead to stochastic processes which arise frequently in management science (see Examples 5A–5C).

Functions of the form

$$w_0(s,y) = yw_1(s), \tag{5.6}$$

in which $w_1(s)$ is a suitable function, generally satisfying the condition

$$w_1(s) = 0 \quad \text{for}\quad s < 0, \tag{5.7}$$

are also frequently considered, especially in models for shot noise.

Another important function is

$$\begin{aligned} w(s) &= 1 &&\text{if}\quad s \geq 0 \\ &= 0 &&\text{if}\quad s < 0, \end{aligned} \tag{5.8}$$

which corresponds to compound Poisson processes (as defined in Section 4–2).

THEOREM 5A

Let $X(t)$ be a filtered Poisson process, defined by Eq. 5.2. Then for any positive number t, and real number u,

$$\varphi_{X(t)}(u) = \exp\left\{\nu \int_0^t E[e^{iuw(t,\tau,Y)} - 1]\, d\tau\right\}, \tag{5.9}$$

and for any $t_2 > t_1 \geq 0$ and real numbers u_1 and u_2

$$\varphi_{X(t_1),X(t_2)}(u_1,u_2) = \exp\left\{\nu \int_0^{t_1} E[e^{i(u_1 w(t_1,\tau,Y)+u_2 w(t_2,\tau,Y))} - 1]\, d\tau \right.$$
$$\left. + \nu \int_{t_1}^{t_2} E[e^{iu_2 w(t_2,\tau,Y)} - 1]\, d\tau \right\} \cdot \tag{5.10}$$

If $E[w^2(t,\tau,Y)] < \infty$ for all τ, then $X(t)$ has finite first and second moments given by

$$E[X(t)] = \nu \int_0^t E[w(t,\tau,Y)]\, d\tau, \tag{5.11}$$

$$\mathrm{Var}[X(t)] = \nu \int_0^t E[w^2(t,\tau,Y)]\, d\tau, \tag{5.12}$$

$$\mathrm{Cov}[X(t_1), X(t_2)] = \nu \int_0^{\min(t_1,t_2)} E[w(t_1,\tau,Y)w(t_2,\tau,Y)]\, d\tau. \tag{5.13}$$

Before proving Theorem 5A, we examine its consequences.

EXAMPLE 5E

The number of busy channels, number of busy servers, etc., is Poisson distributed. Let $X(t)$ denote the number of busy channels in a telephone exchange with an infinite number of channels, the number of busy servers in an infinite-server queue, and similar processes (see Examples 5A–5C). Assume that the arrivals of calls or customers are events of Poisson type with intensity ν, and that the service (or holding) times are independent and identically distributed as a non-negative continuous random variable Y. Then $X(t)$ is a filtered Poisson process whose response function is given by Eqs. 5.3 and 5.4. By Theorem 5A, its one-dimensional characteristic function is given by

$$\varphi_{X(t)}(u) = \exp\left\{\nu \int_0^t E[e^{iu w_0(t-\tau,Y)} - 1]\, d\tau \right\} \cdot \tag{5.14}$$

By a change of variable of integration to $s = t - \tau$, one obtains

$$\varphi_{X(t)}(u) = \exp\left\{\nu \int_0^t E[e^{iu w_0(s,Y)} - 1]\, ds \right\} \cdot \tag{5.15}$$

Now, for fixed s, $\{e^{iu w_0(s,Y)} - 1\}$ is a random variable which is equal either to 0, or to $e^{iu} - 1$; it has the latter value if $Y > s$. Therefore,

$$E[e^{iu w_0(s,Y)} - 1] = (e^{iu} - 1)P[Y > s]$$
$$= (e^{iu} - 1)[1 - F_Y(s)]$$

and

$$\varphi_{X(t)}(u) = \exp\left\{(e^{iu} - 1)\nu \int_0^t [1 - F_Y(s)]\, ds \right\} \cdot \tag{5.16}$$

From Eq. 5.16 it follows that $X(t)$ is Poisson distributed with mean

$$\nu \int_0^t [1 - F_Y(s)]\, ds. \tag{5.17}$$

It may be shown (see *Mod Prob*, p. 211) that for a random variable Y with finite mean,

$$E[Y] = \int_0^\infty [1 - F_Y(s)] \, ds - \int_{-\infty}^0 F_Y(s) \, ds. \tag{5.18}$$

Consequently, it follows that, for large values of t, $X(t)$ is Poisson distributed with mean $\nu E[Y]$. This result can be written somewhat more meaningfully as follows: after an infinite-server queue has been operating for a long time, the number of busy servers is Poisson distributed with parameter

$$\rho = \left\{ \begin{matrix} \text{mean number of} \\ \text{busy servers} \end{matrix} \right\} = \frac{\text{mean service time of a customer}}{\text{mean inter-arrival time between customers}}.$$

The probability that there will be no busy servers is

$$P[X(t) = 0] = e^{-\rho}. \tag{5.19}$$

To obtain the covariance kernel of $X(t)$ we use Eq. 5.13: for $s < t$,

$$\text{Cov}[X(s), X(t)] = \nu \int_0^s E[w_0(s - \tau, Y)w_0(t - \tau, Y)] \, d\tau$$

$$= \nu \int_0^s E[w_0(u, Y)w_0(u + t - s, Y)] \, du.$$

For fixed u, $w_0(u, Y)w_0(u + t - s, Y)$ is equal to 0 or to 1; it has the latter value if $Y > t - s + u$. Therefore, for $s < t$,

$$\text{Cov}[X(s), X(t)] = \nu \int_0^s [1 - F_Y(t - s + u)] \, du. \tag{5.20}$$

For exponentially distributed service times with mean $1/\mu$, $1 - F_Y(y) = e^{-\mu y}$ and

$$\text{Cov}[X(s), X(t)] = \frac{\nu}{\mu} \{e^{-\mu(t-s)} - e^{-\mu t}\} \quad \text{for} \quad s < t. \tag{5.21}$$

It is of especial interest to examine the covariance at two times s and $s + v$ a fixed time v apart:

$$\text{Cov}[X(s), X(s + v)] = \frac{\nu}{\mu} \{e^{-\mu v} - e^{-\mu(s+v)}\}. \tag{5.22}$$

If one lets s tend to ∞ in Eq. 5.22, the limit exists:

$$\lim_{s \to \infty} \text{Cov}[X(s), X(s + v)] = \frac{\nu}{\mu} e^{-\mu v} \quad \text{for} \quad v \geq 0. \tag{5.23}$$

After the queue has been operating a long time, the number of busy servers $X(t)$ can be considered to form a covariance stationary stochastic process (as defined in Chapter 3) with constant mean

$$E[X(t)] = \frac{\nu}{\mu} \qquad (5.24)$$

and covariance function

$$R(v) = \frac{\nu}{\mu} e^{-\mu|v|}, \qquad (5.25)$$

assuming that service times are exponentially distributed with mean $1/\mu$ and arrivals are of Poisson type with intensity ν.

It is left to the reader to show that in the case of a general service time distribution

$$\lim_{s \to \infty} \text{Cov}[X(s), X(s+v)] = \nu E[Y] \left\{ 1 - \int_0^v \left(\frac{1 - F_Y(u)}{E[Y]} \right) du \right\}. \qquad (5.26)$$

EXAMPLE 5F

Shot noise processes and Campbell's theorem. A stochastic process $\{X(t), -\infty < t < \infty\}$ is said to be a *shot noise process* if it may be represented as the superposition of impulses occurring at random times $\cdots, \tau_{-1}, \tau_0, \tau_1, \cdots$. All impulses are assumed to have the same shape $w(s)$ so that one may write

$$X(t) = \sum_{m=-\infty}^{\infty} w(t - \tau_m). \qquad (5.27)$$

More generally, the impulse shapes may be randomly chosen from a family of shapes $w(s,y)$, indexed by a parameter y. At each time τ_m one chooses the parameter y as the value of a random variable Y_m. Consequently $X(t)$ is defined to be the superposition

$$X(t) = \sum_{m=-\infty}^{\infty} w(t - \tau_m, Y_m). \qquad (5.28)$$

The times $\{\tau_m\}$ are assumed to occur in accord with a Poisson process with intensity ν, and $\{Y_n\}$ is assumed to be a sequence of independent identically distributed random variables.

Stochastic processes of the form of Eqs. 5.27 and 5.28 play an important role in the theory of noise in physical devices. In the same way that Theorem 5A is proved one may prove the following results concerning the process $\{X(t), -\infty < t < -\infty\}$ defined by Eq. 5.27:

$$\varphi_{X(t)}(u) = \exp\left[\nu \int_{-\infty}^{\infty} \{e^{\,iuw(s)} - 1\} \, ds\right] \tag{5.29}$$

$$E[X(t)] = \nu \int_{-\infty}^{\infty} w(s) \, ds, \tag{5.30}$$

$$\mathrm{Var}[X(t)] = \nu \int_{-\infty}^{\infty} w^2(s) \, ds, \tag{5.31}$$

$$\mathrm{Cov}[X(t), X(t+v)] = \nu \int_{-\infty}^{\infty} w(s)w(s+v) \, ds. \tag{5.32}$$

Eqs. 5.30–5.32 are usually referred to as *Campbell's theorem* on the superposition of random impulses (for references to the history and literature of Campbell's theorem, see Rice [1944]).

To illustrate the use of Campbell's theorem, let us find the mean, variance, and covariance of the stochastic process of the form of Eq. 5.27 with

$$w(s) = \frac{2e}{T^2} s, \quad 0 \le s \le T \tag{5.33}$$

$$= 0, \qquad \text{otherwise,}$$

where e and T are given constants. (This process describes the total current $X(t)$ flowing through a vacuum tube at time t, due to the superposition of current pulses produced by individual electrons passing from the cathode to the anode; each electron has charge $-e$ and takes a time T to pass from the cathode to the anode [see Davenport and Root, 1958].) From Campbell's theorem it follows that

$$E[X(t)] = \nu \int_0^T \left(\frac{2es}{T^2}\right) ds = \nu e, \tag{5.34}$$

$$\mathrm{Var}[X(t)] = \nu \int_0^T \left(\frac{2es}{T^2}\right)^2 ds = \frac{4\nu e^2}{3T}, \tag{5.35}$$

$$\mathrm{Cov}[X(t), X(t+v)] = \frac{4\nu e^2}{3T}\left(1 - \frac{3}{2}\frac{|v|}{T} + \frac{1}{2}\frac{|v|^3}{T^3}\right) \quad \text{for} \quad |v| \le T$$

$$= 0, \quad \text{otherwise.} \tag{5.36}$$

Methods of obtaining the amplitude distributions of shot noise processes have been extensively investigated by Gilbert and Pollak (1960).

EXAMPLE 5G

Models for income variation and acceleration in chaos. Various stochastic processes which have been introduced recently by Mandelbrot (1960) and Good (1961) as models for income variation and acceleration in chaos, respectively, may be treated as filtered Poisson processes. Let us first state the mathematical results which we will then give an intuitive interpretation. Consider a filtered Poisson process $\{X(t), -\infty < t < \infty\}$ defined by

$$X(t) = \sum_{m=-\infty}^{\infty} w(t - \tau_m), \qquad (5.37)$$

where $\{\tau_m\}$ are the times of occurrence of events of Poisson type with intensity ν and

$$w(s) = c \mid s \mid^{-\beta} \quad \text{if} \quad s \geq 0 \qquad (5.38)$$
$$= -w(-s) \quad \text{if} \quad s < 0,$$

where c and β are positive constants. It will be seen below that β should satisfy the condition $\beta > 1/2$. The characteristic function of $X(t)$ is given by

$$\begin{aligned}
\varphi_{X(t)}(u) &= \exp\left\{\nu \int_{-\infty}^{\infty} [e^{iuw(t-\tau)} - 1]\, d\tau\right\} \\
&= \exp\left\{\nu \int_{-\infty}^{\infty} [e^{iuw(s)} - 1]\, ds\right\} \qquad (5.39) \\
&= \exp\left\{2\nu \int_{0}^{\infty} [\cos uw(s) - 1]\, ds\right\}.
\end{aligned}$$

We evaluate the integral in the exponent as follows. Assume that $u > 0$. Under the change of variable $y = ucs^{-\beta}$ or $s = (y/uc)^{-\alpha}$, where $\alpha = 1/\beta$,

$$\int_{0}^{\infty} \{\cos(ucs^{-\beta}) - 1\}\, ds = -u^{\alpha}ac^{\alpha} \int_{0}^{\infty} y^{-\alpha-1}(1 - \cos y)\, dy.$$

The last integral converges only if $0 < \alpha < 2$ which requires that $\beta > 1/2$. We have thus shown that the characteristic function of the filtered Poisson process, Eq. 5.37, is given by

$$\varphi_{X(t)}(u) = e^{-k|u|^{\alpha}}, \qquad (5.40)$$

where $\alpha = 1/\beta$, and $k = 2\alpha\nu c^{\alpha} \int_{0}^{\infty} y^{-\alpha-1}(1 - \cos y)\,dy$.

A characteristic function of the form of Eq. 5.40, where $0 < \alpha < 2$, is said to be a *real stable* characteristic function (see Loève [1960], p. 328). For $\alpha > 2$, the right-hand side of Eq. 5.40 is not a characteristic function. In the case $\alpha = 1$, Eq. 5.40 is the characteristic function of a Cauchy distribution and in the case $\alpha = 2$, Eq 5.40 is the characteristic function of a normal distribution. Only in the case $\alpha = 2$ does the characteristic function possess finite second moments.

There are many ways in which one can give a physical interpretation to the filtered Poisson process, Eq. 5.37. For example, one can consider particles distributed randomly on a line (in accord with a Poisson process with intensity ν). Let the force between any two particles be one of attraction and of magnitude $cr^{-\beta}$, where c is a positive constant, and r is the distance between the two particles. Then $X(t)$ represents the total force that would be exerted on a particle located at t. Since force and acceleration are equal, up to a constant factor, one can also interpret $X(t)$ as the acceleration of a particle located at t. Good (1961) calls this the acceleration of a particle in a chaotic environment, and Eq. 5.40 then describes the probability law of acceleration in chaos.

The model for acceleration in chaos may be generalized to particles distributed in space. It then includes Holtzmark's treatment (see Chandrasekhar [1943], p. 70) of the force acting on a star due to the gravitational attraction of the neighboring stars.

Extension of the notion of a filtered Poisson process. It is possible to reformulate the notion of a filtered Poisson process as follows. Let $\{W(t,\tau),\, t \geq 0,\, \tau \geq 0\}$ be a stochastic process and let

$$\varphi_{W(t,\tau)}(u) = E[\exp\{iuW(t,\tau)\}] \tag{5.41}$$

be its characteristic function. Next, let

$$\{\,\{W_m(t,\tau),\, t \geq 0,\, \tau \geq 0\},\, m = 1, 2, \cdots\,\}$$

be a sequence of stochastic processes identically distributed as $\{W(t,\tau),\, t \geq 0,\, \tau \geq 0\}$. Let $\tau_1 < \tau_2 < \cdots$ be the times of occurrence of events of Poisson type with intensity ν, and let $N(t)$ be the number of events which have occurred in the interval $(0,t]$. The stochastic process $\{X(t),\, t \geq 0\}$ defined by

$$X(t) = \sum_{m=1}^{N(t)} W_m(t,\tau_m) \tag{5.42}$$

is said to be a filtered Poisson process (in the extended sense) if all the stochastic processes $\{N(t),\, t \geq 0\}$ and $\{W_m(t,\tau),\, t \geq 0,\, \tau \geq 0\}$, $m = 1, 2, \cdots$, are assumed to be independent. In the same way that Theorem 5A is proved one may prove that

$$\varphi_{X(t)}(u) = \exp\left\{-\nu t + \nu \int_0^t \varphi_{W(t,\tau)}(u)\, d\tau\right\}, \tag{5.43}$$

$$E[X(t)] = \nu \int_0^t E[W(t,\tau)]\, d\tau, \tag{5.44}$$

$$\mathrm{Var}[X(t)] = \nu \int_0^t E[W^2(t,\tau)]\, d\tau. \tag{5.45}$$

EXAMPLE 5H

Population processes with immigration. If an animal of a certain species immigrates to a certain region at time τ, then the number of descendants of this animal who will be present in this region at time t is a random variable $W(t,\tau)$. Suppose that at time 0 there are no animals of this species in the region but that animals of this species immigrate into the region at times $\tau_1 < \tau_2 < \cdots$ which are of Poisson type with intensity ν. Let $W_m(t,\tau_m)$ denote the number of descendants at time t of the animal which immigrated to the region at time τ_m. The total number $X(t)$ of animals in the region at time t is then given by Eq. 5.42. Assuming independence of all the population processes under consideration, it follows that $\{X(t), t \geq 0\}$ is a filtered Poisson process (in the extended sense). Instead of writing an expression for its characteristic function, it is more usual to consider its probability generating function:

$$\psi_X(z;t) = \sum_{n=0}^{\infty} z^n P[X(t) = n].$$

If one knows the probability generating function

$$\psi_W(z;t,\tau) = \sum_{n=0}^{\infty} z^n P[W(t,\tau) = n]$$

then

$$\psi_X(z;t) = \exp\left\{-\nu t + \nu \int_0^t \psi_W(z;t,\tau)\, d\tau\right\}. \tag{5.46}$$

An application of this result is given in Example 4B of Chapter 7.

Proof of Theorem 5A. To prove the theorem it suffices to find the characteristic function $E[\exp i\{u_1 X(t_1) + u_2 X(t_2)\}]$. Now, $w(t,\tau,y)$ may be assumed to have the property that for any y and τ,

$$w(t,\tau,y) = 0 \quad \text{if} \quad t < \tau,$$

since a signal occurring at time τ has no effect at an earlier time t. Consequently, $X(t)$ may be represented

$$X(t) = \sum_{m=1}^{\infty} w(t,\tau_m, Y_m),$$

and

$$u_1 X(t_1) + u_2 X(t_2) = \sum_{m=1}^{N(t_2)} g(\tau_m, Y_m),$$

defining

$$g(\tau,y) = u_1 w(t_1,\tau,y) + u_2 w(t_2,\tau,y).$$

To evaluate the characteristic function of $X(t)$ it suffices to evaluate

$$\Phi = E[e^{iZ}], \quad Z = \sum_{m=1}^{N(t_2)} g(\tau_m, Y_m).$$

Define, for any $\tau > 0$,

$$\varphi(\tau) = E[e^{ig(\tau, Y)} - 1]. \tag{5.47}$$

We shall show that

$$\Phi = \exp\left\{\nu \int_0^{t_2} \varphi(\tau) \, d\tau\right\} \tag{5.48}$$

from which Eq. 5.10 follows. To prove Eq. 5.48 there are several possible methods, of which we discuss one. We first write

$$E[e^{iZ}] = \sum_{n=0}^{\infty} E[e^{iZ} \mid N(t_2) - N(0) = n]P[N(t_2) - N(0) = n]. \tag{5.49}$$

Now, the conditional distribution of the times $0 < \tau_1 < \cdots < \tau_n \leq t_2$ at which events have occurred, given that exactly n events occurred, is the same as the distribution of ordered random variables corresponding to n independent random variables uniformly distributed over the interval 0 to t_2. Consequently, it may be inferred (as is shown in detail below) that

$$E[\exp i \sum_{m=1}^{n} g(\tau_m, Y_m) \mid N(t_2) - N(0) = n]$$
$$= \{E[\exp ig(U, Y)]\}^n = \left\{\frac{1}{t_2} \int_0^{t_2} E[\exp ig(\tau, Y)] \, d\tau\right\}^n \tag{5.50}$$

in which U is uniformly distributed over the interval 0 to t_2. Combining Eqs. 5.49 and 5.50 it follows that

$$E[e^{iZ}] = \sum_{n=0}^{\infty} e^{-\nu t_2} \frac{(\nu t_2)^n}{n!} \left\{\frac{1}{t_2} \int_0^{t_2} E[e^{ig(\tau, Y)}] \, d\tau\right\}^n$$
$$= e^{-\nu t_2} \sum_{n=0}^{\infty} \frac{1}{n!} \left\{\nu \int_0^{t_2} E[e^{ig(\tau, Y)}] \, d\tau\right\}^n$$
$$= \exp\left\{\nu \int_0^{t_2} (E[e^{ig(\tau, Y)}] - 1) \, d\tau\right\}$$
$$= \exp\left\{\nu \int_0^{t_2} \varphi(\tau) \, d\tau\right\}$$

and Eq. 5.48 is proved.

It may be worthwhile to give more details about the derivation of Eq. 5.50. Let us define A to be the event that $N(t_2) - N(0) = n$, and

$$\varphi = E\left[\exp i \sum_{m=1}^{n} g(\tau_m, Y_m) \mid A\right].$$

Now, define for any real numbers $s_1, \cdots, s_n$ satisfying $0 \le s_1 < s_2 < \cdots < s_n \le t_2$

$$\varphi(s_1, \ldots, s_n) = E\left[\exp i \sum_{m=1}^{n} g(\tau_m, Y_m) \mid A, \tau_1 = s_1, \ldots, \tau_n = s_n \right].$$

One may verify that

$$\varphi(s_1, \ldots, s_n) = E\left[\exp i \sum_{m=1}^{n} g(s_m, Y_m) \mid A, \tau_1 = s_1, \ldots, \tau_n = s_n \right]$$

$$= \prod_{m=1}^{n} E[\exp ig(s_m, Y)].$$

Next, one may verify that

$$\varphi = \int_0^{t_2} ds_1 \int_{s_1}^{t_2} ds_2 \ldots \int_{s_{n-1}}^{t_2} ds_n \varphi(s_1, \ldots, s_n) f_{\tau_1, \ldots, \tau_n}(s_1, \ldots, s_n)$$

$$= \frac{n!}{(t_2)^n} \int_0^{t_2} ds_1 \int_{s_1}^{t_2} ds_2 \ldots \int_{s_{n-1}}^{t_2} ds_n \varphi(s_1, \ldots, s_n)$$

$$= \frac{1}{(t_2)^n} \int_0^{t_2} ds_1 \ldots \int_0^{t_2} ds_n \varphi(s_1, \ldots, s_n), \qquad (5.51)$$

because $\varphi(s_1, \cdots, s_n)$ is a symmetric function of its arguments. From Eq. 5.51 one immediately obtains Eq. 5.50.

To establish the formulas for the moments of $X(t)$, one uses the facts that

$$iE[X(t)] = \frac{d}{du} \log \varphi_{X(t)}(0),$$

$$i^2 \operatorname{Var}[X(t)] = \frac{d^2}{du^2} \log \varphi_{X(t)}(0),$$

$$i^2 \operatorname{Cov}[X(t_1), X(t_2)] = \frac{\partial^2}{\partial u_1 \, \partial u_2} \log \varphi_{X(t_1), X(t_2)}(0,0). \qquad (5.52)$$

COMPLEMENTS

5A *Generalization of Campbell's theorem.* Consider a stochastic process of the form

$$X(t) = \sum_{-\infty < \tau_n < \infty} Y_n \, w(t - \tau_n),$$

where $\{\tau_n\}$ are the times of occurrence of events happening in accord with a Poisson process at a mean rate ν per unit time, and $\{Y_n\}$ are independent random variables identically distributed as a random variable Y. Show that the kth cumulant of $X(t)$ is given by

$$\frac{1}{i^k} \frac{d^k}{du^k} \log \varphi_{X(t)}(0) = \nu \, E[Y^k] \int_{-\infty}^{\infty} w^k(s) \, ds,$$

assuming absolute integrability of the integral and expectation.

5B *Filtered generalized or compound Poisson processes.* Consider events which happen in groups rather than singly (for example, the production of showers of particles by cosmic rays). For $k = 1, 2, \cdots$ suppose that the event that k particles arrive simultaneously occurs in accordance with a Poisson process $N_k(t)$ at a mean rate λ_k per unit time. Assume independence of the processes $\{N_k(t), k = 1, 2, \cdots\}$. Let $N(t)$ be the total number of particles that have arrived in the interval $(0,t]$. Then

$$N(t) = N_1(t) + 2N_2(t) + \cdots + kN_k(t) + \cdots.$$

(i) Show that $E[N(t)] = \mu_1 t$ and $\mathrm{Var}[N(t)] = \mu_2 t$, where

$$\mu_1 = \sum_{k=1}^{\infty} k\lambda_k, \qquad \mu_2 = \sum_{k=1}^{\infty} k^2\lambda_k.$$

(ii) Show that $N(t)$ may be represented as a compound Poisson process.

(iii) Suppose that the showers consist of particles which communicate equal charges to an electrometer whose deflection at time t is of the form

$$X(t) = \sum_{0 < \tau_n \leq t} w(t - \tau_n),$$

where $\{\tau_n\}$ are the times at which charges are received, and $w(u)$ is an impulse response function. Show that

$$E[X(t)] = \mu_1 \int_0^t w(s) \, ds, \qquad \mathrm{Var}[X(t)] = \mu_2 \int_0^t w^2(s) \, ds.$$

5C *Filtered non-homogeneous Poisson processes.* Let $\{N(t), t \geq 0\}$ be a non-homogeneous Poisson process, whose mean value function $m(t) = E[N(t)]$ possesses a continuous derivative

$$\nu(t) = \frac{d}{dt} m(t).$$

Note that $\nu(t) \, dt$ is approximately the probability that in the time interval $[t, t + dt]$ exactly one jump will occur in the process $N(t)$. Let $\{X(t), t \geq 0\}$ be defined by Eq. 5.2. Show that

$$\log \varphi_{X(t)}(u) = \int_0^t \nu(\tau) \, E[e^{iuw(t,\tau,Y)} - 1] \, d\tau,$$

$$E[X(t)] = \int_0^t \nu(\tau) \, E[w(t,\tau,Y)] \, d\tau,$$

$$\mathrm{Var}[X(t)] = \int_0^t \nu(\tau) \, E[w^2(t,\tau,Y)] \, d\tau,$$

$$\mathrm{Cov}[X(t_1), X(t_2)] = \int_0^{min\ (t_1,t_2)} \nu(\tau) \, E[w(t_1,\tau,Y)w(t_2,\tau,Y)] \, d\tau.$$

Remark. These results provide a model for *non-stationary shot noise.* Consider the shot noise generated in a temperature-limited diode operating on an alternating current filament supply. The probability of electron emission from the cathode then varies periodically. The number $N(t)$ of electrons emitted in time interval $(0,t]$ is then a non-homogeneous Poisson process with $\nu(t)$ a periodic function.

5D *Asymptotic normality of filtered Poisson processes.* Let

$$X(t) = \sum_{-\infty < \tau_n < \infty} w(t, \tau_n, Y_n)$$

be a filtered Poisson process. Let

$$m(t) = \nu \int_{-\infty}^{\infty} E[w(t,\tau,Y)]\, d\tau,$$

$$\sigma^2(t) = \nu \int_{-\infty}^{\infty} E[\,|\,w(t,\tau,Y)\,|^2\,]\, d\tau,$$

$$\rho(s,t) = \frac{\int_{-\infty}^{\infty} E[w(s,\tau,Y)w(t,\tau,Y)]\, d\tau}{\left\{\int_{-\infty}^{\infty} E[\,|\,w(s,\tau,Y)\,|^2\,]\, d\tau \int_{-\infty}^{\infty} E[\,|\,w(t,\tau,Y)\,|^2\,]\, d\tau\right\}^{1/2}},$$

$$K(t) = \nu \int_{-\infty}^{\infty} E[\,|\,w(t,\tau,Y)\,|^3\,]\, d\tau.$$

Suppose that, for each t, $m(t)$, $\sigma(t)$, and $K(t)$ are finite and that

$$\frac{K(t)}{\sigma^3(t)} = \frac{1}{\sqrt{\nu}}\, \frac{\int_{-\infty}^{\infty} E[\,|\,w(t,\tau,Y)\,|^3\,]\, d\tau}{\left\{\int_{-\infty}^{\infty} E[\,|\,w(t,\tau,Y)\,|^2\,]d\tau\right\}^{3/2}}$$

tends to 0 as certain parameters (such as ν) tend to prescribed limits. Then as these parameters tend to the limits given, $\{X(t), -\infty < t < \infty\}$ is asymptotically normal in the sense that Eq. 5.8 of Chapter 3 holds.

Hint. Using the fact that

$$\log \varphi_{X(t_1),\dots,X(t_n)}(u_1, \cdots, u_n) = i\sum_{j=1}^{n} u_j m(t_j)$$

$$-\frac{1}{2}\sum_{j,k=1}^{} u_j u_k \rho(t_j,t_k)\sigma(t_j)\sigma(t_k)$$

$$+ \theta \sum_{j=1}^{} K(t_j)\,|\,u_j\,|^3,$$

where θ denotes a quantity of modulus less than 1, show that the standardized process $\{X^*(t), -\infty < t < \infty\}$ satisfies

$$\log \varphi_{X^*(t_1),\dots,X^*(t_n)}(u_1, \cdots, u_n) = -\frac{1}{2}\sum_{j,k=1}^{n} u_j u_k \rho(t_j,t_k)$$

$$+ \theta \sum_{j=1}^{n} \frac{K(t_j)}{\sigma^3(t_j)}\,|\,u_j\,|^3.$$

EXERCISES

For each of the stochastic processes $X(t)$ described in Exercises 5.1 to 5.7

(i) find for $t \geq 0$ the mean value $m(t) = E[X(t)]$ and the limit $m = \lim_{t \to \infty} m(t)$ if it exists,

(ii) find for $t \geq 0$ the variance $\sigma^2(t) = \text{Var}[X(t)]$ and the limit $\sigma^2 = \lim_{t \to \infty} \sigma^2(t)$ if it exists,

(iii) find the covariance $K(t_1, t_2) = \text{Cov}[X(t_1), X(t_2)]$, and the limit $R(t) = \lim_{s \to \infty} \text{Cov}[X(s), X(s + t)]$ if it exists,

(iv) find for $t \geq 0$ the one-dimensional characteristic function $\varphi_{X(t)}(u)$.

5.1 Consider the housewife in Exercise 2.2 (p. 132). Let $X(t)$ be her total commission from subscriptions ordered during the period of one year immediately preceding t. Assume that we begin to observe the housewife one year after she has begun selling subscriptions.

5.2 Consider the housewife described in Exercise 2.2. Let $X(t)$ be the number of subscriptions in force at time t which were sold by the housewife.

5.3 *A paralyzable counter.* Let electrical impulses (in a typical case, activated by the arrival of radioactive particles such as cosmic rays or x-rays) arrive in accord with a Poisson process at a counting mechanism. In a paralyzable (or type II) counter, a particle arriving at the counter locks the counter for a random time Y; consequently, a particle arriving at the counter is registered (counted) if and only if at the time it arrives the locking times of all previously arrived particles have expired. Let $X(t)$ be the number of particles locking the counter at time t. The probability $P[X(t) = 0]$ is especially important; it represents the probability that the counter is unlocked at time t.

5.4 Telephone calls arrive at a telephone exchange with an infinite number of channels at a rate of 30 a minute. Each conversation is exponentially distributed with a mean of 3 minutes. Let $X(t)$ be the number of conversations going on at time t.

5.5 Orders for a certain item arrive at a factory at a rate of 1 per week. Assume order arrivals are events of Poisson type. The item is produced only to order. The firm has unlimited capacity to produce the item so that production on each item ordered starts immediately on receipt of the order. The length of time it takes to produce the item is a random variable uniformly distributed between 80 and 90 days. Let $X(t)$ be the number of orders in production at time t.

5.6 *Continuation of Exercise 5.5.* Let $X(t)$ be the number of days of production time remaining on the unfilled orders.

5.7   Consider a stochastic process of the form

$$X(t) = \sum_{m=1}^{N(t)} Y_m e^{-(t-\tau_m)/RC}$$

in which $\{\tau_n\}$ are the times of occurrence of events happening in accord with a Poisson process $\{N(t), t \geq 0\}$ with intensity ν, $\{Y_n\}$ are independent random variables identically distributed as an exponentially distributed random variable Y, and R and C are positive constants.

5.8   Consider the filtered Poisson process $\{X(t), t \geq 0\}$ defined in Exercise 5.7. Show that the process is approximately normal if RC is large.

Renewal
counting processes

AN INTEGER-VALUED, or counting, process $\{N(t), t \geq 0\}$ corresponding to a series of points distributed on the interval 0 to ∞ is said to be a *renewal counting process* if the inter-arrival times $T_1, T_2, \cdots$ between successive points are independent identically distributed positive random variables. This chapter discusses some of the basic properties of renewal counting processes.

It should be pointed out that the phrase "a renewal counting process" is not standard terminology. Many writers use the phrase "renewal process" to describe the sequence of independent positive random variables $T_1, T_2, \cdots$ which represent the inter-arrival times. Renewal theory is defined to be the study of renewal processes. It seems to me that the counting process $\{N(t), t \geq 0\}$ representing the number of renewals in an interval deserves an explicit name. Consequently, I have taken the liberty of introducing the phrase "a renewal counting process."

5-1 EXAMPLES OF RENEWAL COUNTING PROCESSES

In this section we describe the wide variety of phenomena that can be described by renewal counting processes. A typical sample function of a renewal counting process is sketched in Figure 5.1.

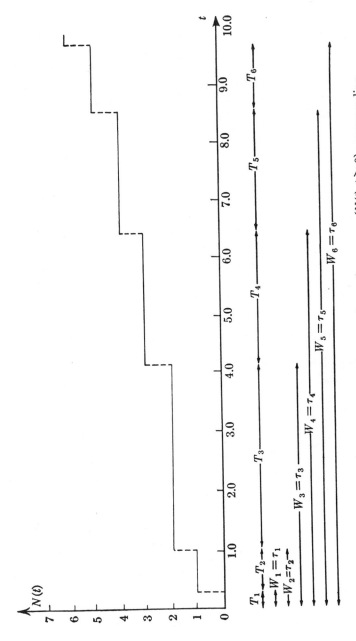

Fig. 5.1. Typical sample function of a renewal counting process $\{N(t), t \geq 0\}$ corresponding to inter-arrival times with probability density function $f(x) = xe^{-x}$, for $x \geq 0$.

EXAMPLE 1A

Failures and replacements. Consider an item (such as an electric light bulb, a vacuum tube, a machine, and so on) which is used until it fails and is then replaced (or renewed) by a new item of similar type. The length of time such an item serves before it fails is assumed to be a random variable T with known probability law. The lifetimes T_1, T_2, $\cdots$ of the successive items put into service are assumed to be independent random variables identically distributed as a random variable T. If $N(t)$ denotes the number of items replaced (or renewed) in the time 0 to t, then $\{N(t), t \geq 0\}$ is a renewal counting process.

EXAMPLE 1B

Traffic flow. Imagine a central mail-order house which receives orders from a number of sources, or a telephone exchange at which calls arrive. Orders or calls arrive at times τ_1, τ_2, $\cdots$ where $0 < \tau_1 < \tau_2 \cdots$. It is often reasonable to assume that the successive inter-arrival times $T_1 = \tau_1$, $T_2 = \tau_2 - \tau_1$, $\cdots$, $T_n = \tau_n - \tau_{n-1}$, $\cdots$ are independent identically distributed random variables. Let $N(t)$ denote the number of orders (or calls) received in the interval $(0, t]$. Then $\{N(t), t \geq 0\}$ is a renewal counting process.

EXAMPLE 1C

Nuclear particle counters—scaling circuits. Renewal counting processes play an important part in the theory of nuclear particle counters. One way in which renewal counting processes arise is as a model for the outputs of *scaling circuits*.

Devices for counting radioactive emissions often have finite resolvability; that is, pulses arriving too soon after a pulse has been recorded are not counted. For most detection instruments, there is a characteristic constant, called the resolving time; after recording one pulse, the counter is unresponsive to successive pulses until a time interval greater than or equal to its resolving time has elapsed. One way of reducing losses due to positive resolving time is to introduce a circuit which makes only one count for a given number (say, s) input events; a device which does this is called a scale of s scaling circuit. Let T_1, T_2, $\cdots$ denote the time intervals between successive counts with a scale of s scaling circuit. It may be shown (see Theorem 3A of Chapter 4) that if pulse arrivals are events of Poisson type with intensity ν, then T_1, T_2, $\cdots$ are independent random variables, each obeying a gamma probability law with parameters s and ν.

Delayed renewal counting processes. It often happens that the first inter-arrival time T_1 does not have the same distribution as the later inter-arrival times T_2, T_3, $\cdots$ which are identically distributed as a random variable T. A counting process $N(t)$ corresponding to a series of points distributed on the interval 0 to ∞ is said to be a *delayed* renewal counting process if the inter-arrival times T_1, T_2, $\cdots$ between successive points are independent random variables such that T_2, T_3, $\cdots$ are identically distributed as a positive random variable T.

EXAMPLE 1D
    Nuclear particle counters with deadtime. Let electrical impulses (in a typical case, actuated by the arrival of radioactive particles such as cosmic rays or x-rays) arrive at a counting mechanism at times τ_1, τ_2, $\cdots$ in which $0 < \tau_1 < \tau_2 < \cdots$. Many counting mechanisms have a positive resolving time so that not all particles arriving are registered (or counted). At any given time, the counter either is or is not in a condition to register particles arriving at that time; it is said to be *locked* at a given time if a particle arriving at that time would not be registered (counted).

We must distinguish between the counting process, denoted by $\{N(t), t \geq 0\}$ say, of the *arriving* particles, and the counting process, denoted by $\{M(t), t \geq 0\}$ say, of the *registered* particles. The experimenter observes only the process $\{M(t), t \geq 0\}$; from this he desires to infer the properties of the process $\{N(t), t \geq 0\}$. It is often assumed that $\{N(t), t \geq 0\}$ is a Poisson process with intensity λ. It is then desired to use the observed process $\{M(t), t \geq 0\}$ to estimate λ.

The process $\{M(t), t \geq 0\}$ can often be treated as a renewal counting process. If the times at which particles arrive at the counter are $0 < \tau_1 < \tau_2 < \cdots$, then the times $0 < \tau_1' < \tau_2' < \cdots$ at which particles are registered clearly form a subsequence of the times at which particles arrive. Let

$$T_1 = \tau_1, \, T_2 = \tau_2 - \tau_1, \, \cdots, \, T_n = \tau_n - \tau_{n-1}, \, \cdots$$

be the successive inter-arrival times between the times at which particles arrive, and let

$$T_1' = \tau_1', \, T_2' = \tau_2' - \tau_1', \, \cdots, \, T_n' = \tau_n' - \tau_{n-1}', \, \cdots$$

be the successive inter-arrival times between the times at which particles are registered.

Let $M(t)$ be the number of particles registered in the interval $(0,t]$. In order to show that $\{M(t), t \geq 0\}$ is a renewal counting process generated by the inter-arrival times T_1', T_2', $\cdots$, it suffices to show that (i) the random variables $\{T_n'\}$ are independent and (ii) T_1' is exponentially distributed with mean $1/\lambda$, while T_2', T_3', $\cdots$ are identically distributed, so that $\{M(t), t \geq 0\}$ is possibly a delayed renewal counting process. In order to show that this is the case one must make certain assumptions about the counting mechanism.

The counter devices in use belong mainly to two types, called *non-paralyzable* (or type I) and *paralyzable* (or type II).

In a non-paralyzable counter, if a particle arriving at the counter is registered (that is, counted) then the counter remains locked for a random time Y called the deadtime, locking time, or holding time of the particle. Particles arriving during the time Y are neither registered nor do they affect the counter in any way. Counters of non-paralyzable type are scintillation counters and self-quenching Geiger-Müller counters connected to sensitive preamplifiers; see Evans (1955), p. 785.

In a paralyzable counter, a particle arriving at the counter locks the counter for a random locking time Y, regardless of whether or not it was registered. Consequently, a particle arriving at the counter is registered if and only if at the time it arrives the locking times of all previously arriving particles have expired. Counters of paralyzable type are electromechanical registers and non-self-quenching Geiger-Müller counters connected to high resistance preamplifiers.

The effect of a non-paralyzable counter and of a paralyzable counter on a series of particles is illustrated in Figure 5.2.

One may conceive (at least mathematically) of a counter which contains both these types as special cases. A particle arriving at the counter when it is unlocked is registered, and in addition locks the counter for a random locking time Y. A particle arriving at the counter when it is locked is not registered, but has probability p (where $0 \le p \le 1$) of locking the counter for a random locking time Y. We call such a counter a *type p counter*. It is clear that a non-paralyzable counter is a type 0 counter, and a paralyzable counter is a type 1 counter.

No matter which counter is considered it is assumed that the successive locking times Y_1, Y_2, $\cdots$ are independent random variables identically distributed as a random variable Y with finite second moment. It then follows that the counting process $\{M(t),\ t \ge 0\}$ of registered particles is a renewal counting process.

It should be pointed out that both the mathematical and physical literature of nuclear particle counters employs the classification of counters into type I and type II. However, what some authors call type I is called type II by other authors (see Feller [1948] and Evans [1955] or Korff [1955]). It seems to me that the physically meaningful terminology of non-paralyzable and paralyzable counters has much to commend it. In order to make it easier for the reader to consult the mathematical literature, the terms type I and type II are used in this book in the sense assigned to them by the probabilists who have written about counters (see Pyke [1958], Smith [1958] and Takács [1958, 1960]).

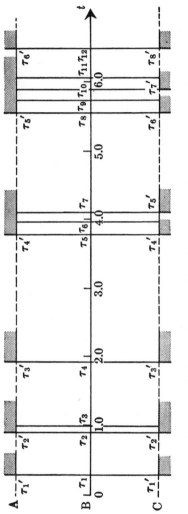

Fig. 5.2. Schematic illustration of the behavior of nuclear particle counters having a constant deadtime L. The time axis is from left to right; arrival times of particles are shown by vertical lines. Paralyzable or type II counters with constant deadtime L respond only to inter-arrival times longer than L. The number of registered particles and the time during which a type II counter is insensitive (locked) are shown by shaded blocks at the top. Nonparalyzable or type I counters (with constant deadtime L) are insensitive for a time L after a pulse is registered, but then respond to the first particle arriving after the deadtime. The number of registered particles and the time during which a type I counter is insensitive are shown by shaded blocks at the bottom. In the hypothetical example shown, 12 particles actually arrive, of which 6 would be recorded by a type II counter and 8 by a type I counter.

A. Times at which particles are registered by paralyzable (type II) counters
B. Times at which particles actually arrive
C. Times at which particles are registered by non-paralyzable (type I) counters

EXAMPLE 1E

Paralyzable (or type II) counter with constant deadtime. Consider a paralyzable (or type II) nuclear particle counter with a constant locking time L. Let $N(t)$ be the number of particles arriving at the counter in the time interval $(0,t]$, and let $M(t)$ be the number of particles registered by the counter during $(0,t]$. An arriving event is registered if and only if no particles arrived during the preceding time interval of length L. Consequently, the probability that a particle is registered is $p = e^{-\nu L}$. According to Theorem 3A of Chapter 4, the time intervals between successive particle arrivals are independent. Therefore, the registering of a particle is independent of the registering of other particles. It might be thought that the process $\{M(t), t \geq 0\}$ is derived by random selection (as defined in Section 2–1) from the Poisson process $\{N(t), t \geq 0\}$ and consequently is a Poisson process. However, $\{M(t), t \geq 0\}$ is *not* a Poisson process, since the event that a particle is registered is not independent of the process $\{N(t), t \geq 0\}$.

Nevertheless, the successive inter-arrival times between particle registrations are independent identically distributed random variables. Consequently, $\{M(t), t \geq 0\}$ is a renewal counting process. To determine its properties, it suffices to find the probability law of the time T between particle registrations. We now show that T has characteristic function, mean, and variance given by

$$\varphi_T(u) = \left\{ 1 - \frac{iu}{\nu} e^{(\nu - iu)L} \right\}^{-1} \tag{1.1}$$

$$E[T] = \frac{1}{\nu} e^{\nu L}, \tag{1.2}$$

$$\text{Var}[T] = \frac{e^{2\nu L}}{\nu^2} - \frac{2Le^{\nu L}}{\nu}. \tag{1.3}$$

In the time T between two particle registrations, a certain number, denoted by M, of particles have arrived at the counter and were not registered. Let us first note that for $m = 0, 1, \cdots$

$$P[M = m] = pq^m, \tag{1.4}$$

where $p = e^{-\nu L}$ and $q = 1 - p$, since p is equal to the probability that an inter-arrival time between particle arrivals is greater than L. The waiting time T between particle registrations can be regarded as the sum of a random number of independent random variables:

$$\begin{aligned} T &= U_1 + \cdots + U_M + V \quad \text{if} \quad M \geq 1 \\ &= V \quad \text{if} \quad M = 0, \end{aligned} \tag{1.5}$$

where U_1 is the time until the arrival of the first unregistered particle, U_2 is the time between the arrival of the first and second unregistered particles, and so on, while V is defined as follows: if $M \geq 1$, V is the time between the arrival of the last unregistered particle and the registered particle; if $M = 0$, V is the time until the arrival of the registered particle. Further, $U_1, \cdots, U_M$ are identically distributed as a random variable U with distribution function

$$F_U(u) = \frac{1 - e^{-\nu u}}{1 - e^{-\nu L}}, \quad 0 \leq u \leq L, \tag{1.6}$$

probability density function

$$f_U(u) = \frac{\nu e^{-\nu u}}{1 - e^{-\nu L}}, \quad 0 < u < L \tag{1.7}$$

and characteristic function

$$\varphi_U(u) = \int_0^L e^{ixu} f_U(x) \, dx = \frac{\nu}{\nu - iu} \frac{1 - e^{-\nu L} e^{iuL}}{1 - e^{-\nu L}} . \tag{1.8}$$

On the other hand, V has distribution function

$$1 - F_V(v) = e^{-\nu(v-L)}, \quad v \geq L, \tag{1.9}$$

probability density function

$$f_V(v) = \nu e^{-\nu(v-L)}, \quad v \geq L \tag{1.10}$$

and characteristic function

$$\varphi_V(u) = \int_L^\infty e^{ixu} f_V(x) \, dx = \frac{\nu}{\nu - iu} e^{iuL}. \tag{1.11}$$

The characteristic function of T is given by

$$\varphi_T(u) = \sum_{m=0}^\infty E[e^{iuT} \mid M = m] P[M = m] \tag{1.12}$$

$$= \frac{p\varphi_V(u)}{1 - q\varphi_U(u)} .$$

Combining Eqs. 1.12, 1.11, and 1.8, one obtains Eq. 1.1. To prove Eqs. 1.2 and 1.3 one evaluates at $u = 0$ the derivatives of the logarithm of the characteristic function:

$$\frac{d}{du} \log \varphi_T(u) = \frac{\{i - uL\}}{\{ve^{-(v-iu)L} - iu\}},$$

$$\frac{d^2}{du^2} \log \varphi_T(u) = \frac{\{1 + 2iuL - (2L + iuL^2)ve^{-(v-iu)L}\}}{\{ve^{-(v-iu)L} - iu\}^2}.$$

For examples of the application of renewal counting processes to the mathematical theory of genetic recombination, see Owen (1949).

COMPLEMENTS

The hazard function and distributions for the length of life of mechanical systems, business enterprises, and so on. A method frequently employed for deriving the form of the distribution of inter-arrival times of events such as system failures is to consider the hazard function. Let T be a random variable representing the service life to failure of a specified system. Let $F(x)$ be the distribution function of T, and let $f(x)$ be its probability density function. We define a function $\mu(x)$, called the *intensity function*, or *hazard function*, or *conditional rate of failure function* of T by

$$\mu(x) = \frac{f(x)}{1 - F(x)}. \tag{i}$$

In words, $\mu(x)\,dx$ is the conditional probability that the item will fail between x and $x + dx$, given that it has survived a time T greater than x.

For a given hazard function $\mu(x)$ the corresponding distribution function is

$$1 - F(x) = \{1 - F(x_0)\} \exp\left[- \int_{x_0}^x \mu(z)\,dz\right]$$

where x_0 is an arbitrary value of x, since (i) can be rewritten as

$$\frac{d}{dx} \log[1 - F(x)] = - \mu(x).$$

1A   Show that, if there is a lower bound ϵ to the value of T (more precisely, if $F(\epsilon) = 0$), then

$$1 - F(x) = \exp\left[- \int_{\epsilon}^x \mu(z)\,dz\right], \qquad x > \epsilon.$$

1B   Show that a constant hazard function

$$\mu(x) = \lambda, \qquad\qquad x > 0,$$

and a lower bound $\epsilon = 0$ give rise to an exponential distribution

$$f(x) = \lambda e^{-\lambda x}, \qquad\qquad x > 0.$$

1C   A random variable T which takes only values greater than some number ϵ, is defined to have a *Weibull* distribution with parameters v and k (where $v \geq \epsilon$ and $k > 1$) if its distribution function is given by

$$F(x) = 1 - \exp\left\{-\left(\frac{x-\epsilon}{v-\epsilon}\right)^k\right\}, \qquad x \geq \epsilon$$
$$= 0, \qquad x < \epsilon,$$

or if its probability density function is given by

$$f(x) = \frac{k}{v-\epsilon}\left(\frac{x-\epsilon}{v-\epsilon}\right)^{k-1} \exp -\left\{\left(\frac{x-\epsilon}{v-\epsilon}\right)^k\right\}, \qquad x > \epsilon$$
$$= 0, \qquad x < \epsilon.$$

Show that the Weibull distribution corresponds to the hazard function

$$\mu(t) = k\left(\frac{t-\epsilon}{v-\epsilon}\right)^{k-1}, \qquad t > \epsilon.$$

1D   A random variable T is defined to have an *extreme value* distribution (of exponential type) with parameters u and β (where u and β are constants such that $-\infty < u < \infty$ and $\beta > 0$) if its distribution function is given by

$$F(x) = 1 - \exp\left\{-e^{-\left(\frac{x-u}{\beta}\right)}\right\}, \qquad -\infty < x < \infty,$$

or if its probability density function is given by

$$f(x) = \exp\left\{-\left(\frac{x-u}{\beta}\right) - e^{-\left(\frac{x-u}{\beta}\right)}\right\}, \qquad -\infty < x < \infty.$$

Show that the extreme value distribution (of exponential type) corresponds to the hazard function

$$\mu(t) = \frac{1}{\beta}\exp\left[-\left(\frac{t-u}{\beta}\right)\right], \qquad -\infty < t < \infty,$$

and a lower bound $\epsilon = -\infty$. The parameters u and β of this extreme value distribution act as location and scale parameters in much the same way that the mean m and variance σ^2 of the normal distribution are respectively measures of location and scale. For a complete discussion of extreme value distributions the reader should consult Gumbel (1958).

1E   Find the distribution functions corresponding to the hazard functions (for positive constants a and b)

(i) $\mu(t) = ae^{-bt}$ \qquad\qquad (ii) $\mu(t) = \dfrac{b}{t+a}$

and a lower bound $\epsilon = 0$. For an example in which these hazard functions arise empirically, see Lomax (1954).

EXERCISES

1.1 Just before the beginning of a concert, audience noises (such as coughs) of Poisson type are heard by the conductor at an average rate of one every 10 seconds. Find the characteristic function, mean and variance of the time that the conductor waits to begin the program if before beginning

(i) he waits until he hears a noise which was preceded by at least 20 seconds of silence;

(ii) he waits until there has been continuous silence from the audience for 20 seconds;

(iii) he waits until 20 seconds have elapsed since the first noise he heard (assume he does not begin without hearing at least one noise).

1.2 Consider a non-paralyzable counter in which successive locking times $Y_1, Y_2, \cdots$ are independent random variables identically distributed as a random variable Y with finite second moments. Suppose that particle *arrivals* are events of Poisson type with intensity λ. In terms of the characteristic function, mean and variance of Y, express the characteristic function, mean and variance of the random variables $\{T_n', n \geq 2\}$, where

$$T_n' = \tau_n' - \tau_{n-1}'$$

is the time between two successive particle registrations.

Hint. The inter-arrival time T_n' may be written as a sum

$$T_n' = Y_n + V_n$$

where Y_n is the deadtime caused by the particle registered at τ_{n-1}' and V_n is the time that elapsed between the time $\tau_{n-1}' + Y_n$ at which the counter became unlocked, and the time τ_n' at which the next registration occurred. Because particle arrivals are of Poisson type, it follows that Y_n and V_n are independent and that V_n is exponentially distributed with mean $1/\lambda$.

5-2 THE RENEWAL EQUATION

Let $f(t)$, $g(t)$, and $h(t)$ be functions defined for $t \geq 0$ satisfying the relation

$$g(t) = h(t) + \int_0^t g(t - s)f(s)\, ds, \quad t \geq 0. \tag{2.1}$$

If $f(t)$ and $h(t)$ are known functions, and $g(t)$ is an unknown function to be determined as the solution of the integral equation 2.1, then we say that $g(t)$ satisfies a renewal equation. The integral equation 2.1 is called a *renewal equation* because many quantities of interest in the theory of renewal counting processes satisfy an integral equation of the form of 2.1.

Renewal equation for the mean value function of a renewal counting process. The mean value function

$$m(t) = E[N(t)] = \sum_{n=0}^{\infty} n p_{N(t)}(n) \qquad (2.2)$$

of a renewal counting process $\{N(t), t \geq 0\}$ corresponding to independent identically distributed inter-arrival times $T_1, T_2, \cdots$ each possessing a probability density function $f(\cdot)$ and distribution function $F(\cdot)$ satisfies the renewal equation

$$m(t) = F(t) + \int_0^t m(t-s)f(s) \, ds, \ t \geq 0. \qquad (2.3)$$

To prove Eq. 2.3 we write

$$m(t) = \int_0^{\infty} E[N(t) \mid T_1 = s] f_{T_1}(s) \, ds. \qquad (2.4)$$

Now,

$$E[N(t) \mid T_1 = s] = 0 \quad \text{for} \quad s > t, \qquad (2.5)$$

since $N(t) = 0$ if the first event occurs at a time s later than t, while

$$E[N(t) \mid T_1 = s] = 1 + m(t-s) \quad \text{if} \quad s \leq t, \qquad (2.6)$$

since given that the first event occurs at a time $s \leq t$, the conditional distribution of $N(t)$ is the same as the distribution of $1 + N(t-s)$. It should be emphasized that in writing Eq. 2.6 we have used intuition more than rigor. A rigorous derivation would require a very careful development of the foundations of the theory of stochastic processes and is beyond the scope of this book. For the purposes of applied probability theory, the argument given here is completely adequate. From Eqs. 2.4, 2.5, and 2.6 one obtains Eq. 2.3.

Excess life distribution. Another quantity of interest when considering mechanisms which are being renewed is the amount of life remaining in the item in service at a given time t. We denote this quantity by $\gamma(t)$ and call it the *excess life* at time t. In symbols,

$$\gamma(t) = W_{N(t)+1} - t. \qquad (2.7)$$

Equivalently, $\gamma(t)$ is the time between t and the next occurring event.

The excess life distribution is of interest not only for its own sake but because of its usefulness in studying the distribution of the increment $N(t+v) - N(t)$ of the renewal counting process (where v is a positive constant). The probability distribution of $N(t+v) - N(t)$ may be expressed in terms of the distributions of $N(t)$ and of $\gamma(t)$ as follows: for any positive numbers t and v,

$$P[N(t + v) - N(t) = 0] = P[\gamma(t) > v], \qquad (2.8)$$

while for any integer $n \geq 1$,

$$P[N(t + v) - N(t) = n] = \int_0^v P[N(v - s) = n - 1] \, dF_{\gamma(t)}(s). \quad (2.9)$$

To prove Eq. 2.9, let A be the event that $N(t + v) - N(t) = n$. One may write

$$P[A] = \int_0^\infty P[A \mid \gamma(t) = s] \, dF_{\gamma(t)}(s).$$

Now, $P[A \mid \gamma(t) = s]$ is equal to $P[N(v - s) = n - 1]$ or 0 according as $s \leq v$ or $s > v$.

To determine the probability distribution of the excess life, one uses the fact that

$$g(t,x) = P[\gamma(t) > x] \qquad (2.10)$$

satisfies the renewal equation

$$g(t,x) = 1 - F(t + x) + \int_0^t g(t - s,x)f(s) \, ds, \qquad (2.11)$$

assuming that the successive inter-arrival times possess a probability density function $f(\cdot)$ and distribution function $F(\cdot)$.

To prove Eq. 2.11 we write

$$P[\gamma(t) > x] = \int_0^\infty P[\gamma(t) > x \mid T_1 = s]f_{T_1}(s) \, ds.$$

Now,

$$
\begin{aligned}
P[\gamma(t) > x \mid T_1 = s] &= 1 && \text{if } s > t + x, \\
&= 0 && \text{if } t < s < t + x, \\
&= P[\gamma(t - s) > x] = g(t - s, x) && \text{if } s < t.
\end{aligned}
$$

We discuss only the last equation; given that the first event occurs at $s < t$, the conditional distribution of the excess life of the item in service at time t is the same as the distribution of the excess life of the item in service at time $t - s$. From these equations it follows that

$$g(t,x) = \int_0^t g(t - s,x)f(s) \, ds + \int_{t+x}^\infty f(s) \, ds$$

from which one obtains Eq. 2.11.

We do not discuss here either general methods of solving the renewal equation (see Feller [1941]) or the properties of the renewal equation (see Karlin [1955]). Rather we show how to solve the renewal equation in certain cases of interest.

THEOREM 2A

Solution of the renewal equation in the case of exponentially distributed inter-arrival times. The solution $g(t)$ of the renewal equation

$$g(t) = h(t) + \nu \int_0^t g(t-s)e^{-\nu s}\, ds \qquad (2.12)$$

is given by

$$g(t) = g(0) + \int_0^t e^{-\nu s} \frac{d}{ds} \{e^{\nu s}h(s)\}\, ds. \qquad (2.13)$$

Proof. Let $G(t) = e^{\nu t}g(t)$, $H(t) = e^{\nu t}h(t)$. From Eq. 2.12 it follows that $G(t)$ satisfies the integral equation

$$G(t) = H(t) + \nu \int_0^t G(s)\, ds,$$

and therefore $G(t)$ satisfies the differential equation

$$G'(t) - \nu G(t) = H'(t)$$

whose solution (see Theorem 4A of Chapter 7) is

$$G(t) = G(0)e^{\nu t} - \int_0^t e^{\nu(t-s)}H'(s)\, ds$$

from which one obtains Eq. 2.13.

From Theorem 2A one can obtain a fact of great importance. *For exponential inter-arrival times, the excess life $\gamma(t)$ is exponentially distributed with the same mean as the inter-arrival times;* in symbols,

$$\text{if } \quad F_T(u) = 1 - e^{-\nu u}, \quad \text{then } \quad P[\gamma(t) \le x] = 1 - e^{-\nu x}. \qquad (2.14)$$

Proof. From Eq. 2.11 it follows that

$$P[\gamma(t) > x] = e^{-\nu(t+x)} + \nu \int_0^t P[\gamma(t-s) > x]e^{-\nu s}\, ds. \qquad (2.15)$$

From Eq. 2.15 and Theorem 2A it follows that

$$P[\gamma(t) > x] = P[\gamma(0) > x] + \int_0^t e^{-\nu s} \frac{d}{ds} \{e^{\nu s}e^{-\nu(s+x)}\}ds$$

$$= P[\gamma(0) > x]$$

$$= e^{-\nu x}.$$

The tools are now at hand to prove the converse of Theorem 3A of Chapter 4.

THEOREM 2B

If the inter-arrival times $\{T_n\}$ are exponentially distributed with mean $1/\nu$, then the renewal counting process $\{N(t), t \geq 0\}$ is a Poisson process with intensity ν.

Proof. To prove the theorem, it suffices to show that for any $t \geq s \geq 0$, the increment $N(t) - N(s)$ is Poisson distributed with mean $\nu(t - s)$, no matter what values $N(t')$ has for $t' \leq s$. Now,

$$\{N(t) - N(s), t \geq s\}$$

is a (possibly delayed) renewal counting process, corresponding to independent random variables $T_1, T_2, \cdots$ where T_1 is the time from s to the first event occurring after time s, and so on. It is clear that $T_2, T_3, \cdots$ are exponentially distributed with mean $1/\nu$. By Eq. 2.14 it follows that T_1 is also exponentially distributed with mean $1/\nu$, no matter what the values of $N(t')$ for $t' \leq s$. Consequently, the conditional distribution of $N(t) - N(s)$, given the values of $N(t')$ for $t' \leq s$, is the same as the unconditional distribution of $N(t - s)$.

We complete the proof of the theorem by showing that, for any $t > 0$, $N(t)$ is Poisson distributed with mean νt. Since the inter-arrival times $\{T_n\}$ are exponentially distributed with mean $1/\nu$, it follows that the waiting time W_n to the nth event obeys a gamma probability law with parameters n and ν;

$$f_{W_n}(x) = \frac{1}{\Gamma(n)} \nu^n x^{n-1} e^{-x\nu}, \quad x > 0$$

$$1 - F_{W_n}(t) = \sum_{m=0}^{n-1} \frac{1}{m!} (\nu t)^m e^{-\nu t}, \quad t > 0.$$

Therefore, by Eq. 3.8 of Chapter 4,

$$p_{N(t)}(n) = \frac{1}{n!} (\nu t)^n e^{-\nu t}$$

and $N(t)$ is Poisson distributed with mean νt.

Gamma distributed inter-arrival times. The gamma distribution is a two parameter family of distributions which can be used to approximate any general distribution of inter-arrival times. Consequently, renewal counting processes corresponding to gamma distributed inter-arrival times are frequently considered, especially since the probability law of such a process may be readily computed.

Let $\{N(t), t \geq 0\}$ be a renewal counting process corresponding to independent identically distributed inter-arrival times which obey a gamma probability law with parameters $\lambda > 0$ and $k = 1, 2, \cdots$. The inter-arrival time then has probability density

$$f(t) = \frac{\lambda}{(k-1)!} (\lambda t)^{k-1} e^{-\lambda t}, \quad t > 0 \tag{2.16}$$
$$= 0, \qquad t < 0.$$

It follows that the waiting time to the nth event has distribution function given by

$$1 - F_{W_n}(t) = \sum_{m=0}^{nk-1} \frac{1}{m!} (\lambda t)^m e^{-\lambda t} = \int_t^\infty \frac{\lambda^{nk} x^{nk-1}}{\Gamma(nk)} e^{-\lambda x} dx \tag{2.17}$$

and that $N(t)$ has probability mass function

$$p_{N(t)}(n) = \sum_{m=nk}^{(n+1)k-1} \frac{1}{m!} (\lambda t)^m e^{-\lambda t}. \tag{2.18}$$

We next compute the probability generating function of $N(t)$:

$$\psi(z,t) = \sum_{n=0}^\infty z^n P[N(t) = n]. \tag{2.19}$$

Define

$$G(z,t) = \sum_{n=1}^\infty z^{n-1} F_{W_n}(t). \tag{2.20}$$

It is easily verified (using Eq. 3.8 of Chapter 4) that

$$\psi(z,t) = 1 + (z-1) G(z,t). \tag{2.21}$$

From Eq. 2.17 it follows that

$$G(z,t) = \sum_{n=1}^\infty z^{n-1} \int_0^t \frac{(\lambda x)^{nk-1}}{(nk-1)!} \lambda e^{-\lambda x} dx$$
$$= y^{1-k} \int_0^t \lambda e^{-\lambda x} \left(\sum_{n=1}^\infty \frac{(\lambda x y)^{nk-1}}{(nk-1)!} \right) dx, \tag{2.22}$$

where we define $y^k = z$. The sum in Eq. 2.22 may be evaluated using the following fact: for any real number u and integer k

$$\sum_{n=1}^\infty \frac{u^{nk-1}}{(nk-1)!} = \frac{1}{k} \sum_{r=0}^{k-1} \epsilon^r e^{u \epsilon^r}, \tag{2.23}$$

where

$$\epsilon = \exp\left(\frac{2\pi i}{k}\right), \quad \epsilon^0 = 1, \quad \epsilon^r = \exp\left(\frac{2\pi i r}{k}\right). \tag{2.24}$$

To prove Eq. 2.23, expand $e^{u\epsilon^r}$ in its Taylor series, and use the easily verified fact (using the formula for the sum of a geometric series) that

$$\frac{1}{k}\sum_{r=0}^{k-1}(\epsilon^r)^\nu = 1 \quad \text{if } \nu \text{ is a multiple of } k$$
$$= 0 \quad \text{if } \nu \text{ is not a multiple of } k.$$

From Eqs. 2.22 and 2.23 one may write

$$G(z,t) = y^{1-k}\int_0^t \lambda e^{-\lambda x}\left(\frac{1}{k}\sum_{r=0}^{k-1}\epsilon^r e^{\lambda xy\epsilon^r}\right)dx$$

$$= y^{1-k}\frac{1}{k}\sum_{r=0}^{k-1}\frac{\epsilon^r}{1-y\epsilon^r}\{1-\exp[-\lambda t(1-y\epsilon^r)]\}.$$

Therefore, the probability generating function of $N(t)$ is given by

$$\psi(z,t) = 1 + \left(\frac{z-1}{z}\right)\frac{1}{k}\sum_{r=0}^{k-1}\frac{z^{1/k}\epsilon^r}{1-z^{1/k}\epsilon^r}\{1-\exp[-\lambda t(1-z^{1/k}\epsilon^r)]\}. \quad (2.25)$$

To illustrate the use of Eq. 2.25, let us consider the cases $k=1$ and $k=2$.

The case $k=1$ corresponds to exponentially distributed inter-arrival times for which it is known that $N(t)$ is Poisson distributed. From Eq. 2.25 it follows that

$$\psi(z,t) = 1 + \left(\frac{z-1}{z}\right)\left(\frac{z}{1-z}\right)\{1-\exp[-\lambda t(1-z)]\}$$
$$= \exp[\lambda t(z-1)],$$

which is the probability generating function of a Poisson distribution.

In the case of $k=2$, $\epsilon^0 = 1$, $\epsilon^1 = -1$, and $\psi(z,t)$ may be shown to reduce to

$$\psi(z,t) = \frac{1}{2}e^{-\lambda t}\left\{\left(1+\frac{1}{\sqrt{z}}\right)e^{\lambda t\sqrt{z}} + \left(1-\frac{1}{\sqrt{z}}\right)e^{-\lambda t\sqrt{z}}\right\}$$

$$= e^{-\lambda t}\left\{\cosh(\lambda t\sqrt{z}) + \frac{1}{\sqrt{z}}\sinh(\lambda t\sqrt{z})\right\}. \quad (2.26)$$

From Eq. 2.26 it follows if $\{N(t), t \geq 0\}$ is the renewal counting process corresponding to independent identically distributed inter-arrival times with probability density function

$$f(x) = \lambda^2 x e^{-\lambda x}, \quad x > 0, \quad (2.27)$$

then

$$E[N(t)] = \frac{\lambda}{2}t - \frac{1}{4} - \frac{1}{4}e^{-2\lambda t}, \tag{2.28}$$

$$\mathrm{Var}[N(t)] = \frac{\lambda}{4}t - \frac{\lambda}{2}te^{-2\lambda t}$$

$$+ \frac{1}{4}e^{-\lambda t}\sinh\lambda t - \frac{1}{4}e^{-2\lambda t}\sinh^2\lambda t. \tag{2.29}$$

One may show either directly, or by differentiating the probability generating function, that for gamma distributed inter-arrival times with probability density Eq. 2.16,

$$E[N(t)] = \frac{\lambda t}{k} + \frac{1}{k}\sum_{r=1}^{k-1}\frac{\epsilon^r}{1-\epsilon^r}\{1 - \exp[-\lambda t(1-\epsilon^r)]\}. \tag{2.30}$$

The mean value function of a renewal counting process determines its probability law. The renewal counting process $\{N(t), t \geq 0\}$ corresponding to exponentially distributed inter-arrivals with mean μ has a mean value function

$$m(t) = E[N(t)] = \frac{t}{\mu}, \tag{2.31}$$

which is a linear function of t. The question naturally arises: does the fact that the mean value function of a renewal counting process is a linear function of t imply that the inter-arrival times are exponentially distributed? We prove that this is the case by showing that the mean value function of a renewal counting process actually determines its probability law.

To show that this is the case, let us first define the moment generating function $\psi_T(\theta)$ of the inter-arrival time T [or equivalently the Laplace-Stieltjes transform of the distribution function $F_T(t)$ of T],

$$\psi_T(\theta) = \int_0^\infty e^{-t\theta}\,dF_T(t),$$

and the Laplace-Stieltjes transform $m^*(\theta)$ of the mean value function

$$m^*(\theta) = \int_0^\infty e^{-t\theta}\,dm(t).$$

For example, if

$$f_T(t) = \nu e^{-\nu t}, \quad t \geq 0$$

then

$$\psi_T(\theta) = \int_0^\infty e^{-\theta t}\nu e^{-\nu t}\,dt = \frac{\nu}{\nu+\theta}.$$

If

$$m(t) = \nu t$$

then

$$m^*(\theta) = \int_0^\infty e^{-\theta t}\, d(\nu t) = \nu \int_0^\infty e^{-\theta t}\, dt = \frac{\nu}{\theta}.$$

Since the waiting time W_n may be written as the sum, $W_n = T_1 + T_2 + \cdots + T_n$, of n independent random variables with moment generating function $\psi_T(\theta)$, it follows that

$$\psi_{W_n}(\theta) = \{\psi_T(\theta)\}^n.$$

In the case of a delayed renewal counting process,

$$\psi_{W_n}(\theta) = \psi_{T_1}(\theta)\{\psi_T(\theta)\}^{n-1}.$$

Now, in view of Eq. 3.8 of Chapter 4,

$$m(t) = \sum_{n=1}^\infty n\{F_{W_n}(t) - F_{W_{n+1}}(t)\} = \sum_{n=1}^\infty F_{W_n}(t). \tag{2.32}$$

Taking the Laplace-Stieltjes transform of both sides of Eq. 2.32 it follows that if the renewal counting process is not delayed, then

$$m^*(\theta) = \sum_{n=1}^\infty \psi_{W_n}(\theta) = \sum_{n=1}^\infty \{\psi_T(\theta)\}^n. \tag{2.33}$$

The sum in Eq. 2.33 is the sum of a geometric series. Consequently,

$$m^*(\theta) = \frac{\psi_T(\theta)}{1 - \psi_T(\theta)}, \tag{2.34}$$

from which it follows that

$$\psi_T(\theta) = \frac{m^*(\theta)}{1 + m^*(\theta)}. \tag{2.35}$$

Since the Laplace-Stieltjes transform of a function uniquely determines the function, it follows from Eq. 2.35 that the mean value function of a renewal counting process determines the probability law of the inter-arrival times, and therefore determines the probability law of the renewal counting process.

It should be noted that Eq. 2.35 can be derived immediately from the renewal equation for $m(t)$ (see Theorem 3C).

COMPLEMENTS

2A *Mean excess life.* Show that the solution $g(t)$ of the renewal equation

$$g(t) = m + \int_0^t g(t - s) \, dF_T(s), \qquad (2.36)$$

(in which m is a constant) is given by

$$g(t) = m\{E[N(t)] + 1\}. \qquad (2.37)$$

Consequently, establish the following identity:

$$E[W_{N(t)+1}] = t + E[\gamma(t)] = E[T]\{E[N(t)] + 1\}. \qquad (2.38)$$

Hint. Show that $g(t) = E[W_{N(t)+1}]$ satisfies Eq. 2.36 with $m = E[T]$. To establish Eq. 2.37, use Eq. 2.32; this proof is due to H. Scarf. For the usefulness of Eq. 2.38 in renewal theory, see Smith (1958), p. 246.

2B *Existence of all moments of a renewal counting process.* Let $\{N(t), t \geq 0\}$ be a renewal counting process corresponding to inter-arrival times identically distributed as a random variable T. Show that for each $t > 0$ (i) there exists a positive number θ_0 such that the moment generating function $E[\exp\{\theta N(t)\}]$ exists for all $\theta \leq \theta_0$, and (ii) $E[\{N(t)\}^m]$ is finite for all positive integers m.

Hint. Choose C so that $P[T > C] > 0$. Let $\{T_n'\}$ be the sequence of independent identically distributed random variables defined as follows: $T_n' = C$ or 0 depending on whether $T_n > C$ or $T_n \leq C$. Let $\{N'(t), t \geq 0\}$ be the renewal counting process corresponding to $\{T_n'\}$. Show that $N(t) \leq N'(t)$ for all t, and $N'(t) - r$, for a suitable integer r depending on t, has a negative binomial distribution.

2C Since knowledge of the mean value function $m(t)$ of a renewal counting process $\{N(t), t \geq 0\}$ implies complete knowledge of its probability law, it is not surprising that *the second moment function* $m_2(t) = E[N^2(t)]$ can be expressed in terms of the mean value function. Show that

$$m_2(t) = m(t) + 2 \int_0^t m(t - s) \, dm(s).$$

EXERCISES

2.1 A certain nuclear particle counter registers only every second particle arriving at the counter (starting with the second particle to arrive). Suppose that particle arrivals are events of Poisson type with intensity ν. Let $N(t)$ be the number of particles registered in time t. Find the probability that $N(t)$ is an odd number.

2.2 Show that for a delayed renewal counting process, instead of Eq. 2.34 one obtains

$$m^*(\theta) = \frac{\psi_{T_1}(\theta)}{1 - \psi_T(\theta)}.$$

2.3 *Finding the mean value function by inverting a Laplace-Stieltjes transform.*
Consider the delayed renewal counting process $\{M(t), t \geq 0\}$ defined in
Example 1E. Show that

$$\psi_T(\theta) = \left\{ 1 + \frac{\theta}{\nu} e^{(\nu+\theta)L} \right\}^{-1}, \qquad \psi_{T_1}(\theta) = \frac{\nu}{\nu + \theta},$$

$$m^*(\theta) = \int_0^\infty e^{-t\theta}\, dE[M(t)] = \frac{\nu}{\theta(\nu + \theta)}\{\theta + \nu\, e^{-(\nu+\theta)L}\}.$$

Consequently, show that

$$
\begin{aligned}
m(t) = E[M(t)] &= 1 - e^{-\nu t} & &\text{for} \quad t \leq L \\
&= 1 - e^{-\nu L} + \nu\, e^{-\nu L}(t - L) & &\text{for} \quad t \geq L
\end{aligned}
$$

by computing its Laplace-Stieltjes transform.

5-3   LIMIT THEOREMS FOR RENEWAL COUNTING PROCESSES

Renewal counting processes $\{N(t), t \geq 0\}$ have the attractive
feature that their asymptotic properties (for large values of t) can be
simply stated.

THEOREM 3A

Let $\{N(t), t \geq 0\}$ be a (possibly delayed) renewal counting process
corresponding to independent inter-arrival times $\{T_n, n = 1, 2, \cdots\}$ such
that $T_2, T_3, \cdots$ are identically distributed as a random variable T.

Asymptotic expression for the mean $m(t) = E[N(t)]$: if $\mu = E[T] < \infty$,
then

$$\lim_{t \to \infty} \frac{m(t)}{t} = \frac{1}{\mu}. \tag{3.1}$$

Indeed, a law of large numbers holds:

$$P\left[\lim_{t \to \infty} \frac{N(t)}{t} = \frac{1}{\mu} \right] = 1. \tag{3.2}$$

Asymptotic expression for the variance $\mathrm{Var}[N(t)]$: if $\mu = E[T]$ and
$\sigma^2 = \mathrm{Var}[T]$ are finite, then

$$\lim_{t \to \infty} \frac{\mathrm{Var}[N(t)]}{t} = \frac{\sigma^2}{\mu^3}. \tag{3.3}$$

Asymptotic normality of the renewal process: if $E[T^2] < \infty$, then
for any real x

$$\lim_{t \to \infty} P\left[\frac{N(t) - (t/\mu)}{\sqrt{t\sigma^2/\mu^3}} \leq x \right] = \frac{1}{\sqrt{2\pi}} \int_{-\infty}^x e^{-(1/2)v^2}\, dy. \tag{3.4}$$

In words, Eq. 3.4 says that $N(t)$ is asymptotically normally distributed with asymptotic mean and variance given by Eqs. 3.1 and 3.3.

Unfortunately, the proofs of these theorems require mathematical techniques beyond the scope of this book and are consequently omitted (for references to the history of these theorems, the reader is referred to the thorough review paper by Smith [1958]). Some idea of the proofs of Eqs. 3.2 and 3.4 can be obtained in Section 6–9.

To illustrate the meaning of Eqs. 3.1 and 3.3, let us consider the case of the renewal counting processes $\{N(t), t \geq 0\}$ corresponding to inter-arrival times which are gamma distributed with parameters λ and $r = 2$. The exact expressions for the mean and variance of $N(t)$ are given by Eqs. 2.28 and 2.29 respectively. Consequently,

$$\lim_{t \to \infty} \frac{E[N(t)]}{t} = \frac{\lambda}{2},$$

$$\lim_{t \to \infty} \frac{\text{Var}[N(t)]}{t} = \frac{\lambda}{4}.$$

These expressions can be obtained directly from Eqs. 3.1 and 3.3, since the probability density function, Eq. 2.27, has mean $\mu = 2/\lambda$ and variance $\sigma^2 = 2/\lambda^2$.

EXAMPLE 3A

Corrections for counting losses. Consider a nuclear particle counter with constant deadtime L at which particle arrivals are events of Poisson type with intensity ν. For long observation times t, the actual number $M(t)$ of particles registered may be regarded by Eq. 3.2 as approximately equal to t/μ, where μ is the mean inter-arrival time between particle registrations. For a non-paralyzable counter one may show (see Exercise 1.2) that

$$\mu = L + \frac{1}{\nu} = \frac{1 + \nu L}{\nu},$$

while for a paralyzable counter it was shown in example 1E that

$$\mu = \frac{1}{\nu} e^{\nu L} \doteq \frac{1 + \nu L}{\nu} \quad \text{if } \nu L \text{ is small.}$$

Consequently, to estimate ν from $M(t)$ one may proceed as follows. Let

$$\hat{\nu} = \frac{M(t)}{t}$$

be the observed intensity of registration of events. One may write approximately by Eq. 3.2

$$\hat{\nu} = \frac{1}{\mu}.$$

For a non-paralyzable counter,

$$\hat{\nu} = \frac{\nu}{1 + \nu L}$$

so that

$$\nu = \frac{\hat{\nu}}{1 - \hat{\nu} L}. \tag{3.5}$$

For a paralyzable counter one would determine ν from the equation

$$\hat{\nu} = \nu e^{-\nu L}.$$

If νL can be assumed to be small, then Eq. 3.5 gives ν in both the non-paralyzable and paralyzable cases. One can consider Eq. 3.5 as the correction to be made in the observed intensity $\hat{\nu}$ of particle registrations in order to obtain the true intensity ν of particle arrivals.

For the behavior of the number of particle registrations in a counter with arbitrary distribution of deadtime and Poisson arrivals, see Smith (1957) and Takács (1958). For a thorough discussion of the application of renewal theory to the theory of counters, see Smith (1958).

In the applications of renewal theory an important role is played by the following limit theorem which we state without proof.

THEOREM 3B

If the inter-arrival time T is not a lattice random variable† and has finite mean μ then for any $h > 0$,

$$\lim_{t \to \infty} m(t+h) - m(t) = \frac{h}{\mu}. \tag{3.6}$$

More generally, for any function $Q(t)$ satisfying the conditions

(i) $Q(t) \geq 0$ for all $t > 0$,

(ii) $\int_0^\infty Q(t)\, dt < \infty$,

(iii) $Q(t)$ is non-increasing (that is, $Q(t_1) \geq Q(t_2)$ if $t_1 \leq t_2$),

†A lattice random variable X is a discrete random variable with the property that all values x which X can assume with positive probability are of the form $x = kh$, for some real number h, and integer k (for example, an integer-valued random variable is a lattice random variable).

it holds that

$$\lim_{t \to \infty} \int_0^t Q(t-s) \, dm(s) = \frac{1}{\mu} \int_0^\infty Q(s) \, ds. \tag{3.7}$$

The result Eq. 3.6 is usually called Blackwell's theorem, since it was first proved in full generality by Blackwell (1948). The result Eq. 3.7 is called the *key renewal theorem* (see Smith [1958], p. 247), since a large number of other results can be deduced from it. For a proof of Eq. 3.7, using Eq. 3.6, see the discussion by Takács of Smith's paper.

THEOREM 3C

Let $m(t)$ be the mean value function of a renewal counting process corresponding to independent identically distributed inter-arrival times with non-lattice distribution function $F(x)$ and finite mean μ. Let $g(t)$ be a function satisfying the renewal equation

$$g(t) = Q(t) + \int_0^t g(t-s) \, dF(s). \tag{3.8}$$

Then $g(t)$ is given by

$$g(t) = Q(t) + \int_0^t Q(t-s) \, dm(s). \tag{3.9}$$

If $Q(t)$ satisfies assumptions (i), (ii), and (iii) of Theorem 3B, then

$$\lim_{t \to \infty} g(t) = \frac{1}{\mu} \int_0^\infty Q(s) \, ds. \tag{3.10}$$

Proof. Taking the Laplace-Stieltjes transform of both sides of Eq. 3.8 it follows that

$$g^*(\theta) = Q^*(\theta) + g^*(\theta) \psi(\theta),$$

where

$$g^*(\theta) = \int_0^\infty e^{-t\theta} \, dg(t), \quad Q^*(\theta) = \int_0^\infty e^{-t\theta} \, dQ(t),$$

$$\psi(\theta) = \int_0^\infty e^{-t\theta} \, dF(t).$$

Therefore,

$$g^*(\theta) = \frac{Q^*(\theta)}{1 - \psi(\theta)} = Q^*(\theta) + Q^*(\theta) \frac{\psi(\theta)}{1 - \psi(\theta)}. \tag{3.11}$$

Since the Laplace-Stieltjes transform $m^*(\theta)$ of $m(t)$ satisfies

$$m^*(\theta) = \frac{\psi(\theta)}{1 - \psi(\theta)},$$

one obtains Eq. 3.9 upon inverting Eq. 3.11. That Eq. 3.10 holds follows from Eq. 3.9 and the key renewal theorem.

EXAMPLE 3B

   The asymptotic distribution of excess life. It is shown in Eq. 2.11 that

$$g(t,x) = P[\gamma(t) > x]$$

satisfies the renewal equation

$$g(t,x) = 1 - F(t + x) + \int_0^t g(t - s,x)\, dF(s).$$

Consequently, by Eq. 3.10,

$$\lim_{t \to \infty} P[\gamma(t) > x] = \frac{1}{\mu} \int_0^\infty \{1 - F(s + x)\}\, ds$$

$$= \frac{1}{\mu} \int_x^\infty \{1 - F(y)\}\, dy$$

and

$$\lim_{t \to \infty} P[\gamma(t) \le x] = \int_0^x \frac{1}{\mu} \{1 - F(y)\}\, dy.$$

Thus, for large values of t, the excess life $\gamma(t)$ approximately obeys the probability law specified by the probability density function

$$\frac{1}{\mu} \{1 - F(x)\}, \quad x \ge 0.$$

EXAMPLE 3C

   The asymptotic probability that a system subject to random breakdown and repair is operative. Consider a system which can be in one of two states, "on" or "off." At time 0, it is "on." It then serves before breakdown for a random time T_{on} with distribution function $F_{on}(t)$. It is then off before being repaired for a random time T_{off} with distribution function $F_{off}(t)$. It then repeats the cycle of being operative for a random time and being inoperative for a random time (for another description of this system see Example 4A of Chapter 1). Successive times to breakdown and to repair are assumed to be independent. We desire to find the probability $g(t)$ that at time t the system is operative.

   One can find a renewal equation that $g(t)$ satisfies. Let $T = T_{on} + T_{off}$ be the time from 0 to the time at which the system begins to be operative after having been repaired, and let $F(t)$ be its distribution function. Then

$$g(t) = \int_0^\infty P[\text{system operative at } t \mid T = s]\, dF(s).$$

Now,

$$P[\text{system operative at } t \mid T = s] = g(t - s) \text{ if } s \leq t$$
$$= P[t < T_{\text{on}} \mid T = s] \text{ if } s > t;$$
$$\int_t^\infty P[t < T_{\text{on}} \mid T = s] \, dF(s) = P[t < T_{\text{on}} \text{ and } t < T]$$
$$= P[t < T_{\text{on}}] = 1 - F_{\text{on}}(t).$$

Therefore, $g(t)$ satisfies the renewal equation

$$g(t) = 1 - F_{\text{on}}(t) + \int_0^t g(t - s) \, dF(s).$$

Assume that T is not a lattice random, and has finite mean μ. Then, by Eq. 3.10,

$$\lim_{t \to \infty} g(t) = \frac{1}{\mu} \int_0^\infty \{1 - F_{\text{on}}(t)\} \, dt = \frac{E[T_{\text{on}}]}{E[T_{\text{on}} + T_{\text{off}}]}.$$

The asymptotic probability that the system is operative is thus equal to the ratio of the mean time the system is on to the mean inter-arrival time between operative periods.

EXERCISES

3.1 Let $N(t)$ denote the number of particles registered in time t by a scale of s scaling circuit (see Example 1C) at which particle arrivals are events of Poisson type with intensity ν. Approximately evaluate the mean and variance of $N(t)$.

3.2 Consider a non-paralyzable counter in which successive locking times $Y_1, Y_2, \cdots$ are independent random variables identically distributed as a random variable Y with finite second moments. Suppose that particle arrivals are events of Poisson type with intensity λ. Let $M(t)$ be the number of particles registered in time t. Show that

$$\lim_{t \to \infty} \frac{E[M(t)]}{t} = \frac{\lambda}{1 + \lambda E[Y]}, \quad \lim_{t \to \infty} \frac{\text{Var}[M(t)]}{t} = \lambda \frac{1 + \lambda^2 \, \text{Var}[Y]}{(1 + \lambda E[Y])^3}.$$

3.3 Let $F(x)$ be the distribution function of a non-negative random variable with finite mean μ and second moment μ_2. Show that

$$g(x) = \frac{1}{\mu} \{1 - F(x)\} \qquad \text{for} \quad x \geq 0$$
$$= 0 \qquad\qquad\qquad \text{for} \quad x < 0$$

is a probability density function with mean

$$\int_0^\infty x g(x) \, dx = \frac{\mu_2}{2\mu}.$$

In view of this result show that it is plausible (although not proved here) that if a renewal counting process corresponds to inter-arrival times which are not lattice random variables and have finite mean $E[T]$ and second moment $E[T^2]$, then its excess life $\gamma(t)$ has mean satisfying

$$\lim_{t \to \infty} E[\gamma(t)] = \frac{E[T^2]}{2E[T]}.$$

3.4   *The next term in the expansion of the mean value function.* The mean value function of a renewal counting process corresponding to inter-arrival times which are not lattice random variables and have finite mean μ and second moment μ_2 satisfies

$$\lim_{t \to \infty} \left\{ m(t) - \frac{t}{\mu} \right\} = \frac{\mu_2}{2\mu^2} - 1.$$

Prove this assertion by applying the key renewal theorem to

$$Q(t) = \int_t^\infty \frac{1}{\mu} \{1 - F(s)\} \, ds = 1 - \int_0^t \frac{1}{\mu} \{1 - F(s)\} \, ds.$$

Hint. Show and use the fact that

$$\int_0^t Q(t - s) \, dm(s) = m(t) - \frac{t}{\mu} + \{1 - Q(t)\}.$$

3.5   *The probability that a paralyzable counter is unlocked.* Consider a paralyzable counter in which successive locking times are independent random variables identically distributed as a random variable Y with finite second moment. Suppose that particle *arrivals* are events of Poisson type with intensity λ. Let $\{X(t), t \geq 0\}$ be a stochastic process defined as follows:

$$X(t) = 1 \text{ if at time } t \text{ the counter is unlocked}$$
$$= 0 \text{ if at time } t \text{ the counter is locked.}$$

Let $M(t)$ denote the number of particles *registered* in the interval 0 to t. Let $m(t) = E[M(t)]$, and let $p(t) = P[X(t) = 1]$. Show that

(i) $\dfrac{d}{dt} m(t) = \lambda p(t), \quad m(t) = \lambda \int_0^t p(s) \, ds;$

(ii) $p(t) = \exp\left\{ -\lambda \int_0^t [1 - F_Y(y)] \, dy \right\};$

(*Hint.* Use the theory of filtered Poisson processes.)

(iii) the inter-arrival time between particle registrations has mean

$$\frac{1}{\lambda} \exp\{\lambda E[Y]\};$$

(iv) the random variable T_{locked}, representing the time the counter is locked (from the moment a particle is registered until the counter becomes unlocked), has mean $\dfrac{1}{\lambda} \exp\{\lambda E[Y]\} - \dfrac{1}{\lambda};$

(v) in a type p counter T_{locked} has mean $\dfrac{1}{\lambda p} [\exp\{\lambda p E[Y]\} - 1].$

Markov chains:
discrete parameter

IN CLASSICAL PHYSICS, a basic role is played by the fundamental principle of scientific determinism: from the state of a physical system at the time t_0, one may deduce its state at a later instant t. As a consequence of this principle one obtains a basic method of analyzing physical systems: the state of a physical system at a given time t_2 may be deduced from a knowledge of its state at any earlier (later) time t_1 and does not depend on the history of the system before (after) time t_1.

For physical systems which obey probabilistic laws rather than deterministic laws, one may enunciate an analogous principle (in which, however, time is no longer reversible): the probability that the physical system will be in a given state at a given time t_2 may be deduced from a knowledge of its state at any earlier time t_1, and does not depend on the history of the system before time t_1. Stochastic processes which represent observations on physical systems satisfying this condition are called Markov processes.

A special kind of Markov process is a Markov chain; it may be defined as a stochastic process whose development may be treated as a series of transitions between certain values (called the "states" of the process) which have the property that the probability law of the future development of the process, once it is in a given state, depends only on the state and not on how the process arrived in that state. The number of possible states is either finite or countably infinite.

This chapter is somewhat more mathematical than the other chapters of this book, since its main aim is to give a development of the central ideas of the theory of discrete parameter Markov chains.

6-1 FORMAL DEFINITION OF A MARKOV PROCESS

A discrete parameter stochastic process $\{X(t), t = 0, 1, 2, \cdots\}$ or a continuous parameter stochastic process $\{X(t), t \geq 0\}$ is said to be a *Markov process* if, for any set of n time points $t_1 < t_2 < \cdots < t_n$ in the index set of the process, the conditional distribution of $X(t_n)$, for given values of $X(t_1), \cdots, X(t_{n-1})$, depends only on $X(t_{n-1})$, the most recent known value; more precisely, for any real numbers $x_1, \cdots, x_n$

$$P[X(t_n) \leq x_n \mid X(t_1) = x_1, \cdots, X(t_{n-1}) = x_{n-1}]$$
$$= P[X(t_n) \leq x_n \mid X(t_{n-1}) = x_{n-1}]. \quad (1.1)$$

Intuitively, one interprets Eq. 1.1 to mean that, given the "present" of the process, the "future" is independent of its "past."

Markov processes are classified according to (i) the nature of the index set of the process (whether discrete parameter or continuous parameter) and (ii) the nature of the *state space* of the process.

A real number x is said to be a possible value, or a *state*, of a stochastic process $\{X(t), t \epsilon T\}$ if there exists a time t in T such that the probability $P[x - h < X(t) < x + h]$ is positive for every $h > 0$. The set of possible values of a stochastic process is called its state space. The state space is called discrete if it contains a finite or countably infinite number of states. A state space which is not discrete is called continuous. A Markov process whose state space is discrete is called a *Markov chain*. We shall often use the set of integers $\{0, 1, \cdots\}$ as the state space of a Markov chain.

TABLE 6.1. **Markov processes classified into four basic types**

		State space	
		Discrete	Continuous
Nature of parameter	Discrete	Discrete parameter Markov chain	Discrete parameter Markov process
	Continuous	Continuous parameter Markov chain	Continuous parameter Markov process

A Markov process is described by a *transition probability function*, often denoted by $P(x,t_0;E,t)$ or $P(E,t \mid x,t_0)$, which represents the conditional probability that the state of the system will at time t belong to the set E, given that at time $t_0(< t)$ the system is in state x. The Markov process is said to have *stationary transition probabilities*, or to be *homogeneous* in time, if $P(x,t_0;E,t)$ depends on t and t_0 only through the difference $(t - t_0)$.

There is an extensive literature on Markov processes, concerned on the one hand with the mathematical foundations of the theory and on the other with the applications of the theory. In this book our aim is to show how Markov processes arise as models for natural phenomena. In order to keep the mathematics simple, only Markov chains are treated. We shall be mainly concerned with examining:

(i) the time dependent (or "transient") behavior of a Markov chain; to find the transition probability function by finding and solving (differential, integral, or other kinds of functional) equations which it satisfies;

(ii) the long run (or "steady state") behavior of a Markov chain; in particular, to determine conditions under which there exists a probability measure $\pi(E)$ such that the limit

$$\lim_{t \to \infty} P(x,t_0;E,t) = \pi(E)$$

exists independently of x and t_0;

(iii) the behavior of various occupation times and first passage times; to study the probability distribution of the amount of time the chain spends in various states (and sets of states) and the length of time it takes the system to pass from one set of states to a second set of states.

EXAMPLE 1A

Discrete parameter Markov chain. Consider a physical system which is observed at a discrete set of times. Let the successive observations be denoted by $X_0, X_1, \cdots, X_n, \cdots$. It is assumed that X_n is a random variable. The value of X_n represents the state at time n of the physical system. The sequence $\{X_n\}$ is called a *chain* if it is assumed that there are only a finite or countably infinite number of states in which the system can be. The sequence $\{X_n\}$ is a Markov chain if each random variable X_n is discrete and if the following condition is satisfied: for any integer $m > 2$ and any set of m points $n_1 < n_2 < \cdots < n_m$ the conditional distribution of X_{n_m}, for given values of $X_{n_1}, \cdots, X_{n_{m-1}}$, depends only on $X_{n_{m-1}}$, the most recent known value; in particular, for any real numbers $x_0, x_1, \cdots, x_m$ it holds that

$$P[X_m = x_m \mid X_0 = x_0, \ldots, X_{m-1} = x_{m-1}] = P[X_m = x_m \mid X_{m-1} = x_{m-1}],$$
(1.2)

whenever the left-hand side of Eq. 1.2 is defined.

EXAMPLE 1B

An imbedded Markov chain. As an example of a Markov chain, consider *the number of persons in a queue* waiting for service. Consider a box office, with a single cashier, at which the arrivals of customers are events of Poisson type with intensity λ. Suppose that the service times of successive customers are independent identically distributed random variables. For $n \geq 1$, let X_n denote the number of persons waiting in line for service at the moment when the nth person to be served (on a given day) has finished being served. The sequence $\{X_n\}$ is a Markov chain; we show this by showing that the conditional distribution of X_{n+1}, given the values of $X_1, X_2, \cdots, X_n$, depends only on the value of X_n. Let U_n denote the number of customers arriving at the box office during the time that the nth customer is being served. We may then write

$$X_{n+1} = X_n - \delta(X_n) + U_{n+1}, \tag{1.3}$$

where we define

$$\begin{aligned} \delta(x) &= 1 && \text{if } x \neq 0 \\ &= 0 && \text{if } x = 0. \end{aligned} \tag{1.4}$$

In words, the number of persons waiting for service when the $(n+1)$st customer leaves depends on whether the $(n+1)$st customer was in the queue when the nth customer departed service. If $\delta(X_n) = 0$, then $X_{n+1} = U_{n+1}$; while if $\delta(X_n) = 1$, then $X_{n+1} = U_{n+1} + X_n - 1$. Since U_{n+1} is independent of $X_1, \cdots, X_n$ it follows that given the value of X_n, one need not know the values of $X_1, \cdots, X_{n-1}$ in order to determine the conditional probability distribution of X_{n+1}.

The Markov chain $\{X_n\}$ defined by Eq. 1.3 is known as an *imbedded Markov chain*, since it corresponds to observing the stochastic process $\{N(t), t \geq 0\}$, where $N(t)$ represents the number of customers in the queue at time t, at a sequence of times $\{t_n\}$ corresponding to the moments when successive customers depart service; in symbols

$$X_n = N(t_n). \tag{1.5}$$

The use of imbedded Markov chains to study the properties of continuous parameter stochastic processes constitutes a very important technique of applied probability theory. In the case that service times are exponentially distributed it may be shown (see Section 7–2) that $\{N(t), t \geq 0\}$ is itself a Markov chain.

EXAMPLE 1C

Continuous parameter Markov chains. Consider a population, such as the molecules present in a certain sub-volume of gas, the particles emitted by a radioactive source, biological organisms of a certain kind present in a certain environment, persons waiting in a line (queue) for service, and so on. For $t \geq 0$, let $X(t)$ denote the size of the population at time t. Consequently, $\{X(t), t \geq 0\}$ is an integer-valued process. For each $t \geq 0$, $X(t)$ has as its possible values the integers $\{0, 1, 2, \cdots\}$. If $\{X(t), t \geq 0\}$ is a Markov process, as is the case for example if $\{X(t), t \geq 0\}$ is a process with independent increments, then it is a Markov chain.

For the sake of giving a formal definition to the notion of a Markov process and of a Markov chain, we have considered only real valued random variables. However, the notion of a Markov process derives its wide applicability from the fact that it applies to stochastic processes $\{X(t), t \geq 0\}$ or $\{X_n, n = 1, 2, \cdots\}$ in which the random elements $X(t)$ or X_n can take other kinds of values than real numbers.

EXAMPLE 1D

Written language as a multiple Markov chain. The letters of the alphabet may be divided into two categories, vowels and consonants. Let us denote a letter by a 0 if it is a vowel and by a 1 if it is a consonant. A page of written text then appears as a sequence of 0's and 1's. The vowels and consonants form a Markov chain if given any string of letters the probability for the next letter to be a vowel or consonant (0 or 1) is the same as the probability that the next letter will be a vowel or consonant knowing only the nature of the last letter of the string. For most languages this would not be the case, although it does seem to be the case for sufficiently simple languages such as Samoan. It has been reported (Newman, as quoted by Miller [1952]) that the vowels and consonants in Samoan form a Markov chain in which a consonant is never followed by a consonant, and a vowel has probability 0.51 of being followed by a vowel.

However, suppose we consider successive blocks of letters. A block of r consecutive letters would be represented by an r-tuple $(z_1, \cdots, z_r)$ where each z_i is either a 0 or a 1. There are 2^r possible states for a block. A sequence of n letters would lead to $n - r + 1$ blocks of r consecutive letters. Thus the phrase of 19 letters:

stochastic processes

would lead to 18 blocks of 2 letters:

(1,1),   (1,0),   (0,1),   (1,1),   (1,0),   (0,1),   (1,1),   (1,0),   (0,1),
(1,1),   (1,1),   (1,0),   (0,1),   (1,0),   (0,1),   (1,1),   (1,0),   (0,1).

The original series of letters is said to form a *multiple Markov chain of order* r if the series of blocks of r consecutive letters forms a Markov chain in the sense that for any string of blocks (r-tuples), the probability for the next block to be of a specified kind is the same as the probability that the next block will be of the specified kind, knowing only the nature of the last block of the string. Various tests are available for determining the order of a multiple Markov chain (see Anderson and Goodman [1957] and Billingsley [1961]). Such tests provide one means of investigating the efficiency of languages as conveyors of information.

EXAMPLE 1E

System maintenance problems. Markov chains provide a powerful means of studying the effects of various rules for the operation, breakdown, and repair of any system such as an electronic computer or a piece of machinery. In order for a sequence of observations on the state of the system to form a Markov chain one must adopt a suitable definition of the notion of a state. One possibility might be to take $\{1, 2\}$ as the state space, where

state 1 denotes that the system is operative,
state 2 denotes that the system is being repaired.

This model usually does not embody enough of the complications of system breakdown and repair to form a Markov chain. For example, the system may be able to break down in two ways, one needing only a single time period to repair, the other needing several time periods to fix. Or a productive system may go through various phases of breakdown, which can be detected by periodic inspections and preventive maintenance. In the sequel it will be seen what assumptions have to be made in order for the behavior of the system to be treated as a Markov chain.

For a systematic discussion of how stochastic processes can be converted into multiple Markov processes by the inclusion of supplementary variables, see Cox (1955).

EXERCISES

In Exercises 1.1 to 1.4 state whether or not the stochastic process described is (i) a Markov process, (ii) a Markov chain. Explain your reasoning.

1.1 For $n = 1, 2, \cdots, X_n = U_1 + U_2 + \cdots + U_n$ where $\{U_n\}$ is a sequence of independent random variables, (i) each normally distributed, (ii) each taking the values 0 and 1 with probabilities p and q, respectively.

1.2 For $n = 1, 2, \cdots, X_n = \{U_1 + U_2 + \cdots + U_n\}^2$ where $\{U_n\}$ is a sequence of independent random variables.

1.3 $\{X(t), t > 0\}$ is (i) the Wiener process, (ii) the Poisson process.

1.4 $\{X_n, n = 1, 2, \cdots\}$ is the solution of the stochastic difference equation $X_n = \rho X_{n-1} + I_n$, where ρ is a known constant, $X_0 = 0$, and $\{I_n, n = 1, 2, \cdots\}$ is a sequence of independent identically distributed random variables.

6-2 TRANSITION PROBABILITIES AND THE CHAPMAN-KOLMOGOROV EQUATION

In order to specify the probability law of a discrete parameter Markov chain $\{X_n\}$ it suffices to state for all times $n \geq m \geq 0$, and states j and k, the probability mass function

$$p_j(n) = P[X_n = j] \tag{2.1}$$

and the conditional probability mass function

$$p_{j,k}(m,n) = P[X_n = k \mid X_m = j]. \tag{2.2}$$

The function $p_{j,k}(m,n)$ is called the *transition probability function* of the Markov chain. The probability law of a Markov chain is determined by the functions in Eqs. 2.1 and 2.2, since for all integers q, and any q time points $n_1 < n_2 < \cdots < n_q$, and states $k_1, \cdots, k_q$

$$P[X_{n_1} = k_1, \ldots, X_{n_q} = k_q]$$
$$= p_{k_1}(n_1)p_{k_1,k_2}(n_1,n_2)p_{k_2,k_3}(n_2,n_3) \ldots p_{k_{q-1},k_q}(n_{q-1},n_q). \tag{2.3}$$

A Markov chain is said to be *homogeneous* (or to be homogeneous in time or to have stationary transition probabilities) if $p_{j,k}(m,n)$ depends only on the difference $n - m$. We then call

$$p_{j,k}(n) = P[X_{n+t} = k \mid X_t = j] \quad \text{for any integer} \quad t \geq 0 \tag{2.4}$$

the *n-step transition probability function* of the homogeneous Markov chain $\{X_n\}$. In words, $p_{j,k}(n)$ is the conditional probability that a homogeneous Markov chain now in state j will move after n steps to state k. The one-step transition probabilities $p_{j,k}(1)$ are usually written simply $p_{j,k}$. In symbols,

$$p_{j,k} = P[X_{t+1} = k \mid X_t = j] \quad \text{for any integer} \quad t \geq 0. \tag{2.5}$$

Similarly, if $\{X(t), t \geq 0\}$ is a continuous parameter Markov chain, then to specify the probability law of $\{X(t), t \geq 0\}$ it suffices to state for all times $t \geq s \geq 0$, and states j and k, the probability mass function

$$p_k(t) = P[X(t) = k] \tag{2.6}$$

and the conditional probability mass function

$$p_{j,k}(s,t) = P[X(t) = k \mid X(s) = j]. \tag{2.7}$$

The function $p_{j,k}(s,t)$ is called the *transition probability function* of the Markov chain. The Markov chain $\{X(t), t \geq 0\}$ is said to be homogeneous (or to have stationary transition probabilities) if $p_{j,k}(s,t)$ depends only on the difference $t - s$. We then call

$$p_{j,k}(t) = P[X(t+u) = k \mid X(u) = j] \quad \text{for any } u \geq 0 \tag{2.8}$$

the transition probability function of the Markov chain $\{X(t), t \geq 0\}$.

A fundamental relation satisfied by the transition probability function of a Markov chain $\{X_n\}$ is the so-called *Chapman-Kolmogorov* equation: for any times $n > u > m \geq 0$ and states j and k,

$$p_{j,k}(m,n) = \sum_{\text{states } i} p_{j,i}(m,u) \; p_{i,k}(u,n). \tag{2.9}$$

Note that the summation in Eq. 2.9 is over all states of the Markov chain. To prove Eq. 2.9 one uses the Markov property and the easily verified fact that

$$P[X_n = k \mid X_m = j] = \sum_{\text{states } i} P[X_n = k \mid X_u = i, X_m = j] \, P[X_u = i \mid X_m = j]. \tag{2.10}$$

Similarly, one may show that the transition probability function of a Markov chain $\{X(t), t \geq 0\}$ satisfies the Chapman-Kolmogorov equation: for any times $t > u > s \geq 0$ and states j and k,

$$p_{j,k}(s,t) = \sum_{\text{states } i} p_{j,i}(s,u) \; p_{i,k}(u,t). \tag{2.11}$$

Transition probability matrices. The transition probabilities of a Markov chain $\{X_n\}$ with state space $\{0, 1, 2, \cdots \}$ are best exhibited in the form of a matrix:

$$P(m,n) = \begin{bmatrix} p_{0,0}(m,n) & p_{0,1}(m,n) & p_{0,2}(m,n) & \cdots & p_{0,k}(m,n) & \cdots \\ p_{1,0}(m,n) & p_{1,1}(m,n) & p_{1,2}(m,n) & \cdots & p_{1,k}(m,n) & \cdots \\ \vdots & \vdots & \vdots & \cdots & \vdots & \cdots \\ p_{j,0}(m,n) & p_{j,1}(m,n) & p_{j,2}(m,n) & \cdots & p_{j,k}(m,n) & \cdots \\ \vdots & \vdots & \vdots & \cdots & \vdots & \cdots \end{bmatrix}. \tag{2.12}$$

Note that the elements of a transition probability matrix $P(m,n)$ satisfy the conditions

$$p_{j,k}(m,n) \geq 0 \qquad \text{for all } j,k, \tag{2.13}$$

$$\sum_{k} p_{j,k}(m,n) = 1 \qquad \text{for all } j. \tag{2.14}$$

Given a $p \times q$ matrix A and a $q \times r$ matrix B,

$$A = \begin{bmatrix} a_{11} & a_{12} & \cdots & a_{1q} \\ a_{21} & a_{22} & \cdots & a_{2q} \\ \vdots & \vdots & \cdots & \vdots \\ a_{p1} & a_{p2} & \cdots & a_{pq} \end{bmatrix}, \quad B = \begin{bmatrix} b_{11} & b_{12} & \cdots & b_{1r} \\ b_{21} & b_{22} & \cdots & b_{2r} \\ \vdots & \vdots & \cdots & \vdots \\ b_{q1} & b_{q2} & \cdots & b_{qr} \end{bmatrix}$$

the product $C = AB$ of the two matrices is defined as the $p \times r$ matrix whose element c_{jk}, lying at the intersection of the jth row and the kth column, is given by

$$c_{jk} = a_{j1} b_{1k} + a_{j2} b_{2k} + \ldots + a_{jq} b_{qk} = \sum_{i=1}^{q} a_{ji} b_{ik}.$$

Similarly, given two infinite matrices A and B, one can define the product AB as the matrix C whose element c_{jk}, lying at the intersection of the jth row and the kth column, is given by

$$c_{jk} = \sum_{i} a_{ji} b_{ik}.$$

In terms of multiplication of transition probability matrices the Chapman-Kolmogorov equations for all times $n > u > m \geq 0$ may be written:

$$P(m,n) = P(m,u)P(u,n). \tag{2.15}$$

We thus see that given a Markov chain $\{X_n\}$ one can define a family $\{P(m,n)\}$ of matrices satisfying Eqs. 2.13, 2.14, and 2.15. The converse may also be shown: given a family $\{P(m,n)\}$ of matrices satisfying Eqs. 2.13, 2.14, and 2.15, one can define a Markov chain $\{X_n\}$ for which $P(m,n)$ is a transition probability matrix whose elements $p_{j,k}(m,n)$ satisfy Eq. 2.2.

Determining the transition probabilities of a Markov chain. From the Chapman-Kolmogorov equation one may derive various recursive relations (in the discrete parameter case) and differential equations (in the continuous parameter case) for the transition probability functions. In this section we discuss the discrete parameter case. We reserve the discussion of the continuous parameter case for Chapter 7.

Let $\{X_n\}$ be a Markov chain with transition probability matrix $\{P(m,n)\}$. From Eq. 2.15 it follows that

$$\begin{aligned} P(m,n) &= P(m,n-1)P(n-1,n) \\ &= P(m,n-2)P(n-2,n-1)P(n-1,n) \\ &= \cdots\cdots\cdots\cdots\cdots\cdots\cdots \\ &= P(m,m+1)P(m+1,m+2) \cdots P(n-1,n). \end{aligned} \tag{2.16}$$

Thus, to know $P(m,n)$ for all $m \leq n$ it suffices to know the sequence of one-step transition probability matrices

$$P(0,1), P(1,2), \cdots, P(n,n+1), \cdots. \tag{2.17}$$

Next, let us define the unconditional probability vectors (for $n = 0, 1, 2, \cdots$)

$$p(n) = \begin{bmatrix} p_0(n) \\ p_1(n) \\ \vdots \\ p_j(n) \\ \vdots \end{bmatrix}, \quad p_j(n) = P[X_n = j]. \tag{2.18}$$

It is easily verified that

$$p(n) = P(0,n)p(0). \tag{2.19}$$

In view of Eqs. 2.19, 2.16, and 2.3, it follows that *the probability law of a Markov chain $\{X_n\}$ is completely determined once one knows the transition probability matrices given in Eq. 2.17 and the unconditional probability vector $p(0)$ at time* 0.

In the case of a homogeneous Markov chain $\{X_n\}$, let

$$P(n) = \{p_{j,k}(n)\}, \quad P = \{p_{j,k}\} \tag{2.20}$$

denote respectively the n-step and the one-step transition probability matrices. From Eqs. 2.16 and 2.19 it follows that

$$P(n) = P^n, \tag{2.21}$$

$$p(n) = p(0)P^n. \tag{2.22}$$

Consequently, the probability law of a homogeneous Markov chain is completely determined once one knows the one-step transition probability matrix $P = \{p_{j,k}\}$ and the unconditional probability vector $p(0) = \{p_j(0)\}$ at time 0.

A Markov chain $\{X_n\}$ is said to be a *finite Markov chain* with K states if the number of possible values of the random variables $\{X_n\}$ is finite and equal to K. The transition probabilities $p_{j,k}$ are then non-zero for only a finite number of values of j and k, and the transition probability matrix P is then a $K \times K$ matrix.

Using the theory of eigenvalues and eigenvectors of finite matrices one can analytically obtain the n-step transition probabilities of a finite Markov chain. (See Feller [1957], Chapter 16, for an exposition of this theory.) For Markov chains with an infinite number of states it is difficult to obtain analytic expressions for the n-step transition probabilities. Consequently, in the remainder of this chapter we consider the problem of determining the asymptotic behavior (as n tends to ∞) of the n-step transition probabilities of homogeneous Markov chains.

EXAMPLE 2A

Two-state Markov chains. Simple as they are, homogeneous Markov chains with two states are nevertheless important. The transition probability matrix of a homogeneous two-state Markov chain (with states denoted by 0 and 1) is of the form

$$P = \begin{bmatrix} p_{0,0} & p_{0,1} \\ p_{1,0} & p_{1,1} \end{bmatrix}.$$

The two-step transition probability matrix is given by

$$P(2) = P^2 = \begin{bmatrix} p_{0,0}^2 + p_{0,1}p_{1,0} & p_{0,1}(p_{0,0} + p_{1,1}) \\ p_{1,0}(p_{0,0} + p_{1,1}) & p_{1,1}^2 + p_{0,1}p_{1,0} \end{bmatrix}.$$

In the case that $|\, p_{0,0} + p_{1,1} - 1 \,| < 1$ it may be shown by mathematical induction that the n-step transition probability matrix is given by

$$P(n) = \frac{1}{2 - p_{0,0} - p_{1,1}} \begin{bmatrix} 1 - p_{1,1} & 1 - p_{0,0} \\ 1 - p_{1,1} & 1 - p_{0,0} \end{bmatrix}$$
$$+ \frac{(p_{0,0} + p_{1,1} - 1)^n}{2 - p_{0,0} - p_{1,1}} \begin{bmatrix} 1 - p_{0,0} & -(1 - p_{0,0}) \\ -(1 - p_{1,1}) & 1 - p_{1,1} \end{bmatrix}. \quad (2.23)$$

From Eq. 2.23 one obtains simple asymptotic expressions for the n-step transition probabilities:

$$\lim_{n \to \infty} p_{0,0}(n) = \lim_{n \to \infty} p_{1,0}(n) = \frac{1 - p_{1,1}}{2 - p_{0,0} - p_{1,1}}, \quad (2.24)$$

$$\lim_{n \to \infty} p_{0,1}(n) = \lim_{n \to \infty} p_{1,1}(n) = \frac{1 - p_{0,0}}{2 - p_{0,0} - p_{1,1}}.$$

An example of a homogeneous two-state Markov chain is provided by a communications system which transmits the digits 0 and 1. Each digit transmitted must pass through several stages, at each of which there is a probability p that the digit which enters will be unchanged when it leaves. Let X_0 denote the digit entering the first stage of the system and, for $n \geq 1$, let X_n denote the digit leaving the nth stage of the communications system. The sequence $X_0, X_1, X_2, \cdots$ is then a homogeneous Markov chain with transition probability matrix (letting $q = 1 - p$)

$$P = \begin{bmatrix} p & q \\ q & p \end{bmatrix}. \tag{2.25}$$

The corresponding n-step transition probability matrix $P(n)$ may be written

$$P(n) = \begin{bmatrix} \frac{1}{2} + \frac{1}{2}(p-q)^n & \frac{1}{2} - \frac{1}{2}(p-q)^n \\ \frac{1}{2} - \frac{1}{2}(p-q)^n & \frac{1}{2} + \frac{1}{2}(p-q)^n \end{bmatrix}. \tag{2.26}$$

As an example of the use of Eq. 2.26, we note that if $p = 2/3$, then

$$\begin{aligned} P[X_2 = 1 \mid X_0 = 1] &= p_{11}^{(2)} = \tfrac{5}{9}, \\ P[X_3 = 1 \mid X_0 = 1] &= p_{11}^{(3)} = \tfrac{14}{27}; \end{aligned} \tag{2.27}$$

in words, if $p = 2/3$, then a digit entering the system as a 1 has probability $5/9$ of being correctly transmitted after 2 stages and probability $14/27$ of being correctly transmitted after 3 stages.

It is also of interest to compute the probability that a digit transmitted by the system as a 1 in fact entered the system as a 1. We leave it to the reader to show that under the transition probability matrix, Eq. 2.25,

$$P[X_0 = 1 \mid X_n = 1] = \frac{\alpha + \alpha(p-q)^n}{1 + (\alpha - \beta)(p-q)^n}, \tag{2.28}$$

where $\alpha = P[X_0 = 1]$ and $\beta = 1 - \alpha$.

EXAMPLE 2B

Transition probabilities for the imbedded Markov chain of a single-server queue. Let $\{X_n\}$ be the imbedded Markov chain defined in Example 1B. The one-step transition probabilities are easily obtained: for $j = 0$

$$p_{0,k}(n, n+1) = P[U_{n+1} = k] = a_k \tag{2.29}$$

while for $j > 0$

$$p_{j,k}(n, n+1) = P[U_{n+1} = k - j + 1] = a_{k-j+1}, \tag{2.30}$$

where a_k is the probability that k customers arrive during the service time of a customer (see Exercise 2.20). The one-step transition probabilities are independent of n. Therefore, the Markov chain $\{X_n\}$ is homogeneous, with transition probability matrix given by

$$P = \begin{bmatrix} a_0 & a_1 & a_2 & \dots \\ a_0 & a_1 & a_2 & \dots \\ 0 & a_0 & a_1 & \dots \\ 0 & 0 & a_0 & \dots \\ \cdot & \cdot & \cdot & \dots \end{bmatrix}. \tag{2.31}$$

An explicit formula for P^n seems difficult to achieve.

Remark on notation used to describe queues. To describe certain kinds of queues, many writers use a notation proposed by Kendall (1953). It is assumed that the successive inter-arrival times $T_1, T_2, \cdots$ between customers are independent identically distributed random variables. Similarly, the service times $S_1, S_2, \cdots$ are assumed to be independent identically distributed random variables. One then writes a symbol of the form $F_T/F_S/Q$, where F_T denotes the distribution function of inter-arrival times, F_S denotes the distribution function of service times, and Q designates the number of servers. The following symbols are used to denote inter-arrival and service time distributions:

D for a deterministic or constant inter-arrival or service time;

M for exponentially distributed inter-arrival or service time;

E_k for Erlangian (or gamma) distributed inter-arrival or service times (the Erlangian distribution of order k has probability density function

$$f(t) = \frac{\lambda}{\Gamma(k)} (\lambda t)^{k-1} e^{-\lambda t}, \qquad t > 0$$
$$= 0 \qquad\qquad , \qquad t < 0);$$

G for a general distribution of service times;

GI for a general distribution of inter-arrival times.

Thus, $M/G/1$ denotes a queue with exponentially distributed inter-arrival times, no special assumption about service times, and one server, while $GI/M/1$ denotes a queue with no special assumption about inter-arrival times, exponentially distributed service times, and one server. The Markov chain defined in Example 1B is the imbedded Markov chain of an $M/G/1$ queue. In Exercise 2.21 we consider the imbedded Markov chain of a $GI/M/1$ queue.

EXAMPLE 2C

Discrete branching processes. Consider a population consisting of individuals able to produce new individuals of like kind. The original individuals are called the zeroth generation. The number of original individuals, denoted by X_0, is called the size of the zeroth generation. All offspring of the zeroth generation constitute the first generation and their number is denoted by X_1. In general, let X_n denote the size of the nth generation (the set of offspring of the $(n-1)$st generation). Let us assume that the number of offspring of different individuals are independent random variables identically distributed as a random variable Z with probability mass function $\{p_j\}$: for $j = 0, 1, \cdots$

$$p_j = P[Z = j] = P[\text{an individual has } j \text{ offspring}]. \qquad (2.32)$$

It is assumed that none of the probabilities $p_0, p_1, \cdots$ is equal to 1 and that $p_0 + p_1 < 1$ and $p_0 > 0$. Now,

$$X_{n+1} = \sum_{i=1}^{X_n} Z_i \tag{2.33}$$

where Z_i is the number of offspring of the ith member of the nth generation. Consequently, the conditional distribution of X_{n+1} given X_n is the same as the distribution of the sum of X_n independent random variables, each identically distributed as Z. In terms of probability generating functions this assertion may be written: for any $j > 0$

$$\sum_{k=0}^{\infty} p_{j,k}\, z^k = \left(\sum_{i=0}^{\infty} p_i\, z^i \right)^j. \tag{2.34}$$

Therefore, the one-step transition probability $p_{j,k}$ is equal to the coefficient of z^k in the expansion of the right-hand side of Eq. 2.34 in powers of z.

It is of particular interest to find the size X_n of the nth generation given that $X_0 = 1$ (the zeroth generation had size 1). The unconditional probability distribution of X_n is given by

$$p_j(1) = P[X_1 = j] = p_j \tag{2.35}$$

for $n = 1$, while for $n > 1$

$$p_j(n) = P[X_n = j] = p_{1,j}(n). \tag{2.36}$$

Let

$$P_n(z) = \sum_{j=0}^{\infty} p_j(n)\, z^j \tag{2.37}$$

be the probability generating function of the size of the nth generation. A *recursive relation* for $P_n(z)$ may be obtained:

$$P_{n+1}(z) = P(P_n(z)) = P_n(P(z)), \tag{2.38}$$

where we define

$$P(z) = P_1(z) = \sum_{j=0}^{\infty} p_j\, z^j. \tag{2.39}$$

To prove Eq. 2.38 note that from Eq. 2.33 it follows that

$$P_{n+1}(z) = E[z^{X_{n+1}}] = \sum_{j=1}^{\infty} E[z^{X_{n+1}} \mid X_n = j] P[X_n = j]$$

$$= \sum_{j=1}^{\infty} \{P(z)\}^j\, P[X_n = j] = P_n(P(z)).$$

From the fact that $P_{n+1}(z) = P_n(P(z))$ one may prove (by induction) that $P_{n+1}(z) = P(P_n(z))$. From Eq. 2.38 we are able to state a formula for the n-step transition probability:

$$p_{j,k}(n) = \text{coefficient of } z^k \text{ in } [P_n(z)]^j. \tag{2.40}$$

The probability of extinction of a population described by a branching process. An important problem, first raised in connection with the extinction of family surnames, is to find the limiting probability of extinction; that is, find

$$\pi_0 = \lim_{n \to \infty} p_0(n), \tag{2.41}$$

the limit of the probability $p_0(n)$ of no individuals in the nth generation. That the limit exists is clear, since $p_0(n)$ is a bounded non-decreasing sequence:

$$0 \leq p_0(n) \leq p_0(n+1) \leq 1. \tag{2.42}$$

Although this problem was first raised at the end of the last century (see Galton [1889]) it was not completely solved until 1930 (see Steffensen [1930]).

THEOREM 2A

The fundamental theorem of branching processes. The probability π_0 of ultimate extinction (given that the zeroth population had size 1) is the smallest positive number p satisfying the relation

$$p = P(p), \tag{2.43}$$

where $P(z)$ is the probability generating function of the number of offspring per individual. Further, the probability of ultimate extinction equals 1 if and only if the mean number μ of offspring per individual is not greater than 1; in symbols,

$$\pi_0 = 1 \text{ if and only if } \mu \leq 1. \tag{2.44}$$

Remark. To illustrate the use of the theorem let us consider an example given by Lotka (1931) who found that for white males in the United States in 1920 the probability generating function of the number of male offspring per male is approximately given by

$$P(z) = \frac{0.482 - 0.041z}{1 - 0.559z}. \tag{2.45}$$

Consequently, the equation $P(p) = p$ leads to the quadratic equation

$$0 = p(1 - 0.559p) - (0.482 - 0.041p). \qquad (2.46)$$

Eq. 2.46 is easily verified to have two roots: $p = 1$ and

$$p = \frac{0.482}{0.559} = 0.86. \qquad (2.47)$$

Consequently, the probability π_0 of extinction of a surname (descended from a single male) is given by 0.86.

Proof. Since $p_0(n) = P_n(0)$, it follows from Eq. 2.38 that $p_n(0)$ satisfies the recursive relation

$$p_0(n + 1) = P(p_0(n)). \qquad (2.48)$$

Letting $n \to \infty$ in Eq. 2.48 it follows that π_0 satisfies

$$\pi_0 = P(\pi_0). \qquad (2.49)$$

To show that π_0 is the smallest positive number satisfying Eq. 2.43, let us show that

$$\pi_0 \leq p \qquad (2.50)$$

for any positive number p satisfying Eq. 2.43. Since $0 \leq a < b$ implies $P(a) \leq P(b)$ it follows that

$$p_0(1) = P(0) \leq P(p) = p$$

and by induction, for any n,

$$p_0(n) \leq p$$

from which Eq. 2.50 follows by letting n tend to ∞.

To prove Eq. 2.44, we first note (i) that $\mu = P'(1)$, the derivative of the generating function at 1, (ii) that $P(1) = 1$, so that 1 is a solution of Eq. 2.43, and (iii) that $P(0) = p_0(1) > 0$. It is clear geometrically (and it can be argued analytically, using the fact that $P(z)$ is a convex function) that either Figure 6.1 or Figure 6.2 holds. Either there is *not* a number p in the interval $0 < p < 1$ satisfying Eq. 2.43 or there is. There is not such a number p if and only if the slope μ, of the tangent to the generating function at $z = 1$, is less than or equal to 1. On the other hand, $\pi_0 = 1$ if and only if there is not a root p in $0 < p < 1$ of the equation $p = P(p)$. Our geometrical proof of Eq. 2.44 is now complete. For an analytic formulation of this proof, see Feller (1957), p. 275, or Bharucha-Reid (1960), pp. 24–26.

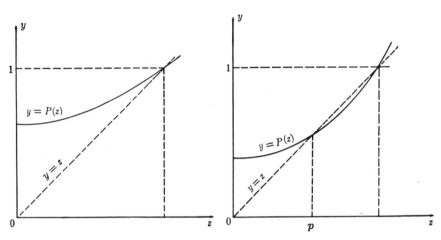

Fig. 6.1. Case when $\mu \leq 1$ **Fig. 6.2.** Case when $\mu > 1$

COMPLEMENT

2A *A non-Markov process satisfying the Chapman-Kolmogorov equation.* One way of studying the general properties of Markov chains (which has been extensively pursued) is to study the properties of solutions of the Chapman-Kolmogorov equations. It should be realized that there exist non-Markovian stochastic processes whose transition probabilities defined by Eq. 2.2 also satisfy the Chapman-Kolmogorov equations; a simple example of such a process is given below. (For further examples, see Feller [1959].) Thus while it is true that the transition probabilities of a Markov chain satisfy the Chapman-Kolmogorov equations, it is not true that a stochastic process is Markov if its transition probabilities satisfy the Chapman-Kolmogorov equations.

Let a sequence $\{X_n\}$ of random variables be defined as follows: Consider a sequence of urns each containing 4 balls, numbered 1 to 4. Let a ball be chosen successively and independently from each urn. For $m = 1, 2, \cdots$ let $A_m^{(1)}$ be the event that the ball drawn from the mth urn is 1 or 4, let $A_m^{(2)}$ be the event that the ball drawn from the mth urn is 2 or 4, and let $A_m^{(3)}$ be the event that the ball drawn from the mth urn is 3 or 4. For $m = 1, 2, \cdots$ and $j = 1, 2, 3$, let

$$X_{3(m-1)+j} = 1 \text{ or } 0 \text{ depending on whether or not } A_m^{(j)} \text{ occurred.}$$

Show that, for k, k_1, k_2 equal to 0 or 1,

$$P[X_n = k] = P[X_n = k_2 \mid X_m = k_1] = \tfrac{1}{2} \qquad \text{for } n > m,$$

while

$$P[X_{3m+3} = 1 \mid X_{3m+2} = 1, X_{3m+1} = 1] = 1 \qquad \text{for } m = 0, 1, \cdots.$$

Consequently, show that the process $\{X_n\}$ satisfies the Chapman-Kolmogorov equations but is not a Markov chain.

EXERCISES

In Exercises 2.1 to 2.10 describe the state space and the one-step and two-step transition probability matrices for the homogeneous Markov chain $\{X_n\}$ described.

2.1　A group of 4 children play a game consisting of throwing a ball to one another. At each stage the child with the ball is equally likely to throw it to any of the other 3 children. Let X_0 denote the child who had the ball originally, and for $n \geq 1$ let X_n denote the child who has the ball after it has been tossed exactly n times.

2.2　A number X_1 is chosen at random from the integers 1 to 6. For $n > 1$, X_n is chosen at random from the integers $1, 2, \cdots, X_{n-1}$.

2.3　Consider independent tosses of a fair die. Let X_n be the maximum of the numbers appearing in the first n throws.

2.4　Consider independent tosses of a coin which has probability p of falling heads. Let X_n be the total number of heads in the first n tosses.

2.5　Two black balls and two white balls are placed in 2 urns so that each urn contains 2 balls. At each step 1 ball is selected at random from each urn. The 2 balls selected are interchanged. Let X_0 denote the number of white balls initially in the first urn. For $n \geq 1$, let X_n denote the number of white balls in the first urn after n interchanges have taken place.

2.6　Consider a particle performing a random walk on a circle on which 4 points (denoted 0, 1, 2, 3) in clockwise order have been marked. The particle has probability p of moving to the point on its right (clockwise) and probability $1 - p$ of moving to the point on its left (counterclockwise). Let X_0 denote the particle's initial position, and for $n \geq 1$ let X_n denote the particle's position after n steps.

2.7　A white rat is put into the maze shown below:

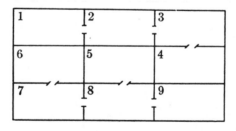

The rat moves through the compartments at random; i.e., if there are k ways to leave a compartment he chooses each of these with probability $1/k$.

He makes one change of compartment at each instant of time. The state of the system is the number of the compartment the rat is in.

2.8 A factory has two machines, only one of which is used at any given time. A machine breaks down on any given day with probability p. There is a single repairman who takes two days to repair a machine and can work on only one machine at a time. When a machine breaks down, it breaks down at the end of the day so that the repair man starts work the next day, and the other machine (if it is available) is not put into use until the next day. The state of the system is the pair (x,y) where x is the number of machines in operating condition at the end of a day and y is one if one day's work has been put in on a broken machine and zero otherwise. *Hint.* The state space is $\{(2,0), (1,0), (1,1), (0,1)\}$.

2.9 Consider a series of independent repeated tosses of a coin that has probability p of falling heads. For $n \geq 2$, let X_n be equal to 0, 1, 2, or 3 depending on whether the $(n-1)$st and nth trials had as their outcomes (heads, heads), (heads, tails), (tails, heads) or (tails, tails) respectively.

2.10 A sequence of experiments is performed, in each of which two fair coins are tossed. Let X_n be equal to the numbers of heads in n repetitions of the experiment.

2.11 *A non-Markov chain.* Consider a series of repeated tosses of a coin that has probability p of falling heads. For $n \geq 2$, let X_n be equal to 0 or 1, depending on whether or not the $(n-1)$st and nth trials both resulted in heads. Show that $\{X_n\}$ is not a Markov chain. (Note that this stochastic process is obtained from the Markov chain of Exercise 2.9 by grouping together three states.) Does $\{X_n\}$ satisfy the Chapman-Kolmogorov equation?

2.12 Consider Exercise 2.1. Find the probability that after 3 throws the ball will be in the hands of the child (i) who had it originally, (ii) who had it after the first throw.

2.13 Consider Exercise 2.2. Find (i) the probability that $X_3 = 3$; (ii) the conditional probability that $X_2 = 4$, given that $X_3 = 3$.

2.14 Consider Exercise 2.5. Find the most probable value of X_n for $n \geq 2$.

2.15 Using Markov chain considerations, solve the following riddle: "If A, B, C, and D each speak the truth once in three times (independently), and A affirms that B denies that C declares that D is a liar, what is the probability that D was telling the truth"? (For further discussion of this riddle, and references to its history, see *Mod Prob*, p. 133.)

2.16 Suppose that it may be assumed that the probability is equal to p that the weather (rain or no rain) on any arbitrary day is the same as on the preceding day. Let p_1 be the probability of rain on the first day of the year. Find the probability p_n of rain on the nth day. Evaluate the limit of p_n as n tends to infinity.

2.17   Suppose you are confronted with two coins, A and B. You are to make n tosses, using whichever coin you prefer at each toss. You will be paid \$1 for each time the coin tossed falls heads. Coin A has probability $1/2$ of falling heads, and coin B has probability $1/4$ of falling heads. Unfortunately, you are not told which of the coins is coin A. Consequently, you decide to toss the coins in accord with the following system. For the first toss, you choose a coin at random. For all succeeding tosses, you use the coin used on the preceding toss if it fell heads, and otherwise switch coins. (i) What is the probability that coin A is the coin tossed on the nth toss if (a) $n = 4$, (b) n is very large? (ii) What is the probability that the coin tossed on the nth toss will fall heads if (a) $n = 4$, (b) n is very large? (*Hint.* Let X_n be equal to 1 or 0 depending on whether the coin used at the nth trial was coin A or coin B.)

2.18   Prove that Eq. 2.23 holds.

2.19   Prove that Eq. 2.28 holds.

2.20   *Continuation of Examples 1B and 2B.* Show that the probability a_k of k arrivals during the service time of a customer is given by

$$a_k = \int_0^\infty e^{-\lambda t} \frac{(\lambda t)^k}{k!} \, dF_S(t), \tag{2.51}$$

where $F_S(\cdot)$ is the distribution function of service times. Find the mean and variance of the probability distribution $\{a_k\}$.

2.21   *The imbedded Markov chain of the queue GI/M/1.* Consider a single-server queue in which the inter-arrival times of successive customers are independent identically distributed random variables, and the service times of successive customers are independent exponentially distributed random variables with mean $1/\mu$; this queueing system is denoted $GI/M/1$ (see Example 2B). For $n \geq 1$, let X_n denote the number of persons waiting in line for service at the moment of arrival of the nth person to arrive. Let U_n denote the number of customers served during the interval between the arrivals of the nth and $(n + 1)$st customers. Then $X_1 = 0$ and, for $n \geq 1$,

$$X_{n+1} = X_n + 1 - U_n.$$

Show that $\{X_n\}$ is a homogeneous Markov chain with state space $\{0, 1, \cdots \}$ and show that its transition probability matrix is given by

$$P = \begin{bmatrix} B_0 & b_0 & 0 & 0 & 0 & \cdots \\ B_1 & b_1 & b_0 & 0 & 0 & \cdots \\ B_2 & b_2 & b_1 & b_0 & 0 & \cdots \\ \cdots & \cdots & \cdots & \cdots & \cdots & \cdots \end{bmatrix}, \tag{2.52}$$

where

$$b_k = \int_0^\infty e^{-\mu t} \frac{(\mu t)^k}{k!} \, dF_T(t), \quad B_k = \sum_{j=k+1}^\infty b_j, \tag{2.53}$$

and $F_T(t)$ is the distribution function of inter-arrival times.

2.22 Prove analytically the following reformulation of the fundamental theorem of branching processes. Let

$$P(z) = \sum_{n=0}^\infty z^n p_n$$

be the generating function of a probability distribution $\{p_n\}$ with $p_0 > 0$. The equation $p = P(p)$ possesses a root p in the interval $p_0 \leq p < 1$ if and only if

$$\mu = \sum_{n=0}^\infty n p_n > 1.$$

2.23 Show that the probability π_0 of eventual extinction of a branching process, given that the zeroth population had size $k \geq 1$, is p^k where p is the smallest positive number satisfying Eq. 2.43.

2.24 Suppose that the probabilities $p_1, p_2, \cdots$ defined by Eq. 2.32 form a geometric series:

$$p_j = b \, r^{j-1}, \qquad j = 1, 2, \cdots,$$

where $0 < r < 1$ and $0 < b < 1 - r$, while

$$p_0 = 1 - \sum_{j=1}^\infty p_j = 1 - \frac{b}{1-r} = \frac{1 - (r+b)}{1-r}.$$

(i) Find the corresponding probability generating function $P(z)$ and mean μ.

(ii) Show that the equation $p = P(p)$ has as its only positive roots 1 and

$$p = \frac{1 - (r+b)}{r(1-r)}.$$

(iii) Show that $p > 1$ if and only if $\mu \leq 1$.

2.25 *Estimating a transition probability matrix.* A two-state Markov chain was observed for 50 transitions. The successive states occupied by the chain were as follows:

```
0  1  0  1  1  1  0  1  0  0  1
0  1  0  1  0  1  1  1  1  0
1  0  0  1  0  1  1  0  0  1
0  1  1  0  1  0  0  0  1  0
1  1  0  0  1  1  0  1  1  0
```

Draw up an estimate of the matrix of one-step transition probabilities, assuming that the chain is homogeneous.

6-3 DECOMPOSITION OF MARKOV CHAINS INTO COMMUNICATING CLASSES

In the remainder of this chapter, we study the evolution in time of a discrete parameter homogeneous Markov chain $\{X_n\}$. It is convenient to begin by classifying the states of the chain according to whether it is possible to go from a given state to another given state.

A state k is said to be *accessible* from a state j if, for some integer $N \geq 1$, $p_{j,k}(N) > 0$. Two states j and k are said to *communicate* if j is accessible from k and k is accessible from j. If k is accessible from j, we write $j \to k$; if j and k communicate, we write $j \leftrightarrow k$.

THEOREM 3A

If $i \to j$ and $j \to k$, then $i \to k$.

Proof. Choose M and N so that $p_{i,j}(M) > 0$ and $p_{j,k}(N) > 0$. Then, by the Chapman-Kolmogorov equation,

$$p_{i,k}(M + N) = \sum_{\text{all states } h} p_{i,h}(M)\, p_{h,k}(N) \geq p_{i,j}(M)\, p_{j,k}(N) > 0.$$

The proof of the theorem is complete.

Using Theorem 3A we obtain the following result.

THEOREM 3B

Communication is symmetric and transitive in the sense that for any states i, j, and k,

$$j \leftrightarrow k \text{ implies } k \leftrightarrow j \tag{3.1}$$

$$i \leftrightarrow j \text{ and } j \leftrightarrow k \text{ implies } i \leftrightarrow k \tag{3.2}$$

Given a state j of a Markov chain, its communicating class $C(j)$ is defined to be the set of all states k in the chain which communicate with j; in symbols,

$$k \in C(j) \text{ if and only if } k \leftrightarrow j. \tag{3.3}$$

It may happen that $C(j)$ is empty (that is, j communicates with no state, not even itself). In this case, we call j a *non-return* state.

If $C(j)$ is non-empty, then j belongs to $C(j)$; to see this, note that by Theorem 3B there exists a state k such that $j \leftrightarrow k$ and $k \leftrightarrow j$ if and only if $j \leftrightarrow j$. A state which communicates with itself is called a *return* state.

A non-empty class C of states in a Markov chain is said to be a *communicating class* if, for some state j, C is equal to $C(j)$.

THEOREM 3C

If C_1 and C_2 are communicating classes, then either $C_1 = C_2$ or $C_1 C_2 = \emptyset$ (that is, C_1 and C_2 are disjoint).

Proof. Suppose C_1 and C_2 have a state (say h) in common. Let j and k be states such that $C_1 = C(j)$ and $C_2 = C(k)$. To prove that $C_1 = C_2$ it suffices to prove that $C(j) \subset C(k)$ since it then follows similarly that $C(k) \subset C(j)$. Let $g \in C(j)$. Since $g \leftrightarrow j$ and $h \leftrightarrow j$, it follows that $g \leftrightarrow h$. But $h \leftrightarrow k$. Therefore $g \leftrightarrow k$ and g belongs to $C(k)$, which was to be proved. It is possible to draw a diagram which proves Theorem 3C at a glance (see Figure 6.3).

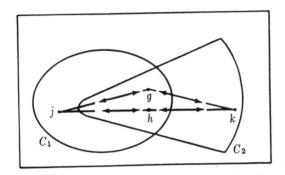

Fig. 6.3.   Diagrammatic proof of Theorem 3C

THEOREM 3D

Decomposition of a Markov chain into disjoint classes. The set S of states of a Markov chain can be written as the union of a finite or countably infinite family $\{C_r\}$ of disjoint sets of states,

$$S = C_1 \cup C_2 \cup \ldots \cup C_r \cup \ldots \quad \text{and} \quad C_i C_j = \emptyset \quad \text{for} \quad i \neq j,$$

where each set C_r is either a communicating class of states, or contains exactly one non-return state.

Proof. Choose a state j_1, and let $C_1 = C(j_1)$ or $\{j_1\}$ according as j_1 is or is not a return state. Then choose a state j_2 not in C_1; let $C_2 = C(j_2)$ or $\{j_2\}$ according as j_2 is or is not a return state. That C_2 and C_1 are disjoint follows from Theorem 3C. Continuing in this manner one obtains the decomposition described.

A non-empty set C of states is said to be *closed* if no state outside the set is accessible from any state inside the set. In symbols, C is closed if and only if, for every j in C and every k not in C, $p_{j,k}(n) = 0$ for every $n = 1, 2, \cdots$. Note that once a Markov chain enters a closed class it remains there. A set C which is not closed is said to be *non-closed.*

EXAMPLE 3A

Consider four Markov chains, each with state space $S = \{1,2,3,4\}$ and with respective transition probability matrices

$$P_1 = \begin{bmatrix} 1 & 0 & 0 & 0 \\ \frac{1}{2} & \frac{1}{2} & 0 & 0 \\ \frac{1}{3} & \frac{1}{3} & 0 & \frac{1}{3} \\ \frac{1}{4} & \frac{1}{4} & \frac{1}{4} & \frac{1}{4} \end{bmatrix}, \quad P_2 = \begin{bmatrix} 1 & 0 & 0 & 0 \\ 1 & 0 & 0 & 0 \\ \frac{1}{2} & \frac{1}{2} & 0 & 0 \\ \frac{1}{3} & \frac{1}{3} & \frac{1}{3} & 0 \end{bmatrix}, \quad P_3 = \begin{bmatrix} 0 & 0 & 1 & 0 \\ 1 & 0 & 0 & 0 \\ \frac{1}{2} & \frac{1}{2} & 0 & 0 \\ \frac{1}{3} & \frac{1}{3} & \frac{1}{3} & 0 \end{bmatrix}, \quad P_4 = \begin{bmatrix} \frac{1}{2} & \frac{1}{2} & 0 & 0 \\ \frac{1}{3} & \frac{1}{3} & \frac{1}{3} & 0 \\ \frac{1}{4} & \frac{1}{4} & \frac{1}{4} & \frac{1}{4} \\ \frac{1}{4} & \frac{1}{4} & \frac{1}{4} & \frac{1}{4} \end{bmatrix}.$$

Let us consider the decomposition of the state space S into disjoint sets having the properties stated in Theorem 3D.

Under P_1, S splits into 1 closed communicating class $\{1\}$, and 2 non-closed communicating classes $\{2\}$, $\{3,4\}$.

Under P_2, S splits into 1 closed communicating class $\{1\}$ and 3 non-return sets $\{2\}$, $\{3\}$, $\{4\}$.

Under P_3, S splits into 1 closed communicating class $\{1,2,3\}$ and 1 non-return set $\{4\}$.

Under P_4, S consists of a single communicating class.

A closed communicating class C of states essentially constitutes a Markov chain which can be extracted and studied independently. If one writes the transition probability matrix P of a Markov chain so that the states in C are written first, then P can be written

$$P = \begin{bmatrix} P_C & 0 \\ * & * \end{bmatrix}$$

where $*$ denotes a possibly non-zero matrix entry and P_C is the submatrix of P whose entries have indices of states in C. The n-step transition probability matrix can be written

$$P(n) = \begin{bmatrix} (P_C)^n & 0 \\ * & * \end{bmatrix}$$

Since a closed communicating class C can be extracted from a Markov chain (by deleting in the transition probability matrix of the chain all rows and columns corresponding to states outside the closed class C), and treated by itself as a Markov chain, it follows that to study the (asymptotic) properties of a Markov chain it suffices to study the (asymptotic) properties of a Markov chain in which there is exactly one closed communicating class and in which all other communicating classes are non-closed classes.

COMPLEMENTS

3A   *A criterion for accessibility.* Show that k is accessible from j if and only if for some integer N, and states $i_1, i_2, \cdots i_{N-1}$

$$p_{j,i_1} \, p_{i_1,i_2} \cdots p_{i_{N-2},i_{N-1}} \, p_{i_{N-1},k} > 0.$$

3B   *A criterion that a state be non-return.* Show that j is a non-return state if and only if $p_{j,j}(n) = 0$ for every integer n.

3C   *A criterion that a state be absorbing.* A state j for which $p_{j,j} = 1$ is called an *absorbing state*. Show that j is an absorbing state if and only if $\{j\}$ is a closed communicating class.

EXERCISES

For $k = 1, 2, \cdots, 10$, Exercise 3.k is to decompose the state space of the Markov chain defined by Exercise 2.k into disjoint sets having the properties stated in Theorem 3D.

In Exercises 3.11 to 3.13, decompose the state space of the Markov chain defined by the transition probability matrix given into disjoint sets having the properties stated in Theorem 3D.

3.11
$$P = \begin{bmatrix} 0 & \frac{1}{3} & \frac{2}{3} \\ \frac{2}{3} & 0 & \frac{1}{3} \\ \frac{1}{3} & \frac{2}{3} & 0 \end{bmatrix}.$$

3.12
$$P = \begin{bmatrix} 0 & 0 & 0 & 1 \\ 0 & 0 & 0 & 1 \\ .5 & .5 & 0 & 0 \\ 0 & 0 & 1 & 0 \end{bmatrix}.$$

3.13
$$P = \begin{bmatrix} .6 & 0 & .4 & 0 & 0 \\ .2 & .6 & .2 & 0 & 0 \\ .2 & 0 & .8 & 0 & 0 \\ 0 & 0 & 0 & .6 & .4 \\ 0 & 0 & 0 & .4 & .6 \end{bmatrix}.$$

6-4   OCCUPATION TIMES AND FIRST PASSAGE TIMES

In order to study the evolution in time of a Markov chain, one must classify the states of the chain into two categories: those which the system visits infinitely often and those which the system visits only finitely often. In the long run the chain will not be at any of the latter states and we need only study the evolution of the chain as it moves among the states which it visits infinitely often. In this section we develop criteria for classifying the states of a Markov chain in these categories. For this purpose we introduce the notion of the *occupation times of a state*.

Consider a Markov chain $\{X_n, n = 0, 1, \cdots\}$. For any state k and $n = 1, 2, \cdots$, define $N_k(n)$ to be the number of times that the state k is visited in the first n transitions; more precisely $N_k(n)$ is equal to the number of integers v satisfying $1 \leq v \leq n$ and $X_v = k$. We call $N_k(n)$ the *occupation time* of the state k in the first n transitions. The random variable

$$N_k(\infty) = \lim_{n \to \infty} N_k(n) \qquad (4.1)$$

is called the *total occupation time of k.*

Occupation times may be usefully represented as sums of random variables. Define for any state k and $n = 1, 2, \cdots$

$$\begin{aligned} Z_k(n) &= 1 \quad \text{if} \quad X_n = k, \\ &= 0 \quad \text{if} \quad X_n \neq k. \end{aligned} \qquad (4.2)$$

In words, $Z_k(n)$ is equal to 1 or 0 depending on whether or not at time n the chain is in state k. We may write

$$N_k(n) = \sum_{m=1}^{n} Z_k(m), \qquad (4.3)$$

$$N_k(\infty) = \sum_{m=1}^{\infty} Z_k(m). \qquad (4.4)$$

We next define the following probabilities, for any states j and k:

$$f_{j,k} = P[N_k(\infty) > 0 \mid X_0 = j], \qquad (4.5)$$
$$g_{j,k} = P[N_k(\infty) = \infty \mid X_0 = j]. \qquad (4.6)$$

In words, $f_{j,k}$ is the conditional probability of ever visiting the state k, given that the Markov chain was initially at state j, while $g_{j,k}$ is the conditional probability of infinitely many visits to the state k, given that the Markov chain was initially at state j.

It is to be noted that the probabilities in Eqs. 4.5 and 4.6 may be undefined since $P[X_0 = j] = 0$. However, by Eqs. 4.5 and 4.6 we mean really

$$f_{j,k} = P[N_k(\infty) - N_k(m) > 0 \mid X_m = j], \qquad (4.7)$$
$$g_{j,k} = P[N_k(\infty) - N_k(m) = \infty \mid X_m = j], \qquad (4.8)$$

for all values of m for which the conditional probabilities are defined. For Markov chains with homogeneous transition probabilities, the quantities in Eqs. 4.7 and 4.8 are independent of the values of m for which they are defined. All definitions given hereafter will be written in the form of Eqs. 4.5 and 4.6 but should be interpreted as in Eqs. 4.7 and 4.8.

We denote by $f_{j,k}(n)$ the *conditional probability that the first passage from j to k occurs in exactly n steps*; more precisely,

$$f_{j,k}(n) = P[V_k(n) \mid X_0 = j], \tag{4.9}$$

where for any state k and integer $n = 1, 2, \cdots$ we define

$$V_k(n) = [X_n = k, X_m \neq k \quad \text{for} \quad m = 1, 2, \ldots, n-1]. \tag{4.10}$$

In words, $V_k(n)$ is the event that the first time among the times $1, 2, \cdots$ at which state k is visited is time n. We call $f_{j,k}(n)$ the probability of first passage from state j to state k at time n, while $f_{j,k}$ is the probability of a first passage from j to k.

EXAMPLE 4A

Two-state Markov chain. In a Markov chain with two states 0 and 1, the first passage probabilities $f_{j,k}(n)$ are clearly given by

$$f_{0,0}(1) = p_{0,0},$$
$$f_{0,0}(n) = p_{0,1} \{p_{1,1}\}^{n-2} p_{1,0}, n \geq 2,$$
$$f_{0,1}(n) = \{p_{0,0}\}^{n-1} p_{0,1}, n \geq 1,$$
$$f_{1,0}(n) = \{p_{1,1}\}^{n-1} p_{1,0}, n \geq 1,$$
$$f_{1,1}(1) = p_{1,1},$$
$$f_{1,1}(n) = p_{1,0} \{p_{0,0}\}^{n-2} p_{0,1}, n \geq 2.$$

THEOREM 4A

For any states j and k

$$f_{j,k} = \sum_{n=1}^{\infty} f_{j,k}(n). \tag{4.11}$$

Proof. To prove Eq. 4.11, define

$$V_k = [X_n = k \text{ for some integer } n > 0] = [N_k(\infty) > 0] \tag{4.12}$$

Since V_k is the union of the disjoint sequence of sets $\{V_k(n), n = 1, 2, \cdots\}$ it follows that

$$f_{j,k} = P[V_k \mid X_0 = j] = \sum_{n=1}^{\infty} P[V_k(n) \mid X_0 = j] = \sum_{n=1}^{\infty} f_{j,k}(n). \tag{4.13}$$

The transition probabilities $p_{j,k}(n)$ and the first passage probabilities are related by the following fundamental formula.

THEOREM 4B

For any states j and k, and integer $n \geq 1$,

$$p_{j,k}(n) = \sum_{m=1}^{n} f_{j,k}(m) \, p_{k,k}(n-m) \tag{4.14}$$

defining

$$p_{j,k}(0) = \delta_{j,k} = \begin{cases} 1 & \text{if } j = k \\ 0 & \text{if } j \neq k. \end{cases} \tag{4.15}$$

Proof. To prove Eq. 4.14, one writes

$$P[X_n = k \mid X_0 = j]$$

$$= \sum_{m=1}^{n} P[X_n = k, X_m = k, X_q \neq k \text{ for } q = 1, \ldots, m-1 \mid X_0 = j]$$

$$= \sum_{m=1}^{n} P[X_n = k \mid X_m = k] P[X_m = k,$$

$$\qquad X_q \neq k \text{ for } q = 1, \ldots, m-1 \mid X_0 = j] \tag{4.16}$$

$$= \sum_{m=1}^{n} f_{j,k}(m) \, p_{k,k}(n-m).$$

The method used in Eq. 4.16 is frequently employed in the study of Markov chains and is referred to as the *method of first entrance*. Its general principle may be described as follows. For any event A (such as the event $A = [X_n = k]$) one may write

$$P[A V_k \mid X_0 = j] = \sum_{m=1}^{\infty} P[A V_k(m) \mid X_0 = j]. \tag{4.17}$$

Now, suppose that A is a subset of V_k and that for each integer m there exists an event A_m depending only on the random variables $\{X_t, t > m\}$ such that

$$A \, V_k(m) = A_m \, V_k(m). \tag{4.18}$$

By the Markov property

$$P[A_m V_k(m) \mid X_0 = j] = P[A_m \mid V_k(m), X_0 = j] \, P[V_k(m) \mid X_0 = j]$$
$$= P[A_m \mid X_m = k] \, f_{j,k}(m). \tag{4.19}$$

From Eqs. 4.17 and 4.19 one obtains the basic formula of the method of first entrance:

$$P[A \mid X_0 = j] = \sum_{m=1}^{\infty} f_{j,k}(m) \, P[A_m \mid X_m = k]. \tag{4.20}$$

In the instances in which use is made of the method of first entrance it will be obvious how to define the events A_m and it will be simple to evaluate $P[A_m \mid X_m = k]$.

As another illustration of the method of first entrance, we prove the following useful formulas.

THEOREM 4C

For any states j and k

$$g_{k,k} = \lim_{n \to \infty} (f_{k,k})^n, \qquad (4.21)$$

$$g_{j,k} = f_{j,k} g_{k,k}. \qquad (4.22)$$

Proof. To prove Eq. 4.22 we use Eq. 4.20 with $A = [N_k(\infty) = \infty]$, $A_m = [N_k(\infty) - N_k(m) = \infty]$; then

$$P[N_k(\infty) = \infty \mid X_0 = j] = \sum_{m=1}^{\infty} f_{j,k}(m) g_{k,k}, \qquad (4.23)$$

since $P[N_k(\infty) - N_k(m) = \infty \mid X_m = k] = g_{k,k}$. From Eqs. 4.23 and 4.11 one obtains Eq. 4.22.

To prove Eq. 4.21 we first note that by the method of first entrance, for any integer $n \geq 1$,

$$P[N_k(\infty) \geq n \mid X_0 = k]$$
$$= \sum_{m=1}^{\infty} f_{k,k}(m) P[N_k(\infty) - N_k(m) \geq n - 1 \mid X_m = k]. \quad (4.24)$$

Since $P[N_k(\infty) - N_k(m) \geq n - 1 \mid X_m = k] = P[N_k(\infty) \geq n - 1 \mid X_0 = k]$, it follows from Eqs. 4.24 and 4.11 that for $n \geq 1$

$$P[N_k(\infty) \geq n \mid X_0 = k] = P[N_k(\infty) \geq n - 1 \mid X_0 = k] f_{k,k}. \quad (4.25)$$

From Eq. 4.25 it follows by induction that

$$P[N_k(\infty) \geq n \mid X_0 = k] = (f_{k,k})^n, \, n \geq 1.$$

Consequently,

$$P[N_k(\infty) = \infty \mid X_0 = k] = \lim_{n \to \infty} P[N_k(\infty) \geq n \mid X_0 = k]$$
$$= \lim_{n \to \infty} (f_{k,k})^n. \qquad (4.26)$$

The proof of Eq. 4.21 is now complete.

In the classification of states, Theorems 4D and 4E play a basic role.

THEOREM 4D

A zero-one law. For any state k, either $g_{k,k} = 1$ or $g_{k,k} = 0$. Further,

$$g_{k,k} = 1 \quad \text{if and only if} \quad f_{k,k} = 1, \qquad (4.27)$$
$$g_{k,k} = 0 \quad \text{if and only if} \quad f_{k,k} < 1. \qquad (4.28)$$

In words, Theorem 4D says that if, starting at k, a Markov chain has probability one of ever returning to k, then with probability one, it will make infinitely many visits to k. On the other hand it can only expect to return to k a finite number of times if there is a positive probability that it will fail to return.

Theorem 4D is an immediate consequence of Eq. 4.21; from Eq. 4.21 it follows that

$$f_{k,k} < 1 \quad \text{implies} \quad g_{k,k} = 0,$$
$$f_{k,k} = 1 \quad \text{implies} \quad g_{k,k} = 1.$$

Consequently, $g_{k,k} = 1$ implies $f_{k,k} = 1$ since if $f_{k,k}$ were unequal to 1, then $g_{k,k}$ would be unequal to 1.

EXAMPLE 4B

A branching process cannot be moderately large. Let X_n be the size of the nth generation in the branching process considered in Example 2C. The extinction probability π_0, that $X_n = 0$ in the limit as n tends to ∞, was shown to depend on the mean μ of the number of offspring of an individual. Let us now show that, no matter what the value of μ and no matter what the initial population size, either the population dies out or it becomes infinitely large. To prove this assertion it suffices to prove that

$$f_{k,k} < 1 \quad \text{for} \quad k = 1, 2, \ldots \tag{4.29}$$

It then follows that

$$g_{k,k} = 0 \quad \text{for} \quad k = 1, 2, \ldots ;$$

in words, no positive value k can be assumed infinitely often by the population size X_n. Any finite set of positive values $1, \cdots, M$ can contain the population size X_n only for a finite (although random) number of generations n. Consequently, Eq. 4.29 implies that the population size goes eventually to 0 or to ∞. To prove Eq. 4.29 note that if each of k individuals in a population have no offspring, then the population is extinguished, so that

$$f_{k,k} < 1 - (p_0)^k < 1 \quad \text{since} \quad p_0 > 0.$$

THEOREM 4E

For any state k in a Markov chain

$$f_{k,k} < 1 \quad \text{if and only if} \quad \sum_{n=1}^{\infty} p_{k,k}(n) < \infty,$$
$$f_{k,k} = 1 \quad \text{if and only if} \quad \sum_{n=1}^{\infty} p_{k,k}(n) = \infty. \tag{4.30}$$

To prove Theorem 4E we use the method of generating functions. The generating functions

$$P_{j,k}(z) = \sum_{n=0}^{\infty} z^n p_{j,k}(n) = \delta_{j,k} + \sum_{n=1}^{\infty} z^n p_{j,k}(n),$$

$$F_{j,k}(z) = \sum_{n=1}^{\infty} z^n f_{j,k}(n) \tag{4.31}$$

are defined for all real numbers z such that $|z| < 1$.

THEOREM 4F

For any two states j and k in a Markov chain the generating functions $P_{j,k}(z)$ and $F_{j,k}(z)$ are related to each other as follows, for $|z| < 1$:

$$P_{j,k}(z) = F_{j,k}(z) \, P_{k,k}(z) \quad \text{if} \quad j \neq k, \tag{4.32}$$

$$P_{k,k}(z) - 1 = F_{k,k}(z) \, P_{k,k}(z), \tag{4.33}$$

$$P_{k,k}(z) = \frac{1}{1 - F_{k,k}(z)}, \quad F_{k,k}(z) = 1 - \frac{1}{P_{k,k}(z)}. \tag{4.34}$$

Proof. If one multiplies both sides of Eq. 4.14 by z^n, and sums over $n = 1, 2, \cdots$ one obtains

$$\sum_{n=1}^{\infty} z^n p_{j,k}(n) = \sum_{n=1}^{\infty} z^n \sum_{m=1}^{n} f_{j,k}(m) p_{k,k}(n - m). \tag{4.35}$$

If one interchanges the order of summation, one obtains

$$\sum_{n=1}^{\infty} z^n p_{j,k}(n) = \sum_{m=1}^{\infty} \sum_{n=m}^{\infty} z^n f_{j,k}(m) p_{k,k}(n - m)$$

$$= \sum_{m=1}^{\infty} z^m f_{j,k}(m) \sum_{n=m}^{\infty} z^{n-m} p_{k,k}(n - m)$$

$$= \sum_{m=1}^{\infty} z^m f_{j,k}(m) \sum_{\nu=0}^{\infty} z^\nu p_{k,k}(\nu),$$

which may be written in terms of generating functions;

$$P_{j,k}(z) - \delta_{j,k} = F_{j,k}(z) \, P_{k,k}(z) \tag{4.36}$$

The proof of Theorem 4F is now complete.

The usefulness of generating functions derives from the following two properties which they possess.

THEOREM 4G

Limit theorems for generating functions. Let $\{a_n\}$ be a *non-negative* sequence of real numbers with generating function

$$A(z) = \sum_{n=0}^{\infty} a_n z^n, \; | z | < 1. \tag{4.37}$$

In order that there exist a finite number S such that

$$\lim_{n \to \infty} \sum_{m=0}^{n} a_m = S \tag{4.38}$$

it is *necessary and sufficient* that there exist a finite number S such that

$$\lim_{z \to 1-} A(z) = S. \tag{4.39}$$

(By $\lim_{z \to 1-}$ we mean the limit as z approaches 1 through values of z less than 1.) In order that there exist a finite number L such that

$$\lim_{n \to \infty} \frac{1}{n} \sum_{m=1}^{n} a_m = L \tag{4.40}$$

it is *necessary and sufficient* that there exist a finite number L such that

$$\lim_{z \to 1-} (1 - z) A(z) = L. \tag{4.41}$$

The fact that Eqs. 4.40 and 4.41 are equivalent is a deep fact of mathematical analysis whose proof is beyond the scope of this book (for a proof, see Hardy [1949], Theorem 96, p. 155). To prove that Eqs. 4.38 and 4.39 are equivalent, we argue as follows.

If $\{a_n\}$ is an absolutely summable sequence, then by the dominated convergence theorem (see Appendix to this chapter)

$$\sum_{n=1}^{\infty} a_n = S \quad \text{implies} \quad \lim_{z \to 1-} A(z) = S. \tag{4.42}$$

Conversely, let us show that if $a_n \geq 0$ for all n, then

$$\lim_{z \to 1-} A(z) = S < \infty \quad \text{implies} \quad \sum_{n=0}^{\infty} a_n = S. \tag{4.43}$$

To prove Eq. 4.43 note that for any integer N and $0 < z < 1$

$$\sum_{n=1}^{N} a_n z^n \leq A(z). \tag{4.44}$$

Taking the limit of Eq. 4.44 as z tends to 1 it follows that

$$\sum_{n=1}^{N} a_n = \lim_{z \to 1-} \sum_{n=1}^{N} a_n z^n \leq S. \tag{4.45}$$

From Eq. 4.45, the sequence of consecutive sums

$$\left\{ \sum_{n=0}^{N} a_n, N = 1, 2, \ldots \right\}$$

is a bounded monotone sequence. Consequently, the infinite series

$$\sum_{n=1}^{\infty} a_n$$

converges. That its value is S follows from Eq. 4.42.

The tools are now at hand to prove Theorem 4E. It suffices to prove Eq. 4.30. From Eqs. 4.34, 4.38, and 4.39 we obtain that (writing "iff" for if and only if)

$$f_{k,k} < 1 \quad \text{iff} \quad \sum_{n=1}^{\infty} f_{k,k}(n) < 1$$

$$\text{iff} \quad \lim_{z \to 1-} F_{k,k}(z) < 1 \quad \text{iff} \quad \lim_{z \to 1-} P_{k,k}(z) < \infty$$

$$\text{iff} \sum_{n=0}^{\infty} p_{k,k}(n) < \infty.$$

A limit theorem for transition probabilities. In the sequel the following theorem will play an important role. For any state k, define

$$m_{k,k} = \sum_{n=1}^{\infty} n f_{k,k}(n). \tag{4.46}$$

THEOREM 4H
Let k be a state such that $f_{k,k} = 1$ and $m_{k,k} < \infty$. Then

$$\lim_{n \to \infty} \frac{1}{n} \sum_{m=1}^{n} p_{k,k}(m) = \frac{1}{m_{k,k}}, \tag{4.47}$$

and for any state j

$$\lim_{n \to \infty} \frac{1}{n} \sum_{m=1}^{n} p_{j,k}(m) = f_{j,k} \frac{1}{m_{k,k}}. \tag{4.48}$$

In view of the equivalence of Eqs. 4.40 and 4.41, to prove Eq. 4.47 it suffices to prove that

$$\lim_{z \to 1} (1 - z) P_{k,k}(z) = \frac{1}{m_{k,k}}. \tag{4.49}$$

In view of Eq. 4.34, to prove Eq. 4.49 it suffices to prove that

$$\lim_{z \to 1} \frac{1 - F_{k,k}(z)}{1 - z} = m_{k,k}. \tag{4.50}$$

However Eq. 4.50 is easily verified since

$$\lim_{z \to 1} \frac{1 - F_{k,k}(z)}{1 - z} = \frac{d}{dz} F_{k,k}(z) \bigg|_{z=1} = \sum_{n=1}^{\infty} n f_{k,k}(n) = m_{k,k}. \tag{4.51}$$

Similarly Eq. 4.48 is an immediate consequence of Eqs. 4.32, 4.50, and the equivalence of Eqs. 4.40 and 4.41.

COMPLEMENTS

In complements 4A to 4F show that the relations given hold for any states j and k in a Markov chain.

4A
$$\sup_n p_{j,k}(n) \le f_{j,k} \le \sum_{n=1}^{\infty} p_{j,k}(n).$$

Consequently, (i) $j \to k$ if and only if $f_{j,k} > 0$, (ii) $j \leftrightarrow k$ if and only if $f_{j,k} f_{k,j} > 0$.

4B
$$\sum_{n=1}^{\infty} p_{j,k}(n) = f_{j,k} \sum_{n=0}^{\infty} p_{k,k}(n)$$

4C If
$$\lim_{n \to \infty} p_{k,k}(n) = \pi_k \text{ exists,}$$

then
$$\lim_{n \to \infty} p_{j,k}(n) = f_{j,k} \pi_k.$$

Hint. Show that for any integer $N \ge 1$,

$$p_{j,k}(n) - f_{j,k} \pi_k = \sum_{m=1}^{n} f_{j,k}(m) \{p_{k,k}(n) - \pi_k\} - \sum_{m=n+1}^{\infty} \pi_k f_{j,k}(m),$$

$$|p_{j,k}(n) - f_{j,k} \pi_k| \le \sum_{m=1}^{N} f_{j,k}(m) |p_{k,k}(n) - \pi_k| + 2 \sum_{m=N+1}^{\infty} f_{j,k}(m).$$

Let first $n \to \infty$, and then $N \to \infty$.

4D If
$$\lim_{N \to \infty} \frac{1}{N} \sum_{m=1}^{N} p_{k,k}(m) = \pi_k \text{ exists,}$$

then
$$\lim_{N \to \infty} \frac{1}{N} \sum_{m=1}^{N} p_{j,k}(m) = f_{j,k} \pi_k.$$

4E   If $f_{k,k} < 1$, then for any state j

(i)
$$\sum_{n=1}^{\infty} p_{j,k}(n) < \infty,$$

(ii)
$$\lim_{n \to \infty} p_{j,k}(n) = 0,$$

(iii)
$$\sum_{n=1}^{\infty} p_{k,k}(n) = \frac{f_{k,k}}{1 - f_{k,k}}.$$

4F   In terms of the transition probabilities $p_{j,k}(n)$ one may give an expression for the conditional mean total occupation time. Show that

$$E[N_k(\infty) \mid X_0 = j] = \sum_{n=1}^{\infty} p_{j,k}(n).$$

4G   Show that if a sequence $\{a_n, n = 1, 2, \cdots\}$ satisfies Eq. 4.40, then it satisfies Eq. 4.41.

Hint. Verify that (i) $(1 - z)A(z) - Lz = (1 - z)^2 \sum_{n=1}^{\infty} (b_n - L)n \, z^n$,

where $b_n = (a_1 + \cdots + a_n)/n$; (ii) for any N,

$$| (1 - z)A(z) - Lz | \leq (1 - z)^2 \sum_{n=1}^{N} | (b_n - L)n \, z^n | + \sup_{n > N} | b_n - L |.$$

Let $z \to 1$ and then let $N \to \infty$.

6-5   RECURRENT AND NON-RECURRENT STATES AND CLASSES

In order to study the asymptotic behavior of Markov chains it is necessary to be able to identify those communicating classes which are closed and those which are not closed. For this purpose we introduce the notion of a recurrent class.

A state k is said to be *recurrent* if $f_{k,k} = 1$; in words, k is recurrent if the probability is one that the Markov chain will eventually return to k, having started at k. A state k is said to be *non-recurrent* if $f_{k,k} < 1$.

Remark. Some writers (in particular, Feller [1957]) call a recurrent state *persistent* and a non-recurrent state *transient*. We have chosen to follow the terminology employed by Chung (1960) and Kendall (1959).

A class C of states is said to be recurrent if all states in C are recurrent. Similarly a class C is said to be non-recurrent if all states in C are non-recurrent.

THEOREM 5A

Let C be a communicating class of states in a Markov chain. Then C is either recurrent or non-recurrent. More precisely, (i) if any state in C

is recurrent, then all states in C are recurrent, (ii) if any state in C is non-recurrent, then all states in C are non-recurrent.

Proof. It suffices to prove that for any states j and k in a Markov chain

$$\text{if } f_{k,k} = 1, \text{ and if } j \leftrightarrow k, \text{ then } f_{j,j} = 1. \tag{5.1}$$

In words, Eq. 5.1 says that only recurrent states communicate with a recurrent state. Consequently, only non-recurrent states communicate with a non-recurrent state. To prove Eq. 5.1 we prove that

$$\text{if } \sum_n p_{k,k}(n) = \infty \quad \text{and if} \quad j \leftrightarrow k \quad \text{then} \quad \sum_n p_{j,j}(n) = \infty.$$

From the Chapman-Kolmogorov equation it follows that for any integers $N, n,$ and M,

$$p_{j,j}(N + n + M) = \sum_{a,b} p_{j,a}(N)\, p_{a,b}(n)\, p_{b,j}(M) \geq p_{j,k}(N)\, p_{k,k}(n)\, p_{k,j}(M),$$
$$\tag{5.2}$$

so that

$$\sum_{n=0}^{\infty} p_{j,j}(N + n + M) \geq p_{j,k}(N)\, p_{k,j}(M) \sum_{n=0}^{\infty} p_{k,k}(n). \tag{5.3}$$

Now, choose M and N so that $p_{j,k}(N) > 0$ and $p_{k,j}(M) > 0$. Then the divergence of the infinite series on the right side of Eq. 5.3 implies the divergence of the infinite series on the left side. The proof of Theorem 5A is complete.

We next prove the following basic fact.

THEOREM 5B

A recurrent communicating class is closed. A closed non-recurrent communicating class possesses infinitely many states.

From Theorems 5A and 5B, the decomposition theorem for Markov chains given by Theorem 3D can be strengthened as follows.

Decomposition theorem. The set S of return states of a Markov chain can be written as the union of disjoint communicating classes,

$$S = C_1 \cup C_2 \cup \ldots \cup C_r \cup \ldots$$

where each class C_r is either (i) a closed recurrent class, (ii) a closed non-recurrent class, or (iii) a non-closed non-recurrent class. In a finite Markov chain there are no closed non-recurrent communicating classes. These assertions are summarized in Table 6.2.

TABLE 6.2. **Classification of communicating classes**

	Closed	Non-closed
Recurrent		Does not exist
Non-recurrent	Does not exist in a finite chain	

Since a closed communicating class C can be extracted from a Markov chain and treated by itself as a Markov chain, it follows that to study the asymptotic properties of a Markov chain it suffices to study the asymptotic properties of a Markov chain in which there is exactly one closed communicating class (which is either recurrent or non-recurrent), and in which all other communicating classes are non-closed non-recurrent classes.

Proof of Theorem 5B. We prove that a recurrent communicating class is closed by showing that

$$\text{if } f_{k,k} = 1, \text{ and } k \to j, \text{ then } f_{j,k} = 1 \text{ and } k \leftrightarrow j; \qquad (5.4)$$

in words, the only states accessible from a recurrent state are states which communicate with it. To prove Eq. 5.4 verify that for any state k and integer n,

$$g_{k,k} = \sum_{\text{states } i} p_{k,i}(n)\, g_{i,k},$$

$$1 - g_{k,k} = \sum_{\text{states } i} p_{k,i}(n)\{1 - g_{i,k}\}. \qquad (5.5)$$

For a recurrent state k, the second sum in Eq. 5.5 vanishes. Consequently, for any integer n and state i

$$0 = p_{k,i}(n)\, \{1 - g_{i,k}\}. \qquad (5.6)$$

If j is accessible from k, there is an integer N such that $p_{k,j}(N) > 0$. Letting $i = j$ and $n = N$ in Eq. 5.6 it follows that $g_{j,k} = 1$. Therefore, $f_{j,k} = 1$, since from Eq. 4.22 it follows that

$$g_{j,k} = f_{j,k} \text{ if } k \text{ is recurrent.} \qquad (5.7)$$

To prove that a non-recurrent closed communicating class is infinite, we first note that from Eq. 4.22 it follows that for any state j,

$$g_{j,k} = 0 \text{ if } k \text{ is non-recurrent}; \qquad (5.8)$$

in words, if a state k is non-recurrent, then the probability is one that k will be visited only a finite number of times, starting at any state j. Consequently, if C is a closed non-recurrent communicating class, then C must possess infinitely many states, since with probability one the chain remains only a finite number of steps in any finite set of non-recurrent states. The proof of Theorem 5B is now complete.

A problem of great interest is to develop criteria that a communicating class be recurrent or non-recurrent. The following criterion will be found helpful.

THEOREM 5C

Let C be a closed communicating class of states, and let k be a fixed state in C. C is recurrent if and only if for every state j in C such that $j \neq k$

$$f_{j,k} = 1. \tag{5.9}$$

Proof. If C is recurrent, then Eq. 5.9 holds by Eq. 5.4. Conversely, one may verify that

$$f_{k,k} = p_{k,k} + \sum_{\substack{j \in C \\ j \neq k}} p_{k,j}\, f_{j,k}. \tag{5.10}$$

From Eqs. 5.9 and 5.10 it follows that

$$f_{k,k} = p_{k,k} + \sum_{\substack{j \in C \\ j \neq k}} p_{k,j} = 1. \tag{5.11}$$

The proof of Theorem 5C is now complete.

Essential and inessential states. A state k is said to be essential if it communicates with every state which is accessible from it (in symbols, $k \to j$ implies $k \leftrightarrow j$); otherwise it is inessential. One may reformulate Eq. 5.4: a recurrent state is essential. From this it follows that an inessential state is non-recurrent. It should be noted that an essential state may also be non-recurrent. In Example 6B we consider a Markov chain in which each state is essential (since all pairs of states communicate) and in which each state is non-recurrent.

COMPLEMENTS

5A *An inequality on first passage probabilities.* Show that for any states i, j, k in a Markov chain

$$f_{i,k} \geq f_{i,j}\, f_{j,k}.$$

Hint. Let $f_{i,j,k}$ be the probability that starting at i, one will at some time pass to j and then at a later time pass to k. Show that

$$f_{i,k} \geq f_{i,j,k} = f_{i,j}\, f_{j,k}.$$

5B   Prove that if k_1 and k_2 are communicating recurrent states, then for any state j

$$f_{j,k_1} = f_{j,k_2}.$$

Hint. Use Complement 5A.

5C   *Criterion for recurrence.* Show that a communicating class C is non-recurrent (recurrent) if and only if there exists a state k in C for which

$$\sum_{n=1}^{\infty} p_{k,k}(n) < \infty\, (= \infty).$$

5D   *Recurrent and non-recurrent random walks.* Consider a one-dimensional random walk on the integers $\{0, \pm 1, \pm 2, \cdots\}$. At each stage one moves to the right with probability p, and moves to the left with probability $q = 1 - p$. Show that for any integer m

$$p_{0,0}(2m) = \binom{2m}{m} p^m\, q^m.$$

Using Stirling's formula, show that

$$\binom{2m}{m} \sim \frac{4^m}{\sqrt{\pi m}},$$

$$p_{0,0}(2m) \sim \frac{(4pq)^m}{\sqrt{\pi m}}.$$

Consequently,

$$\sum_{n=1}^{\infty} p_{0,0}(n) < \infty \quad \text{if} \quad p \neq \tfrac{1}{2}$$
$$= \infty \quad \text{if} \quad p = \tfrac{1}{2}.$$

Using Complement 5C, show that return to any state is certain in a symmetric random walk $(p = \tfrac{1}{2})$ and is uncertain in an unsymmetric random walk.

5E   Let Z denote a random variable whose possible values are the integers $0, \pm 1, \cdots$; that is,

$$p_k \equiv P[Z = k] > 0 \quad \text{for} \quad k = 0, \pm 1, \cdots; \quad \sum_{k=-\infty}^{\infty} p_k = 1.$$

Assume that $E[\,|Z|\,] < \infty$. Let $Z_1, Z_2, \cdots$ be a sequence of independent random variables identically distributed as Z. For $n = 1, 2, \cdots$, let $X_n = Z_1 + Z_2 + \cdots + Z_n$ denote the nth consecutive sum. The sequence $\{X_n, n = 1, 2, \cdots\}$ is a Markov chain. (i) Describe its transition probability matrix. (ii) Show that the Markov chain $\{X_n, n = 1, 2, \cdots\}$ is recurrent if and only if $E[Z] = 0$.

6-6 FIRST PASSAGE AND ABSORPTION PROBABILITIES

In order to determine the probability $f_{j,k}$ (that a Markov chain starting at j will ever visit k) one must distinguish four cases as illustrated in Table 6.3.

TABLE 6.3. **Methods for determining first passage probabilities**

		k	
		Recurrent	Non-recurrent
j	Recurrent	$f_{j,k} = 1$ iff $j \leftrightarrow k$ $f_{j,k} = 0$ otherwise	$f_{j,k} = 0$
	Non-recurrent	Satisfies system of linear equations given in Theorem 6A	$f_{j,k} = \dfrac{\sum_{n=1}^{\infty} p_{j,k}(n)}{\sum_{n=0}^{\infty} p_{k,k}(n)}$

The aim of this section is to show that in the case that j is non-recurrent and k is recurrent, the first passage probability $f_{j,k}$ may be determined as the solution of a system of linear equations involving all such first passage probabilities.

Given a Markov chain, we let S denote the set of all states and T denote the set of all non-recurrent states. The use of the letter T to denote the set of non-recurrent states originates from the terminology employed by probabilists who call non-recurrent states *transient*.

THEOREM 6A

If k is a recurrent state, then the set of first passage probabilities $\{f_{j,k}, j \epsilon T\}$ satisfies the system of equations

$$f_{j,k} = \sum_{i \epsilon T} p_{j,i} f_{i,k} + \sum_{i \epsilon C} p_{j,i}, \quad j \epsilon T, \qquad (6.1)$$

where C denotes the set of all states communicating with k.

Proof. Recall that S denotes the set of states in the Markov chain. It is easily verified that

$$f_{j,k} = \sum_{i \epsilon S} P[N_k(\infty) > 0 \mid X_1 = i] P[X_1 = i \mid X_0 = j].$$

Therefore,

$$f_{j,k} = \sum_{i \epsilon S} p_{j,i} f_{i,k}.$$

Now, $f_{i,k} = 1$ if i belongs to C and $f_{i,k} = 0$ if i belongs to neither C nor T. The proof of Theorem 6A is complete.

Remark. If k_1 and k_2 are communicating recurrent states, then for any state j

$$f_{j,k_1} = f_{j,k_2}.$$

This may be proved either by using Eq. 6.1 or by using Complement 5A. Consequently, we may define the probability $f_{j,c}$ that a Markov chain starting at a state j will eventually be absorbed in the recurrent class C to be equal to $f_{j,k}$ for any state k in C. One may then rewrite Eq. 6.1

$$f_{j,c} = \sum_{i \epsilon T} p_{j,i} f_{i,c} + \sum_{i \epsilon C} p_{j,i}, \quad j \epsilon T. \tag{6.1'}$$

In order for Theorem 6A to be a useful tool in finding the first passage probabilities it must be shown that the system of equations 6.1 has a unique solution. Now, Eq. 6.1 may be rewritten as follows, if one defines $v_j = f_{j,k}$ and $b_j = \sum_{i \epsilon C} p_{j,i}$:

$$v_j = \sum_{i \epsilon T} p_{j,i} v_i + b_j, \quad j \epsilon T. \tag{6.2}$$

Note that Eq. 6.2 is an inhomogeneous system of equations in variables $\{v_j, j \epsilon T\}$ and that $p_{j,i}$ and b_j are known constants. A necessary and sufficient condition that the inhomogeneous system Eq. 6.2 possess a unique bounded solution $\{v_j, j \epsilon T\}$ is that the homogeneous system of equations

$$v_j = \sum_{i \epsilon T} p_{j,i} v_i, \quad j \epsilon T \tag{6.3}$$

possess as its only bounded solution the null solution

$$v_j = 0 \qquad \text{for all } j \text{ in } T.$$

To prove this assertion, let $a_{ji} = p_{j,i} - \delta_{j,i}$. For ease of exposition suppose that T consists of only two states, $T = \{1, 2\}$. Then Eq. 6.3 may be written

$$\begin{aligned} a_{11} v_1 + a_{12} v_2 &= 0 \\ a_{21} v_1 + a_{22} v_2 &= 0, \end{aligned} \tag{6.3'}$$

while Eq. 6.2 may be written

$$\begin{aligned} a_{11} v_1 + a_{12} v_2 &= b_1 \\ a_{21} v_1 + a_{22} v_2 &= b_2. \end{aligned} \tag{6.2'}$$

Suppose there exist two (bounded) solutions (v_1, v_2) and (v_1', v_2') to Eq. 6.2′. Then $(v_1 - v_1', v_2 - v_2')$ is a (bounded) solution to Eq. 6.3′. In order for any two (bounded) solutions (v_1, v_2) and (v_1', v_2') of Eq. 6.2′ to be identical it is necessary and sufficient that the only (bounded) solution to Eq. 6.3′ be the null solution.

It is possible to give a probabilistic meaning to the validity of Eq. 6.3. Define, for j in T,

$$y_j = P[X_n \text{ belongs to } T \text{ for all } n \mid X_0 = j]. \tag{6.4}$$

In words, y_j is the probability that the Markov chain remains forever in the set T of non-recurrent states, given that it started at the non-recurrent state j. It is easily verified that $\{y_j, j \in T\}$ satisfies Eq. 6.3:

$$y_j = \sum_{i \in T} p_{j,i} y_i, \quad j \in T, \tag{6.5}$$

since if A denotes the event that, for all n, X_n belongs to T, then

$$y_j = P[A \mid X_0 = j] = \sum_{i \in S} p_{j,i} P[A \mid X_1 = i].$$

We next show that $\{y_j, j \in T\}$ is the maximal solution of Eq. 6.5 which is bounded by 1; in symbols,

$$|v_j| \le 1 \quad \text{and} \quad v_j = \sum_{i \in T} p_{j,i} v_i \quad \text{for all} \quad j \text{ in } T,$$

$$\text{implies} \quad |v_j| \le y_j \quad \text{for all} \quad j \text{ in } T. \tag{6.6}$$

Proof of Eq. 6.6. For $n = 1, 2, \cdots$ and j in T, define

$$y_j(n) = P[X_n \text{ belongs to } T \mid X_0 = j].$$

One may verify that

$$y_j = \lim_{n \to \infty} y_j(n).$$

Consequently, to prove Eq. 6.6 it suffices to prove that for $n = 1, 2, \cdots$

$$|v_j| \le y_j(n). \tag{6.7}$$

We prove Eq. 6.7 by induction. Now,

$$y_j(1) = P[X_1 \text{ belongs to } T \mid X_0 = j]$$
$$= \sum_{i \in T} p_{j,i} \ge \left| \sum_{i \in T} p_{j,i} v_i \right| = |v_j|.$$

Next, suppose that Eq. 6.7 holds for n. Then

$$y_j(n+1) = \sum_{i \epsilon T} p_{j,i} \, y_i(n) \geq \sum_{i \epsilon T} p_{j,i} \, |v_i| \geq \left| \sum_{i \epsilon T} p_{j,i} v_i \right| = |v_j|.$$

The proof of Eq. 6.7, and therefore of Eq. 6.6, is now complete. From Eqs. 6.5 and 6.6 one obtains immediately the following theorem.

THEOREM 6B

In order that the homogeneous system of equations 6.3 possess as its only bounded solution the null solution, it is necessary and sufficient that

$$y_j = 0 \qquad \text{for all } j \text{ in } T. \tag{6.8}$$

Remark. In a finite Markov chain, Eq. 6.8 holds, since a finite set of non-recurrent states is visited only a finite number of times. Consequently, in a finite Markov chain the first passage probabilities $\{f_{j,k}, j \epsilon T\}$ are the unique bounded solution of the system of equations 6.1.

A state k is called an *absorbing* state if $p_{k,k} = 1$, so that once the chain visits k it remains there forever. An absorbing state is clearly recurrent. In the case that k is an absorbing state, the first passage probability $f_{j,k}$ is called the probability of absorption into k, having started at j.

From Theorem 6A, one readily obtains the following assertion.

System of equations for absorption probabilities. If S is the state space of a Markov chain, and k is an absorbing state, then the set of absorption probabilities $\{f_{j,k}, j \epsilon S\}$ satisfy the system of equations

$$f_{j,k} = \sum_{i \epsilon S} p_{j,i} \, f_{i,k}, \quad j \epsilon S, \tag{6.9}$$

subject to the conditions

$$\begin{aligned} f_{k,k} &= 1, \\ f_{j,k} &= 0 \qquad \text{if } j \text{ is recurrent and } j \neq k. \end{aligned} \tag{6.9'}$$

EXAMPLE 6A

Absorption probabilities in random walks (the problem of gambler's ruin). A random walk is a Markov chain $\{X_n, n = 0, 1, \cdots\}$ on a state space consisting of integers (or a lattice of points in n-dimensional Euclidean space) with the property that if the system is in a given state k then in a single transition the system either remains at k or moves to one of the states immediately adjacent to k (in other words, the system can only move to a nearest neighbor).

The general random walk on the state space $\{0, 1, 2, \cdots\}$ has a transition probability matrix of the form

$$P = \begin{bmatrix} r_0 & p_0 & 0 & 0 & 0 & \cdots \\ q_1 & r_1 & p_1 & 0 & 0 & \cdots \\ 0 & q_2 & r_2 & p_2 & 0 & \cdots \\ 0 & 0 & q_3 & r_3 & p_3 & \cdots \\ \cdots\cdots\cdots\cdots\cdots\cdots\cdots\cdots \\ \cdots\cdots\cdots\cdots\cdots\cdots\cdots\cdots \end{bmatrix}. \tag{6.10}$$

where (for $k = 0, 1, \cdots$) p_k, r_k, and q_k are non-negative real numbers such that

$$r_0 + p_0 = 1, \quad q_k + r_k + p_k = 1 \qquad \text{for } k = 1, 2, \cdots.$$

The name "random walk" is motivated by the idea of regarding the random variable X_n as being the position at time n of a particle moving on a straight line in such a manner that at each step the particle either remains where it is or moves one step to the left or right. Random walks serve as reasonable approximations to physical processes involving diffusing particles. Random walks also provide a model useful in the analysis of nuclear reactions. One is then interested in studying such problems as the history of a particle (such as a neutron) which is performing a random walk inside a container whose walls are made of a material, such as cadmium, which absorbs neutrons. The walk stops as soon as the neutron hits one of the boundaries of the container.

The fortune of a player engaged in a series of contests can be represented by a random walk. Consider an individual (denoted Peter) playing a game against a possibly infinitely rich adversary (say a gambling casino). Suppose that when Peter's fortune is k, he has probability p_k of winning one unit, probability q_k of losing one unit, and probability r_k of no change in his fortune.

An important special case is the case of repeated independent plays of a game in which Peter has probability p of winning; then

$$p_k = p, q_k = 1 - p, r_k = 0 \qquad \text{for } k = 1, 2, \cdots. \tag{6.11}$$

The quantities p and q may be interpreted as follows: if $p > q$, then the game is advantageous for Peter, either because he is more skillful or because the rules of the game favor him; if $p = q$, the game is fair; if $p < q$, the game is disadvantageous to Peter. If the random walk represents the position of a particle, the case $p > q$ corresponds to a drift to the right, because the particle is more likely to undergo shocks from the left than from the right; when $p = q = 1/2$, the random walk is called symmetric.

In regard to r_0 and p_0 there are several possible assumptions of which we consider two:

(i) random walk with absorbing barrier at 0 (gambler's ruin problem),

$$r_0 = 1, \ p_0 = 0; \tag{6.12}$$

(ii) random walk with reflecting barrier at 0,

$$r_0 > 0, \ p_0 > 0. \tag{6.13}$$

In case (i), Peter's adversary is playing to obtain his fortune and stops playing when Peter's fortune is gone; in case (ii), Peter's adversary is playing for sport, and permits Peter to continue playing when his fortune is gone.

In case (i), 0 is an absorbing state. The first passage (or absorption) probability $f_{j,0}$ represents the probability that Peter is ruined, given that his initial fortune was j. By Eq. 6.9 $\{f_{j,0}, j = 0, 1, 2, \cdots\}$ satisfies the system of equations

$$f_{j,0} = \sum_{i=0}^{\infty} p_{j,i} \, f_{i,0}, \ j = 1, 2, \cdots$$
$$f_{0,0} = 1. \tag{6.14}$$

Let us assume, for $j \geq 1$, that

$$p_{j,i} = \begin{cases} p_j > 0 & \text{if} \ i = j+1 \\ r_j \geq 0 & \text{if} \ i = j \\ q_j > 0 & \text{if} \ i = j-1 \\ 0, & \text{otherwise.} \end{cases} \tag{6.15}$$

Therefore Eq. 6.14 may be written

$$f_{j,0} = q_j \, f_{j-1,0} + r_j \, f_{j,0} + p_j \, f_{j+1,0}, \ j = 1, 2, 3, \cdots. \tag{6.16}$$

The system of equations 6.16 may be solved recursively. Since $r_j = 1 - p_j - q_j$, one may rewrite Eq. 6.16

$$p_j(f_{j+1,0} - f_{j,0}) = q_j(f_{j,0} - f_{j-1,0}), \ j = 1, 2, \cdots. \tag{6.17}$$

From Eq. 6.17 it follows that for any integers $m > j \geq 1$

$$f_{m+1,0} - f_{m,0} = \frac{q_m \cdots q_j}{p_m \cdots p_j} (f_{j,0} - f_{j-1,0}). \tag{6.18}$$

Let us define

$$\rho_0 = 1, \ \rho_m = \frac{q_m \cdots q_1}{p_m \cdots p_1} \ \text{for} \ m = 1, 2, \cdots. \tag{6.19}$$

Then for $m = 0, 1, \cdots$

Now,
$$f_{m+1,0} - f_{m,0} = \rho_m (f_{1,0} - 1). \tag{6.20}$$

$$f_{k+1,0} - 1 = \sum_{m=0}^{k} (f_{m+1,0} - f_{m,0}). \tag{6.21}$$

Consequently, for $k = 0, 1, \cdots$,

$$f_{k+1,0} - 1 = (f_{1,0} - 1) \sum_{m=0}^{k} \rho_m. \tag{6.22}$$

We have thus determined $f_{k,0}$ up to the undetermined constant $f_{1,0}$.

To proceed further, let us first consider the case in which the game stops whenever Peter's fortune reaches a preassigned amount K; in symbols

$$q_K = 0, r_K = 1, p_K = 0 \tag{6.23}$$

The random walk is then a finite Markov chain on the state space $\{0, 1, \cdots, K\}$ with transition probability matrix

$$\begin{bmatrix} 1 & 0 & 0 & 0 & \cdots & 0 & 0 & 0 \\ q_1 & r_1 & p_1 & 0 & \cdots & 0 & 0 & 0 \\ 0 & q_2 & r_2 & p_2 & \cdots & 0 & 0 & 0 \\ \cdot & \cdot & \cdot & \cdot & \cdots & \cdot & \cdot & \cdot \\ 0 & 0 & 0 & 0 & \cdots & q_{K-1} & r_{K-1} & p_{K-1} \\ 0 & 0 & 0 & 0 & \cdots & 0 & 0 & 1 \end{bmatrix}. \tag{6.24}$$

The absorption probabilities $\{f_{j,0}, j = 0, 1, \cdots, K\}$ satisfy the system of equations

$$f_{j,0} = \sum_{i=0}^{K} p_{j,i} f_{i,0}, j = 1, 2, \cdots, K - 1 \tag{6.25}$$
$$f_{0,0} = 1$$
$$f_{K,0} = 0.$$

One may show that Eq. 6.22 now holds for $k = 0, 1, \cdots, K - 1$. In the case of a finite Markov chain, we are able to determine $f_{1,0}$ without difficulty. From the fact that $f_{K,0} = 0$, it follows from Eq. 6.22 with $k = K - 1$ that

$$-1 = (f_{1,0} - 1) \sum_{m=0}^{K-1} \rho_m.$$

Therefore,

$$f_{1,0} = 1 - \left\{ \sum_{m=0}^{K-1} \rho_m \right\}^{-1} = \frac{\sum_{m=1}^{K-1} \rho_m}{\sum_{m=0}^{K-1} \rho_m}. \tag{6.26}$$

From Eqs. 6.26 and 6.22 it follows that for the general random walk with state space $\{0, 1, \cdots, K\}$ and transition probability matrix 6.24 the probabilities of absorption in state 0 satisfy

$$1 - f_{j,0} = \frac{\sum\limits_{m=0}^{j-1} \rho_m}{\sum\limits_{m=0}^{K-1} \rho_m}, j = 1, 2, \cdots, K. \tag{6.27}$$

It seems reasonable that for the general random walk with state space $\{0, 1, 2, \cdots\}$ and transition probability matrix 6.10 the probabilities $f_{j,0}$ may be obtained by taking the limit of Eq. 6.27 as K tends to ∞ :

$$1 - f_{j,0} = \frac{\sum\limits_{m=0}^{j-1} \rho_m}{\sum\limits_{m=0}^{\infty} \rho_m}, j = 1, 2, \cdots \tag{6.28}$$

if

$$\sum_{m=0}^{\infty} \rho_m < \infty; \tag{6.29}$$

$$f_{j,0} = 1 \quad \text{for} \quad j = 1, 2, \cdots. \tag{6.30}$$

if

$$\sum_{m=0}^{\infty} \rho_m = \infty. \tag{6.31}$$

It is beyond our scope to prove rigorously that Eq. 6.28 holds under Eq. 6.29; for a proof, see Chung (1960), p. 69. However, it will be proved in Example 6B that Eqs. 6.30 and 6.31 are equivalent.

In order to understand the significance of these results, let us consider in detail the case of repeated plays of a single game, so that Eq. 6.11 holds. Then

$$\rho_m = \left(\frac{q}{p}\right)^m. \tag{6.32}$$

In the case that the total fortune available to Peter and his adversary is K, it follows from Eq. 6.27 that the probability of Peter's ruin, given an initial fortune j, is (for $j = 1, 2, \cdots, K$)

$$f_{j,0} = 1 - \frac{1 - (q/p)^j}{1 - (q/p)^K} = \frac{(q/p)^j - (q/p)^K}{1 - (q/p)^K} \quad \text{if} \quad p \neq q \tag{6.32}$$

$$= 1 - \frac{j}{K} = \frac{K-j}{K} \quad \text{if} \quad p = q$$

since

$$\sum_{m=0}^{j-1} (q/p)^m = \frac{1 - (q/p)^j}{1 - (q/p)} \quad \text{if} \quad q \neq p \tag{6.33}$$

$$= j \quad \text{if} \quad q = p.$$

As K tends to infinity, one obtains from Eq. 6.32 that

$$f_{j,0} = \left(\frac{q}{p}\right)^j \quad \text{if} \quad p > q \tag{6.34}$$

$$= 1 \quad \text{if} \quad p \leq q.$$

It should be noted that Eq. 6.34 also follows from Eqs. 6.28–6.31.

In words, Eq. 6.34 may be stated as follows. If the game is fair or disadvantageous to Peter, and if Peter's opponent has a much larger fortune than Peter, then Peter is sure to be ruined. However, if the game is advantageous to Peter, there is a positive probability that his opponent will be ruined, even though he has a much larger fortune.

A gambler in a casino is in the following situation. He is playing a disadvantageous game against an infinitely rich adversary, who is always willing to play, while the gambler has the privilege of stopping at his pleasure. If the player has an initial fortune of j coins, bets 1 coin at each play, and plays until he either loses his fortune or increases it by $K - j$ coins, then his probability $f_{j,0}$ of being ruined is given by Eq. 6.32 where p denotes the gambler's probability of winning at each play of the game.

Let it be noted, however, that even though the gambler has positive probability of increasing his fortune to K before being ruined, the expected gain from the strategy described is negative. His gain G is equal to $K - j$ with probability $1 - f_{j,0}$ and is equal to $- j$ with probability $f_{j,0}$. Therefore the expected gain

$$E[G] = (K - j)(1 - f_{j,0}) - j f_{j,0} = K (1 - f_{j,0}) - j, \tag{6.35}$$

which is negative if $p < q$.

A gambler's probability of ruin in various typical situations is given in Table 6.4.

TABLE 6.4. **The fate of a gambler**

Initial fortune j	Desired fortune K	Probability p of winning at each play	Probability of ruin	Mean gain	Mean duration of the game
9	10	.50	0.1	0	9
9	10	.45	0.21	− 1.1	11
99	100	.45	0.182	− 17.2	171.8
90	100	.50	0.1	0	900
90	100	.45	0.866	− 76.6	756.6
90	100	.40	0.983	− 88.3	441.3

For a discussion of the effect of changing stakes on the probabilities of ruin the reader should consult Feller (1957), pp. 315–317. It may be shown that increasing stakes decreases the probability of a player's ruin in a game disadvantageous to him.

Recurrent and non-recurrent irreducible Markov chains. A Markov chain is said to be *irreducible* if all pairs of states of the chain communicate, so that the chain consists of exactly one communicating class. A closed communicating class can be treated as an irreducible Markov chain. An irreducible chain is said to be recurrent (non-recurrent) if each state in the chain is recurrent (non-recurrent). Using Theorem 6B we may obtain a criterion that an irreducible Markov chain be recurrent or non-recurrent.

THEOREM 6C

An irreducible Markov chain with state space C is non-recurrent if and only if there exists a state k in C such that the system of equations

$$v_j = \sum_{\text{states } i \neq k} p_{j,i} v_i, \quad \text{for all states} \quad j \neq k, \tag{6.36}$$

possesses a bounded solution $\{v_j\}$ which is not identically equal to 0.

Proof. For ease of writing suppose that the states of C are $\{0, 1, 2, \cdots\}$ and that the state chosen is $k = 0$. By Theorem 5C, C is recurrent if and only if $f_{j,0} = 1$ for $j = 1, 2, \cdots$. Now, consider a new Markov chain C' with the same state space $\{0, 1, \cdots\}$ as C and whose transition probability matrix P' is related to the transition probability matrix P of C as follows:

$$P = \begin{bmatrix} p_{0,0} & p_{0,1} & p_{0,2} & \cdots \\ p_{1,0} & p_{1,1} & p_{1,2} & \cdots \\ p_{2,0} & p_{2,1} & p_{2,2} & \cdots \\ \cdots\cdots\cdots\cdots\cdots\cdots \end{bmatrix}, \quad P' = \begin{bmatrix} 1 & 0 & 0 & \cdots \\ p_{1,0} & p_{1,1} & p_{1,2} & \cdots \\ p_{2,0} & p_{2,1} & p_{2,2} & \cdots \\ \cdots\cdots\cdots\cdots\cdots\cdots \end{bmatrix}. \tag{6.37}$$

In words, we have changed the original Markov chain into a new chain in which 0 is an absorbing state and the remaining states $\{1, 2, \cdots\}$ are non-recurrent, since the original chain was assumed to be irreducible. Further $f_{j,0}$ represents the probability of absorption in 0, starting at j, and $1 - f_{j,0}$ is the probability that the chain C' will remain infinitely often among the non-recurrent states $\{1, 2, \cdots\}$ starting at j. By Theorem 6B a necessary and sufficient condition that $1 - f_{j,0} > 0$ for some $j = 1$, $2, \cdots$ is that there exist a bounded solution to Eq. 6.36 which is not identically equal to 0. On the other hand, the original chain C is non-recurrent if and only if $1 - f_{j,0} > 0$ for some j. The proof of Theorem 6C is now complete.

EXAMPLE 6B

Recurrent and non-recurrent random walks. Consider the general random walk with state space $\{0, 1, 2, \cdots\}$ and transition probability matrix 6.10 satisfying Eqs. 6.13 and 6.15 so that no state is absorbing. The random walk is clearly an irreducible chain. By Theorem 6C, the random walk is non-recurrent if and only if the system of equations

$$v_j = \sum_{i=1}^{\infty} p_{j,i}\, v_i,\, j = 1, 2, \cdots \tag{6.38}$$

possesses a bounded solution which is not identically equal to 0. In view of Eq. 6.15, Eq. 6.38 may be written

$$v_1 = r_1 v_1 + p_1 v_2$$
$$v_j = q_j v_{j-1} + r_j v_j + p_j v_{j+1}, j = 2, 3, \cdots. \tag{6.39}$$

The second equation in 6.39 may be written

$$p_j(v_{j+1} - v_j) = q_j(v_j - v_{j-1}), j = 2, 3, \cdots. \tag{6.40}$$

Consequently, as in the derivation of Eq. 6.22, for $k = 2, \cdots$

$$v_{k+1} - v_1 = (v_2 - v_1)\left(1 + \sum_{m=2}^{k} \left\{\frac{q_m \cdots q_2}{p_m \cdots p_2}\right\}\right). \tag{6.41}$$

From the first equation in 6.39 it follows that

$$v_2 - v_1 = \frac{q_1}{p_1} v_1. \tag{6.42}$$

From Eqs. 6.41 and 6.42 it follows that the general solution $\{v_k, k = 1, 2, \cdots\}$ of the system of equations 6.38 is given up to an undetermined constant v_1 by

$$v_k = v_1 \sum_{m=0}^{k-1} \rho_m,\ k = 1, 2, \cdots, \tag{6.43}$$

where ρ_m is defined by Eq. 6.19. The sequence $\{v_k\}$ is bounded if and only if Eq. 6.29 holds. Consequently, Eq. 6.29 is a necessary and sufficient condition for the general random walk to be non-recurrent. Equivalently, Eq. 6.31 is a necessary and sufficient condition for the general random walk to be recurrent.

EXERCISES

In Exercises 6.1 to 6.6 find the probabilities $f_{j,0}$ of absorption at 0 for the Markov chain described.

6.1 Consider a random walk on $\{0, 1, \cdots, K\}$ with $r_0 = r_K = 1$, and for $k = 1, 2, \cdots, K - 1$

(i) $p_k = q_k = 0.5$;

(ii) $p_k = 0.4$, $q_k = 0.6$;

(iii) $p_k = 0.4$, $p_k = 0.4$, $r_k = 0.2$.

6.2 Consider a random walk on $\{0, 1, \cdots\}$ with $r_0 = 1, r_K = 1$, and $p_k = p$, $q_k = q, r_k = r$ for other values of k, where p, q, r are positive constants summing to 1.

6.3 Consider a random walk on $\{0, 1, \cdots\}$ with $r_0 = 1$, $p_k = p > 0$, $q_k = 1 - p$ for $k \geq 2$, and $p_1 = p$, $q_1 = (1 - \delta)(1 - p)$, $r_1 = \delta(1 - p)$, where $0 < \delta < 1$. (This random walk corresponds to a contest in which if Peter's fortune is 1 unit and he loses, with probability δ he is permitted to continue in the contest.)

6.4 Let X_n denote the position after n steps of a particle moving among the integers $\{0, 1, 2, \cdots\}$ according to the following rules: the particle moves at each step either two units to the right or one unit to the left, with respective probabilities p and $1 - p$. The particle remains at 0 once there.

6.5 Consider the branching process described in Example 2C. Let X_n denote the size of the nth generation.

6.6 *Success runs.* Consider a sequence of independent tosses of a coin which has probability p of falling heads. Define a sequence of random variables $X_1, X_2, \cdots$ as follows: for $n \geq 1$, $X_n = k$ if the outcomes in the $(n - k)$th trial was tails and the outcomes of the $(n - k + 1)$st, $(n - k + 2)$nd, $\cdots$ nth trials were heads. In words, X_n is the length of the success run underway at the time of the nth trial. In particular, $X_n = 0$ if the outcome of the nth trial was tails.

Hint. $\{X_n\}$ form a Markov chain with transition probability matrix

$$P = \begin{bmatrix} q & p & 0 & 0 & \cdots \\ q & 0 & p & 0 & \cdots \\ q & 0 & 0 & p & \cdots \\ \cdots & \cdots & \cdots & \cdots & \cdots \end{bmatrix}.$$

6.7 Consider a Markov chain whose transition probability matrix P is defined in terms of a sequence $\{q_0, q_1, \cdots\}$ as follows: for $k = 0, 1, 2, \cdots$, $p_{k,0} = q_k$ and $p_{k,k+1} = 1 - q_k$. Prove that the Markov chain is non-recurrent if and only if $\sum_{k=0}^{\infty} q_k < \infty$.

Hint. Use the fact that if $0 < q_k < 1$ then $\lim_{n \to \infty} \prod_{k=0}^{n} (1 - q_k) = 0$ if and only if $\sum_{k=0}^{\infty} q_k = \infty$ which may be proved using the inequalities

$$1 - \sum_{k=m}^{n} q_k < \prod_{k=m}^{n} (1 - q_k) < \exp[- \sum_{k=m}^{n} q_k].$$

6-7 MEAN ABSORPTION, FIRST PASSAGE, AND RECURRENCE TIMES

Consider a Markov chain with state space S, and let T denote the set of non-recurrent states in the chain. It is of interest to determine the probability law of the random variable N', called the *time before absorption*, representing the length of time the chain spends among non-recurrent states before being absorbed. In symbols

$$N' = \sum_{j \in T} N_j(\infty). \tag{7.1}$$

The *time to absorption*, denoted by N, is given by

$$N = N' + 1. \tag{7.2}$$

Define

$$m_j = E[N \mid X_0 = j] = 1 + E[N' \mid X_0 = j] \tag{7.3}$$

to be the *mean time to absorption*, given that the chain started at state j. More generally, define for $r \geq 0$

$$m_j^{(r)} = E[N^r \mid X_0 = j] \tag{7.4}$$

to be the rth moment of the time to absorption.

One may give an expression for m_j in terms of the transition probabilities: for all j in T,

$$m_j = 1 + \sum_{i \in T} \sum_{n=1}^{\infty} p_{j,i}(n) \tag{7.5}$$

$$= \sum_{i \in T} n_{j,i}$$

defining for any states i and j in T

$$n_{j,i} = \sum_{n=0}^{\infty} p_{j,i}(n). \tag{7.6}$$

Proof of Eq. 7.5. From Eqs. 7.1 and 7.2 it follows that

$$m_j = 1 + \sum_{i \in T} E[N_i(\infty) \mid X_0 = j].$$

It is easily verified that

$$E[N_i(\infty) \mid X_0 = j] = \sum_{n=1}^{\infty} p_{j,i}(n).$$

It is usually more convenient to obtain the mean absorption times $\{m_j, j \in T\}$ as the solution of the system of linear equations:

$$m_j = 1 + \sum_{k \in T} p_{j,k} \, m_k, \; j \in T. \tag{7.7}$$

To prove that m_j satisfies Eq. 7.7 one may either use Eq. 7.5 or argue as follows:

$$
\begin{aligned}
m_j = E[N \mid X_0 = j] &= \sum_{k \in S} p_{j,k} \, E[N \mid X_1 = k] \\
&= \sum_{k \in T} p_{j,k}\{1 + E[N \mid X_0 = k]\} + \sum_{k \in T^c} p_{j,k} \\
&= 1 + \sum_{k \in T} p_{j,k} \, m_k.
\end{aligned} \tag{7.8}
$$

In a finite Markov chain, the mean times to absorption are finite, and the system of equations 7.7 has a unique solution. In a Markov chain with an infinite number of non-recurrent states, the mean times to absorption may be infinite. However, it may be shown that if the times to absorption are finite with probability one (that is, the chain is sure not to be eventually in T) then the mean absorption times are the unique solution of the system of equations 7.7.

EXAMPLE 7A

Duration of random walks. Consider two players, Peter and his opponent, playing a contest for unit stakes. At each play, Peter has probability p of winning. Suppose Peter's initial fortune is j, and the combined fortune of Peter and his opponent is K. Let X_n denote Peter's fortune after n plays of the contest. Then $\{X_n, n = 0, 1, \cdots\}$ is a random walk with state space $\{0, 1, \cdots, K\}$ and transition probability matrix 6.24. In this Markov chain, states $1, 2, \cdots, K - 1$ are non-recurrent while states 0 and K are absorbing. The time N to absorption then represents the *duration* of the contest; that is, N is the number of times the contest is repeated until one of the players is ruined (has exhausted his fortune). The mean absorption time $m_j = E[N \mid X_0 = j]$ represents the mean duration of the contest, given that Peter's initial fortune was j. The mean absorption times $\{m_j, j = 1, 2, \cdots, K - 1\}$ are the unique solution of the system of equations

$$m_j = 1 + \sum_{k=1}^{K-1} p_{j,k} \, m_k, \; j = 1, \cdots, K - 1. \tag{7.9}$$

If one defines $m_0 = m_K = 0$, then using Eq. 6.24 one may rewrite Eq. 7.9

$$m_j = 1 + q_j \, m_{j-1} + r_j \, m_j + p_j \, m_{j+1}, j = 1, \cdots, K - 1. \tag{7.10}$$

In turn one may rewrite Eq. 7.10 as a first-order difference equation

$$p_j \, M_{j+1} = q_j \, M_j - 1, \quad j = 1, 2, \cdots, K - 1, \tag{7.11}$$

where

$$M_j = m_j - m_{j-1}, \quad j = 1, 2, \cdots, K. \tag{7.12}$$

Solving Eq. 7.11 recursively, one obtains

$$\begin{aligned}
M_{j+1} &= \frac{q_j}{p_j} \frac{q_{j-1}}{p_{j-1}} \cdots \frac{q_{j-m}}{p_{j-m}} M_{j-m} \\
&\quad - \frac{1}{p_j}\left(1 + \frac{q_j}{p_{j-1}} + \cdots + \frac{q_j \cdots q_{j-m+1}}{p_{j-1} \cdots p_{j-m}}\right).
\end{aligned} \tag{7.13}$$

From Eqs. 7.12 and 7.13 one could obtain an expression for m_j. To illustrate the procedure, let us consider the case of repeated trials. Then, for $j = 1, 2, \cdots, K - 1$,

$$\begin{aligned}
M_{j+1} &= \left(\frac{q}{p}\right)^j M_1 - \frac{1}{p}\left\{1 + \frac{q}{p} + \cdots + \left(\frac{q}{p}\right)^{j-1}\right\} \\
&= \left(\frac{q}{p}\right)^j M_1 - \left(\frac{1}{p-q}\right)\left\{1 - \left(\frac{q}{p}\right)^j\right\} \quad \text{if} \quad q \neq p \\
&= M_1 - \frac{1}{p}j \quad\quad\quad\quad\quad\quad\quad \text{if} \quad p = q. \tag{7.14}
\end{aligned}$$

Now, for $k = 1, 2, \cdots, K$

$$\begin{aligned}
m_k = m_k - m_0 &= \sum_{j=0}^{k-1} M_{j+1} \\
&= kM_1 - k(k-1)\left(\frac{1}{2p}\right) \quad\quad\quad \text{if} \quad p = q \\
&= \left\{M_1 + \left(\frac{1}{p-q}\right)\right\} \frac{1 - (q/p)^k}{1 - (q/p)} - \frac{k}{p-q} \quad \text{if} \quad p \neq q. \tag{7.15}
\end{aligned}$$

Thus, $\{m_k, \ k = 1, 2, \cdots, K\}$ is given by Eq. 7.15 up to an undetermined constant M_1. To determine M_1 we use the fact that $m_K = 0$. Thus, if $p = q$,

$$0 = m_K = K M_1 - K(K-1)\left(\frac{1}{2p}\right),$$

so that $M_1 = (K - 1)/2p$ if $p = q$. One thus obtains that the mean time to absorption, having started at k, is given by

$$\begin{aligned}
m_k &= \frac{k(K - k)}{2p} \quad\quad\quad\quad\quad \text{if} \quad p = q \\
&= \frac{k}{q-p} - \frac{K}{q-p} \frac{1 - (q/p)^k}{1 - (q/p)^K} \quad \text{if} \quad p \neq q. \tag{7.16}
\end{aligned}$$

If one lets K tend to ∞ in Eq. 7.16 one obtains the mean time to absorption at 0 in the random walk on $\{0, 1, \cdots\}$:

$$m_k = \frac{k}{q - p} \quad \text{if} \quad q > p$$
$$= \infty \quad \text{if} \quad q \leq p. \tag{7.17}$$

Matrix formulation of the equations for mean absorption time in the case of finite Markov chains. Consider a finite Markov chain with state space S, and let T be the set of non-recurrent states. Let Q be the matrix

$$Q = \{p_{j,k} : j, k \in T\} \tag{7.18}$$

of transition probabilities of states in T to states in T. Let m be a column vector whose components are $m_j, j \in T$. The system of equations 7.7 may then be written in matrix form

$$Im = 1 + Q\, m, \tag{7.19}$$

where 1 is the column vector each of whose components are 1, and I is the identity matrix. One may rewrite Eq. 7.19

$$(I - Q)\, m = 1. \tag{7.20}$$

Let N be the inverse matrix of $I - Q$:

$$N = (I - Q)^{-1}. \tag{7.21}$$

It may be verified that $I - Q$ possesses an inverse; in fact

$$(I - Q)^{-1} = I + Q + Q^2 + \cdots + Q^n + \cdots. \tag{7.22}$$

To prove that Eq. 7.22 holds prove that the matrix defined as an infinite series converges and has the property that when multiplied by $I - Q$ the product is I.

We may now write the vector m of mean absorption times in terms of N. From Eq. 7.20

$$m = N\, 1. \tag{7.23}$$

Kemeny and Snell (1960) call N the fundamental matrix of an absorbing Markov chain and show that a number of interesting quantities, in addition to the mean absorption times, can be expressed in terms of N (see Complement 7A).

EXAMPLE 7B

Consider the random walk in the integers $\{0, 1, 2, 3, 4\}$ with transition probability matrix

$$P = \begin{bmatrix} 1 & 0 & 0 & 0 & 0 \\ q & 0 & p & 0 & 0 \\ 0 & q & 0 & p & 0 \\ 0 & 0 & q & 0 & p \\ 0 & 0 & 0 & 0 & 1 \end{bmatrix}.$$

Then

$$Q = \begin{bmatrix} 0 & p & 0 \\ q & 0 & p \\ 0 & q & 0 \end{bmatrix},$$

$$I - Q = \begin{bmatrix} 1 & -p & 0 \\ -q & 1 & -p \\ 0 & -q & 1 \end{bmatrix}.$$

If one computes the inverse $(I - Q)^{-1}$ one finds

$$N = (I - Q)^{-1} = \begin{bmatrix} \dfrac{p + q^2}{p^2 + q^2} & \dfrac{p}{p^2 + q^2} & \dfrac{p^2}{p^2 + q^2} \\[2mm] \dfrac{q}{p^2 + q^2} & \dfrac{1}{p^2 + q^2} & \dfrac{p}{p^2 + q^2} \\[2mm] \dfrac{q^2}{p^2 + q^2} & \dfrac{q}{p^2 + q^2} & \dfrac{q + p^2}{p^2 + q^2} \end{bmatrix}.$$

In particular, if $p = 2/3$, then

$$N = \begin{bmatrix} \frac{7}{5} & \frac{6}{5} & \frac{4}{5} \\[1mm] \frac{3}{5} & \frac{9}{5} & \frac{6}{5} \\[1mm] \frac{1}{5} & \frac{3}{5} & \frac{7}{5} \end{bmatrix}, \quad m = \begin{bmatrix} \frac{17}{5} \\[1mm] \frac{18}{5} \\[1mm] \frac{11}{5} \end{bmatrix}.$$

First passage and recurrence times. Consider an irreducible recurrent Markov chain with state space C. For each pair of states j and k in C, the sequence of first passage probabilities $\{f_{j,k}(n), n = 1, 2, \cdots\}$ is a probability distribution. The mean

$$m_{j,k} = \sum_{n=1}^{\infty} n f_{j,k}(n) \tag{7.24}$$

is called the *mean first passage time* from j to k in the case that $j \neq k$ and is called the *mean recurrence time* of the state k if $j = k$.

Given a fixed state k, the mean first passage times $\{m_{j,k}, j \neq k\}$ can be regarded as mean absorption times in a new Markov chain obtained from the original Markov chain by making k an absorbing state. (For an example of how this is done, see the proof of Theorem 6C.) More precisely if $P = \{p_{j,k}\}$ is the transition probability matrix of the original Markov chain, define a new transition probability matrix $P' = \{p'_{i,j}\}$ as follows:

$$p'_{i,j} = 1 \quad \text{if} \quad i = k, j = k$$
$$p'_{i,j} = 0 \quad \text{if} \quad i = k, j \neq k$$
$$p'_{i,j} = p_{i,j} \quad \text{if} \quad i \neq k, j \text{ any state.}$$

The resulting Markov chain contains a single absorbing state k, while all other states are non-recurrent since they are inessential. The behavior of the new Markov chain before absorption is the same as the behavior of the original Markov chain before visiting state k for the first time. In particular, the mean first passage time from j to k in the original Markov chain is the same as the mean time to absorption in the new chain. Consequently from Eq. 7.7 we obtain the following theorem.

THEOREM 7A

Let C be an irreducible recurrent Markov chain. Let k be a fixed state in C. The set of mean first passage times $\{m_{j,k}, j \neq k\}$ satisfy a system of linear equations

$$m_{j,k} = 1 + \sum_{i \neq k} p_{j,i} m_{i,k}, \quad j \neq k. \tag{7.25}$$

The system 7.25 has a unique solution.

To compute the mean recurrence times in an irreducible recurrent Markov chain there are two methods available. The mean recurrence times may be obtained from the mean first passage times:

$$m_{k,k} = 1 + \sum_{i \neq k} p_{k,i} m_{i,k}, \quad k \in C, \tag{7.26}$$

since

$$m_{k,k} = p_{k,k} + \sum_{i \neq k} p_{k,i} \{1 + m_{i,k}\}.$$

A different method of finding mean recurrence times is given by Eq. 8.35.

EXAMPLE 7C

Mean first passage times in a random walk. Consider the random walk on the integers $\{0, 1, 2, \cdots\}$ with transition probability matrix (in which $p > 0$ and $q = 1 - p$)

$$P = \begin{bmatrix} q & p & 0 & 0 & 0 & \cdots \\ q & 0 & p & 0 & 0 & \cdots \\ 0 & q & 0 & p & 0 & \cdots \\ 0 & 0 & q & 0 & p & \cdots \\ \cdots\cdots\cdots\cdots\cdots\cdots\cdots \\ \cdots\cdots\cdots\cdots\cdots\cdots\cdots \end{bmatrix} \qquad (7.27)$$

The Markov chain with transition probability matrix P is clearly irreducible (all states communicate). It is recurrent if and only if $q \geq p$ (for proof see Example 6C). To obtain the mean first passage times $\{m_{j,k}, j \neq k\}$ one uses Eq. 7.25. The results can be readily obtained by suitably adapting the results of Example 7A. Consider the random walk on the integers $\{0, 1, \cdots, K\}$ with transition probability matrix

$$P = \begin{bmatrix} q & p & 0 & 0 & \cdots & 0 & 0 & 0 \\ q & 0 & p & 0 & \cdots & 0 & 0 & 0 \\ 0 & q & 0 & p & \cdots & 0 & 0 & 0 \\ \cdot & \cdot & \cdot & \cdot & \cdots & \cdot & \cdot & \cdot \\ 0 & 0 & 0 & 0 & \cdots & q & 0 & p \\ 0 & 0 & 0 & 0 & \cdots & 0 & 0 & 1 \end{bmatrix},$$

so that K is an absorbing state. It may be shown that in this random walk the mean absorption time m_j from state j to state K is

$$m_j = \frac{q}{(q-p)^2}\left\{\left(\frac{q}{p}\right)^K - \left(\frac{q}{p}\right)^j\right\} - \left(\frac{K-j}{q-p}\right). \qquad (7.28)$$

From Eq. 7.28 it follows that for $j < k$ the mean first passage times $m_{j,k}$ of the random walk 7.27 are given by

$$m_{j,k} = \frac{q}{(q-p)^2}\left\{\left(\frac{q}{p}\right)^k - \left(\frac{q}{p}\right)^j\right\} - \frac{k-j}{q-p} \quad \text{if} \quad j < k. \qquad (7.29)$$

To find $m_{j,k}$ for $j > k$ we use Eq. 7.17. It follows that

$$\begin{aligned} m_{j,k} &= \frac{j-k}{q-p} \quad \text{if} \quad k < j \ \text{ and } \ q > p \\ &= \infty \qquad \text{if} \quad k < j \ \text{ and } \ q = p. \end{aligned} \qquad (7.30)$$

The mean recurrence times may be obtained from the mean first passage times by Eq. 7.26. For $k > 0$,

$$\begin{aligned} m_{k,k} &= 1 + q\, m_{k-1,k} + p\, m_{k+1,k} \\ &= \frac{q}{q-p}\left(\frac{q}{p}\right)^k \quad \text{if} \quad q > p \\ &= \infty \qquad\qquad \text{if} \quad q = p, \end{aligned} \qquad (7.31)$$

while

$$m_{0,0} = 1 + p\, m_{1,0}$$
$$= \frac{q}{q-p} \quad \text{if} \quad q > p$$
$$= \infty \quad \text{if} \quad q = p. \tag{7.32}$$

Positive and null recurrent states. While a recurrent state k has the property that the Markov chain is sure to return eventually to k, the mean time of recurrence of k may be infinite. This is true, for example, of any state in a symmetric random walk on $\{0, 1, 2, \cdots\}$ in which all states communicate.

A recurrent state k is said to be *positive* if its mean recurrence time $m_{k,k}$ is finite and is said to be *null* if its mean recurrence time is infinite.

EXAMPLE 7D

Positive recurrent random walks. The random walk with transition probability matrix 7.27 is recurrent if and only if $q \geq p$. However, it is positive recurrent if and only if $q > p$.

It will be shown in the next section that positive recurrence is a class property in the sense that if C is a communicating class of *recurrent* states, then either all states in C are positive or all states in C are null.

An irreducible Markov chain is said to be *positive recurrent* if all its states are positive recurrent. An irreducible positive recurrent Markov chain is called by many authors an *ergodic* Markov chain.

Characteristic function and higher moments of the time to absorption. To obtain the variance and higher moments of the time N to absorption in a Markov chain, one may obtain a set of linear equations which these moments satisfy. The conditional characteristic function of the time N to absorption, given that the Markov chain started in the non-recurrent state j, is defined (letting $i = \sqrt{-1}$) by

$$\varphi_j(u) = E[e^{iuN} \mid X_0 = j], \; j \in T. \tag{7.33}$$

It satisfies the system of equations

$$e^{-iu}\varphi_j(u) = 1 + \sum_{k \in T} p_{j,k}\{\varphi_k(u) - 1\}, \; j \in T, \tag{7.34}$$

since

$$\varphi_j(u) = \sum_{k \in C} p_{j,k} E[e^{iuN} \mid X_1 = k]$$
$$= \sum_{k \in T^c} p_{j,k}\, e^{iu} + \sum_{k \in T} p_{j,k}\{e^{iu}\varphi_k(u)\}.$$

Differentiating Eq. 7.34 with respect to u one obtains

$$e^{-iu}\{\varphi_j{}'(u) - i\,\varphi_j(u)\} = \sum_{k\,\epsilon\,T} p_{j,k}\,\varphi_k{}'(u).$$ (7.35)

Letting $u = 0$ in Eq. 7.35, one obtains

$$i\,m_j - i = i \sum_{k\,\epsilon\,T} p_{j,k}\,m_k,$$

which leads to Eq. 7.7.

Differentiating Eq. 7.35 with respect to u one obtains

$$e^{-iu}\{\varphi_j{}''(u) - 2\,i\,\varphi_j{}'(u) - \varphi_j(u)\} = \sum_{k\,\epsilon\,T} p_{j,k}\,\varphi_k{}''(u).$$ (7.36)

Letting $u = 0$ in Eq. 7.36 one obtains a system of linear equations in the second moments $m_j{}^{(2)}$ of the time to absorption:

$$m_j{}^{(2)} = 2m_j - 1 + \sum_{k\,\epsilon\,T} p_{j,k}\,m_k{}^{(2)},\ j\,\epsilon\,T.$$ (7.37)

In the same way that Eq. 7.25 is derived from Eq. 7.7 one may derive from Eq. 7.37 a system of linear equations in the second moments

$$m_{j,k}{}^{(2)} = \sum_{n=1}^{\infty} n^2 f_{j,k}(n)$$ (7.38)

of the first passage times in an irreducible recurrent Markov chain:

$$m_{j,k}{}^{(2)} = 2m_{j,k} - 1 + \sum_{i\neq k} p_{j,i}\,m_{i,k}{}^{(2)},\ j \neq k.$$ (7.39)

COMPLEMENTS

7A *Uses of the fundamental matrix of a finite absorbing chain.* Consider a finite Markov chain, and let T be the set of non-recurrent states. Assume that all recurrent states are absorbing. The transition probability matrix P can then be written

$$P = \begin{pmatrix} I & 0 \\ R & Q \end{pmatrix},$$

where Q is defined by Eq. 7.18, I is an identity matrix, and R is the (rectangular) matrix of transition probabilities from non-recurrent states to absorbing states. Let $F = \{f_{j,k}\}$, where $f_{j,k}$ is the absorption probability from a non-recurrent state j to a recurrent state k. Let $\boldsymbol{m}^{(2)} = \{m_j{}^{(2)}\}$, where $m_j{}^{(2)}$ is the second moment of the absorption time from j to the set of recurrent states. Show that

$$F = NR,$$
$$\boldsymbol{m}^{(2)} = N(2\boldsymbol{m} - 1).$$

Find F and $\boldsymbol{m}^{(2)}$ for the transition probability matrix given in Example 7B.

7B   Let S be the set of states of a Markov chain and let T denote the set of non-recurrent states in the chain. Let $\{b_j, j \in T\}$ be a set of known constants, and $\{v_j, j \in T\}$ a set of variables satisfying the system of equations

$$v_j = b_j + \sum_{i \in T} p_{j,i}\, v_i, \qquad j \in T.$$

Show that

$$v_j = \sum_{i \in T} n_{j,i}\, b_i, \qquad j \in T,$$

where $n_{j,i}$ is defined by Eq. 7.6.

EXERCISE

7.1   Consider the Markov chain described in Exercise 2.8.

(i) Find the mean number of days between days when no machine is working.

(ii) If both machines are in working condition at the end of a particular day, what is the mean number of days before the first day that no machine is working?

6-8   LONG-RUN AND STATIONARY DISTRIBUTIONS

The tools are now at hand to study the asymptotic behavior of the transition probabilities of a Markov chain.

If k is non-recurrent, then for any state j

$$\lim_{n \to \infty} p_{j,k}(n) = 0, \tag{8.1}$$

since in fact

$$\sum_{n=1}^{\infty} p_{j,k}(n) < \infty. \tag{8.2}$$

If k is recurrent, the situation is more complicated. In general the transition probabilities $\{p_{j,k}(n), n = 1, 2, \cdots\}$ only possess a limit in Cesàro mean, given by

$$\lim_{n \to \infty} \frac{1}{n} \sum_{m=1}^{n} p_{j,k}(m) = f_{j,k} \frac{1}{m_{k,k}} ; \tag{8.3}$$

recall that $f_{j,k}$ is the probability of a first passage from j to k, while $m_{k,k}$ is the mean recurrence time of k. If $m_{k,k}$ is infinite, then $1/m_{k,k}$ is to be interpreted as 0.

In order to prove Eq. 8.3 it suffices to prove that if k is a recurrent state then

$$\lim_{n \to \infty} \frac{1}{n} \sum_{m=1}^{n} p_{k,k}(n) = \frac{1}{m_{k,k}} . \tag{8.4}$$

In view of Complement 4D, Eq. 8.3 is a consequence of Eq. 8.4. Similarly if one shows that the limit in Eq. 8.4 holds in the ordinary sense, so that

$$\lim_{n \to \infty} p_{k,k}(n) = \frac{1}{m_{k,k}} \tag{8.5}$$

it follows from Complement 4C that Eq. 8.3 can be strengthened to

$$\lim_{n \to \infty} p_{j,k}(n) = f_{j,k} \frac{1}{m_{k,k}} . \tag{8.6}$$

An analytic proof of Eq. 8.4 was sketched in Theorem 4H. Conditions under which Eq. 8.5 holds are stated below. Probabilistic proofs of Eqs. 8.4 and 8.5 are discussed in Sections 6–9 and 6–10, respectively. In this section we discuss their consequences.

A Markov chain with state space C is said to possess a *long-run distribution* if there exists a probability distribution $\{\pi_k, k \in C\}$, having the property that for every j and k in C

$$\lim_{n \to \infty} p_{j,k}(n) = \pi_k. \tag{8.7}$$

The name "long-run distribution" derives from the fact that no matter what the initial unconditional probability distribution $\{p_k(0), k \in C\}$, the unconditional probability $p_k(n)$ tends to π_k as n tends to ∞:

$$\lim_{n \to \infty} p_k(n) = \lim_{n \to \infty} \sum_{j \in C} p_j(0) \, p_{j,k}(n) \tag{8.8}$$

$$= \sum_{j \in C} p_j(0) \, \{\lim_{n \to \infty} p_{j,k}(n)\}$$

$$= \pi_k.$$

A Markov chain with state space C is said to possess a *stationary distribution* if there exists a probability distribution $\{\pi_k, k \in C\}$ having the property that for every k in C

$$\pi_k = \sum_{j \in C} \pi_j \, p_{j,k}. \tag{8.9}$$

The name "stationary distribution" derives from the fact (shown by Eq. 8.28) that if Eq. 8.9 holds, then for every integer n

$$\pi_k = \sum_{j \in C} \pi_j \, p_{j,k}(n). \tag{8.10}$$

Consequently, if the initial unconditional distribution $\{p_k(0), k \in C\}$ is taken to be $\{\pi_k, k \in C\}$ then for every n, $p_k(n) = \pi_k$. Therefore, the Markov chain $\{X_n\}$ has stationary unconditional distributions; indeed, it is a strictly stationary stochastic process.

In order to determine conditions that a Markov chain possesses a stationary distribution we make use of the following consequence of Eqs. 8.1 and 8.3.

THEOREM 8A

To every irreducible Markov chain with state space C there is a sequence $\{\pi_k, k \in C\}$ such that for every j and k in C

$$\lim_{n \to \infty} \frac{1}{n} \sum_{m=1}^{n} p_{j,k}(m) = \pi_k. \tag{8.11}$$

In words, Eq. 8.11 says that the sequence $p_{j,k}(n)$ converges in Cesàro mean to a limit π_k which is independent of j. Note that a sequence $\{\pi_k, k \in C\}$ which satisfies Eq. 8.7 also satisfies Eq. 8.11.

The sequence $\{\pi_k, k \in C\}$ satisfying Eq. 8.11 is non-negative:

$$\pi_k \geq 0 \quad \text{for all } k \text{ in } C. \tag{8.12}$$

If C is a finite Markov chain, then $\{\pi_k, k \in C\}$ is a probability distribution; that is,

$$\sum_{k \in C} \pi_k = 1. \tag{8.13}$$

If C is infinite, Eq. 8.13 does not necessarily hold. In particular, Eq. 8.13 does not hold if C is an irreducible non-recurrent chain since then $\pi_k = 0$ for all k in C. In order to determine conditions under which Eq. 8.13 holds, we first determine various relations which are satisfied by a sequence $\{\pi_k, k \in C\}$ satisfying Eq. 8.11.

THEOREM 8B

Let C be the state space of an irreducible Markov chain and let $\{\pi_k, k \in C\}$ be a sequence satisfying Eq. 8.11. Then

$$\sum_{k \in C} \pi_k \leq 1, \tag{8.14}$$

and the sequence $\{\pi_k, k \in C\}$ satisfies the system of linear equations

$$\pi_k = \sum_{j \in C} \pi_j \, p_{j,k}, \quad k \in C. \tag{8.15}$$

Further, if there exists a sequence $\{u_k, k \in C\}$ such that

$$\sum_k |u_k| < \infty, \tag{8.16}$$

which satisfies the system of linear equations

$$u_k = \sum_{j \epsilon C} u_j \, p_{j,k}, \quad k \epsilon C, \tag{8.17}$$

then

$$u_k = \pi_k \left(\sum_{j \epsilon C} u_j \right). \tag{8.18}$$

Remark. From Eq. 8.15 it follows that if a Markov chain possesses a long-run distribution, then it possesses a stationary distribution.

From Eq. 8.18 it follows that, *if it exists*, the stationary distribution $\{\pi_k, \, k \, \epsilon \, C\}$ is the unique solution of the system of equations 8.15 which satisfies

$$\sum_{k \epsilon C} \pi_k = 1. \tag{8.19}$$

Proof. For ease of writing let us assume that the state space C is the set of integers $\{0, 1, 2, \cdots \}$. Next, let us define

$$p_{j,k}{}^*(n) = \frac{1}{n} \sum_{m=1}^{n} p_{j,k}(n). \tag{8.20}$$

To prove Eq. 8.14 we note that

$$\sum_{k=0}^{\infty} p_{j,k}{}^*(n) = 1 \quad \text{for all } n. \tag{8.21}$$

Consequently, by Fatou's lemma (see Appendix to this chapter),

$$\sum_{k=0}^{\infty} \pi_k = \sum_{k=0}^{\infty} \lim_{n \to \infty} p_{j,k}{}^*(n) \le \lim_{n \to \infty} \sum_{k=0}^{\infty} p_{j,k}{}^*(n) = 1. \tag{8.22}$$

We next prove Eq. 8.15. By the Chapman-Kolmogorov equation

$$p_{j,k}(n+1) = \sum_{i=0}^{\infty} p_{j,i}(n) \, p_{i,k}. \tag{8.23}$$

Consequently,

$$\left(1 + \frac{1}{n}\right) p_{j,k}{}^*(n+1) - \frac{1}{n} p_{j,k}(1) = \sum_{i=0}^{\infty} p_{j,i}{}^*(n) \, p_{i,k}. \tag{8.24}$$

Taking the limit of Eq. 8.24 as n tends to ∞, one obtains

$$\pi_k = \lim_{n \to \infty} \sum_{i=0}^{\infty} p_{j,i}{}^*(n) \, p_{i,k} \ge \sum_{i=0}^{\infty} \{ \lim_{n \to \infty} p_{j,i}{}^*(n) \} \, p_{i,k} = \sum_{i=0}^{\infty} \pi_i \, p_{i,k}. \tag{8.25}$$

To prove that Eq 8.15 holds we argue by contradiction. If Eq. 8.15 does not hold, it follows from Eq. 8.25 that for some k

$$\pi_k > \sum_{i=0}^{\infty} \pi_i \, p_{i,k}. \tag{8.26}$$

From Eqs. 8.25 and 8.26 one obtains that

$$\sum_{k=0}^{\infty} \pi_k > \sum_{k=0}^{\infty} \sum_{i=0}^{\infty} \pi_i \, p_{i,k} = \sum_{i=0}^{\infty} \pi_i \sum_{k=0}^{\infty} p_{i,k} = \sum_{i=0}^{\infty} \pi_i, \tag{8.27}$$

which is impossible. The proof of Eq. 8.15 is complete.

To prove Eq. 8.18, let us first show that if Eq. 8.17 holds, then

$$u_k = \sum_{j=0}^{\infty} u_j \, p_{j,k}(n) \quad \text{for} \quad n = 1, 2, \cdots. \tag{8.28}$$

To prove Eq. 8.28 we use the principle of mathematical induction. By hypothesis, Eq. 8.28 holds for $n = 1$. Next, let us show that if Eq. 8.28 holds for $n - 1$ then it holds for n. Now,

$$\sum_{j=0}^{\infty} u_j \, p_{j,k}(n) = \sum_{j=0}^{\infty} u_j \sum_{h=0}^{\infty} p_{j,h} \, p_{h,k}(n - 1)$$

$$= \sum_{h=0}^{\infty} \left(\sum_{j=0}^{\infty} u_j p_{j,h} \right) p_{h,k}(n - 1)$$

$$= \sum_{h=0}^{\infty} u_h p_{h,k}(n - 1) = u_k.$$

The proof of Eq. 8.28 is now complete. From Eq. 8.28 it follows that

$$u_k = \sum_{j=0}^{\infty} u_j \, p_{j,k}^*(n) \quad \text{for} \quad n = 1, 2, \cdots. \tag{8.29}$$

Now, let n tend to ∞ in Eq. 8.29; by the dominated convergence theorem (see Appendix to this chapter) one obtains Eq. 8.18.

We are now in a position to determine conditions under which an irreducible Markov chain possesses a stationary distribution.

THEOREM 8C

Let C be the state space of an irreducible Markov chain. If C is finite, then the Markov chain possesses a unique stationary distribution. If C consists of infinitely many states, then in order that the Markov chain possess a unique stationary distribution it is necessary and sufficient that the sequence $\{\pi_k, k \in C\}$ satisfying Eq. 8.11 have the property that

$$\pi_k > 0 \text{ for some } k \text{ in } C. \tag{8.30}$$

In order that Eq. 8.30 hold it is necessary and sufficient that there exist an absolutely convergent sequence $\{u_k, k \,\epsilon\, C\}$, not identically equal to zero, satisfying Eq. 8.17. If Eq. 8.30 holds, then

$$\pi_k > 0 \text{ for every } k \text{ in } C. \tag{8.31}$$

Proof. From Theorem 8B it follows that, for every k in C, the sequence $\{\pi_k, k \,\epsilon\, C\}$ satisfies

$$\pi_k = \pi_k \Big(\sum_{j \,\epsilon\, C} \pi_j \Big). \tag{8.32}$$

From Eqs. 8.30 and 8.32 it follows that the sequence $\{\pi_k, k \,\epsilon\, C\}$ satisfies Eq. 8.13. Next, a convergent sequence $\{u_k\}$, not identically equal to zero, which satisfies Eq. 8.17 also satisfies Eq. 8.18, which implies that Eq. 8.30 holds. To prove Eq. 8.31, let j be a state in C. Choose M and N so that $p_{j,k}(N) > 0$ and $p_{k,j}(M) > 0$. By Eq. 5.2

$$\pi_j \geq p_{j,k}(N) \; \pi_k \, p_{k,j}(M) > 0.$$

THEOREM 8D

Let C be the state space of an irreducible Markov chain. The following statements are equivalent:

(i) The Markov chain possesses a stationary distribution.

(ii) C is positive recurrent (that is, every state in C is recurrent and possesses a finite mean recurrence time).

(iii) There exists an absolutely convergent sequence $\{\pi_k, k \,\epsilon\, C\}$, not identically zero, satisfying Eq. 8.17.

Proof. To prove the theorem it suffices to prove the equivalence of (i) and (ii). If the Markov chain possesses a stationary distribution, then the sequence $\{\pi_k, k \,\epsilon\, C\}$ satisfying Eq. 8.11 satisfies Eq. 8.31. Clearly, C is recurrent, since $\pi_k = 0$ if k is non-recurrent. If k is recurrent,

$$\pi_k = \frac{1}{m_{k,k}}. \tag{8.33}$$

Consequently, Eq. 8.31 implies that

$$m_{k,k} < \infty \quad \text{for every } k \text{ in } C. \tag{8.34}$$

Conversely, if C is recurrent and Eq. 8.34 holds, then from Eq. 8.33 it follows that Eq. 8.31 holds.

Remark. From Theorem 8C we obtain a proof of the assertion made in the last section that if any state in an irreducible chain is positive recurrent, then they are all positive recurrent. If, for some k in C, k is recurrent and $m_{k,k} < \infty$, then from Eq. 8.33 it follows that Eq. 8.30 holds. Therefore, Eq. 8.31 holds and again from Eq. 8.33 it follows that Eq. 8.34 holds.

COROLLARY

In terms of the solution $\{u_k, k \in C\}$ of Eq. 8.17 the mean recurrence times $\{m_{k,k}, k \in C\}$ are given by

$$m_{k,k} = \frac{1}{u_k}\left(\sum_{j \in C} u_j\right). \tag{8.35}$$

Proof. By Eq. 8.18

$$\frac{1}{m_{k,k}} = \pi_k = \frac{u_k}{\sum_{j \in C} u_j}.$$

An alternative proof of Eq. 8.35, which is rigorously valid only for finite Markov chains, is as follows. From Theorem 7B it follows that the mean first passage times $m_{j,k}$ satisfy

$$m_{j,k} = 1 + \sum_{i \neq k} p_{j,i}\, m_{i,k}$$
$$= 1 + \sum_{i \in C} p_{j,i}\, m_{i,k} - p_{j,k}\, m_{k,k}.$$

Now, multiply by u_j and sum over j in C:

$$\sum_{j \in C} u_j m_{j,k} = \sum_{j \in C} u_j + \sum_{j \in C} u_j \sum_{i \in C} p_{j,i}\, m_{i,k} - m_{k,k}\sum_{j \in C} u_j p_{j,k}. \tag{8.36}$$

From Eqs. 8.36 and 8.17 it follows that

$$\sum_{j \in C} u_j m_{j,k} = \sum_{j \in C} u_j + \sum_{i \in C} u_i m_{i,k} - m_{k,k}\, u_k. \tag{8.37}$$

From Eq. 8.37 one obtains Eq. 8.35, since in a finite Markov chain $\sum_{j \in C} u_j m_{j,k}$ is finite and may be subtracted from both sides of Eq. 8.37.

EXAMPLE 8A

Positive recurrent random walks. The general random walk on the state space $\{0, 1, 2, \cdots\}$ and transition probability matrix given by Eq. 6.10 is an irreducible Markov chain. It was shown (in Example 6C) to be non-recurrent if and only if Eq. 6.29 holds. To obtain conditions under which it is positive recurrent, we need to examine the conditions under which there exists an absolutely convergent solution to Eq. 8.17. For a general random walk, the system of equations Eq. 8.17 may be written

$$u_0 = u_0\, r_0 + u_1\, q_1,$$
$$u_j = u_{j-1}\, p_{j-1} + u_j\, r_j + u_{j+1}\, q_{j+1} \qquad \text{for } j \geq 1. \tag{8.38}$$

Since for $j \geq 0$, $r_j = 1 - p_j - q_j$ and $r_0 = 1 - p_0$ one may rewrite these equations

$$u_1\, q_1 - u_0\, p_0 = 0,$$
$$u_{j+1}\, q_{j+1} - u_j\, p_j = u_j\, q_j - u_{j-1}\, p_{j-1} \qquad \text{for } j \geq 1. \tag{8.39}$$

From Eq. 8.39 it follows that for any $k \geq 0$

$$u_{k+1}\, q_{k+1} = u_k\, p_k, \tag{8.40}$$

since

$$u_{j+1}\, q_{j+1} - u_j\, p_j = u_j\, q_j - u_{j-1}\, p_{j-1} = \cdots = u_1\, q_1 - u_0\, p_0 = 0.$$

The general solution $\{u_k,\, k = 0, 1, \cdots\}$ of Eq. 8.38 is, therefore, given up to an undetermined constant u_0 by

$$u_k = u_0\, \frac{p_0 \cdots p_{k-1}}{q_1 \cdots q_k}. \tag{8.41}$$

In order for the sequence $\{u_k\}$ defined by Eq. 8.41 to be absolutely convergent it is necessary and sufficient that

$$\sum_{k=1}^{\infty} \frac{p_0 \cdots p_{k-1}}{q_1 \cdots q_k} < \infty. \tag{8.42}$$

Thus Eq. 8.42 is a necessary and sufficient condition that the general random walk be positive recurrent. In the case of a random walk of repeated trials, so that $p_k = p$, $q_k = q$, Eq. 8.42 holds if and only if $p < q$. The stationary distribution $\{\pi_k,\, k = 0, 1, \cdots\}$ is given by

$$\pi_k = \frac{u_k}{\displaystyle\sum_{k=0}^{\infty} u_k} = \frac{u_0 (p/q)^k}{\displaystyle\sum_{k=0}^{\infty} u_0 (p/q)^k} \tag{8.43}$$

$$= \frac{q - p}{q} \left(\frac{p}{q}\right)^k.$$

From Eqs. 8.43 and 8.33 it follows that the mean recurrence times are given by

$$m_{k,k} = \frac{1}{\pi_k} = \frac{q}{q - p} \left(\frac{q}{p}\right)^k \qquad \text{if } p < q. \tag{8.44}$$

Although Eq. 8.44 was previously obtained in Example 7C, the present derivation is much simpler.

Doubly stochastic transition probability matrices. For certain irreducible Markov chains, it is possible to determine by inspection the solution $\{\pi_k, k \in C\}$ of Eq. 8.15. A transition probability matrix $P = \{p_{j,k}\}$ is said to be *doubly* stochastic if the sum over any *column* equals 1:

$$\sum_j p_{j,k} = 1 \quad \text{for all } k. \tag{8.45}$$

For an irreducible Markov chain with doubly stochastic transition probability matrix and whose state space C is finite and contains K states, the solution of Eq. 8.15 is given by

$$\pi_k = \frac{1}{K}, \quad k \in C. \tag{8.46}$$

To prove this assertion one need only verify that the probability distribution on C defined by Eq. 8.46 satisfies Eq. 8.15. Consequently, the mean recurrence times are given by

$$m_{k,k} = K, \quad k \in C. \tag{8.47}$$

EXAMPLE 8B
For the Markov chain with transition probability matrix

$$P = \begin{bmatrix} 0.3 & 0.4 & 0.3 \\ 0.4 & 0.3 & 0.3 \\ 0.3 & 0.3 & 0.4 \end{bmatrix},$$

$$\pi_k = \tfrac{1}{3}, \ m_{k,k} = 3 \qquad \text{for all } k.$$

Periodic and aperiodic states. In order to state conditions under which an irreducible Markov chain possesses a long-run distribution, we need to introduce the notion of the *period* of a state.

The period $d(k)$ of a return state k of a Markov chain is defined to be the greatest common divisor of the set of integers n for which $p_{k,k}(n) > 0$; in symbols

$$d(k) = g.c.d. \ \{n : p_{k,k}(n) > 0\}. \tag{8.48}$$

A state is said to be *aperiodic* if it has period 1.

The period $d(k)$ of a state is the largest integer such that any integer n which has the property that it is possible to return to k in exactly n steps is necessarily a multiple of $d(k)$. It is natural to ask: does every integer n which is a multiple of $d(k)$ have the property that $p_{k,k}(n) > 0$? In Complement 8D it is shown that to every state k there is an integer M_k such that $p_{k,k}(n) > 0$ for every integer n which is a multiple of $d(k)$ and is greater than M_k.

In Complement 8C, it is shown that two states which communicate have the same period. Consequently, we define the *period of a communicating class* to be the common period of the states in the class.

In Section 6–10, the following theorem is proved.

THEOREM 8E

For an irreducible aperiodic Markov chain, with state space C, Eq. 8.5 holds for every k in C. Consequently, there is a sequence $\{\pi_k, k \in C\}$ such that Eq. 8.7 holds for every j and k in C.

It is easy to give an example which shows that Eq. 8.5 does not hold for a periodic Markov chain. Consider a homogeneous Markov chain with state space $\{0, 1\}$ and transition probability matrix

$$P = \begin{bmatrix} 0 & 1 \\ 1 & 0 \end{bmatrix}.$$

It is easily verified that the chain is irreducible and each state has period 2. Further,

$$\begin{aligned} p_{0,0}(n) &= 1 && \text{if } n \text{ is even} \\ &= 0 && \text{if } n \text{ is odd}, \end{aligned}$$

so that Eq. 8.5 does not hold for $k = 0$. The transition probability matrix P is doubly stochastic so that Eq. 8.11 holds with $\pi_0 = \pi_1 = \frac{1}{2}$.

Using Theorems 8B, 8C, and 8E one obtains the following extremely important theorem, which plays a central role in applications of Markov chains.

THEOREM 8F

An irreducible aperiodic positive recurrent Markov chain possesses a unique long-run distribution. A necessary and sufficient condition for an irreducible aperiodic Markov chain to possess a long-run distribution is that there exist a convergent sequence $\{u_k, k \in C\}$ not identically equal to zero, satisfying Eq. 8.17. The long-run distribution $\{\pi_k, k \in C\}$ of an irreducible aperiodic positive recurrent Markov chain is the unique solution of the system of equations 8.15 satisfying Eq. 8.13.

There are several ways in which an irreducible Markov chain can be proved to be aperiodic. One way is to exhibit a state k for which $p_{k,k} > 0$; such a state is clearly aperiodic. Another way, available for finite Markov chains, is to exhibit an integer n such that $p_{j,k}(n) > 0$ for all j and k in C. If such an n exists, it usually can be quickly found by successively computing the transition probability matrices P^2, P^4, P^8, $\cdots P^{2^n}$, $\cdots$.

EXAMPLE 8C

Social mobility. A problem of interest in sociology is the following: to what extent does the social class of the father, grandfather, etc., affect the social class of the son? One method of determining a person's class is by his occupation. One may then determine the probability that a son will have an upper, middle, or lower class occupation given that his father has an upper, middle, or lower class occupation. One study of social mobility (for references and further discussion see Prais [1955]) reported the following table of conditional probabilities:

Son's social class

		Upper	Middle	Lower
Father's	Upper	.448	.484	.068
social class	Middle	.054	.699	.247
	Lower	.011	.503	.486

Now suppose that transitions between social classes of the successive generations in a family can be regarded as transitions of a Markov chain (with states 1, 2, 3, respectively representing upper, middle, lower class) with transition probability matrix

$$P = \begin{bmatrix} .448 & .484 & .068 \\ .054 & .699 & .247 \\ .011 & .503 & .486 \end{bmatrix}. \tag{8.49}$$

This chain is finite, irreducible, and aperiodic. It therefore possesses a long-run distribution (π_1, π_2, π_3) which is the unique solution of the system of equations (written in matrix form)

$$(\pi_1, \pi_2, \pi_3)P = (\pi_1, \pi_2, \pi_3). \tag{8.50}$$

For P given by Eq. 8.49

$$\begin{pmatrix} \pi_1 \\ \pi_2 \\ \pi_3 \end{pmatrix} = \begin{pmatrix} .067 \\ .624 \\ .309 \end{pmatrix}. \tag{8.51}$$

In words one may interpret Eq. 8.51 as follows: a society in which social mobility between classes is a Markov chain with transition probability matrix P given by Eq. 8.49 would, after many generations, consist of the following proportions in each social class: 6.7% in the upper class, 62.4% in the middle class, 30.9% in the lower class.

We have now completed our survey of the basic theory of discrete parameter Markov chains. In Table 6.5 is summarized the procedure to be followed in classifying an irreducible Markov chain. In Table 6.6 the basic properties of general random walks are summarized. In Table 6.7 we list some criteria available for classifying irreducible Markov chains. For the proofs of these theorems, the reader is referred to Foster (1953). Example 8D illustrates the use of these criteria.

TABLE 6.5. **Classification of an irreducible Markov chain**

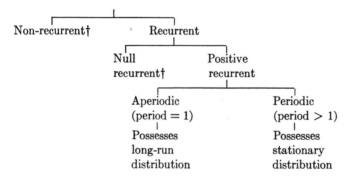

†A property which can be possessed only by an infinite Markov chain.

EXAMPLE 8D

Stationary probabilities for the imbedded Markov chain of an $M/G/1$ **queue.** One important application of the theory of stationary and long-run distributions of irreducible Markov chains is to queueing theory. Consider the imbedded Markov chain of an $M/G/1$ queue. Its transition probability matrix is given in Example 2B. Let

$$\rho = \sum_{n=0}^{\infty} n \, a_n$$

be the mean number of customers that arrive during the service time of a customer. It may be shown that

$$\rho = \lambda \, E[S] = \frac{\text{mean service time}}{\text{mean inter-arrival time}}.$$

TABLE 6.6. **Properties of general random walk on {0, 1, 2, $\cdots$} with transition probability matrix satisfying Eqs. 6.13 and 6.15**

Define $\rho_0 = 1, \rho_m = \dfrac{q_1 \cdots q_m}{p_1 \cdots p_m}, P_m = \dfrac{p_0 \cdots p_{m-1}}{q_1 \cdots q_m} = \dfrac{p_0}{p_m \rho_m}$ for $m \geq 1$.

Random walk is	if and only if		Proved in
Recurrent	$\displaystyle\sum_{m=1}^{\infty} \rho_m = \infty$		Example 6B
Non-recurrent	$\displaystyle\sum_{m=1}^{\infty} \rho_m < \infty$		Example 6B
Positive recurrent	$\displaystyle\sum_{m=1}^{\infty} P_m < \infty$ and	$\displaystyle\sum_{m=1}^{\infty} \rho_m = \infty$	Example 8A
Null recurrent	$\displaystyle\sum_{m=1}^{\infty} \rho_m = \infty$ and	$\displaystyle\sum_{m=1}^{\infty} \dfrac{1}{p_m \rho_m} = \infty$	Example 8A

First passage probabilities in non-recurrent case (Example 6A):

$$f_{j,j} = 1 - \left\{ \frac{p_j \rho_j}{\displaystyle\sum_{m=j}^{\infty} \rho_m} \right\},$$

$$f_{j,k} = \frac{\displaystyle\sum_{m=j}^{\infty} \rho_m}{\displaystyle\sum_{m=k}^{\infty} \rho_m}, \; j < k.$$

Mean first passage times in recurrent case (Examples 7A and 7C):

$$m_{0,0} = 1 + \sum_{m=1}^{\infty} \frac{p_0}{p_m \rho_m} = 1 + \sum_{m=1}^{\infty} P_m;$$

$$m_{k,k} = m_{0,0} \frac{1}{P_k}, \; k > 1;$$

$$m_{j,k} = \sum_{s=j}^{k-1} \rho_s \left\{ 1 + \sum_{m=1}^{s} \frac{1}{p_m \rho_m} \right\}, \; j < k,$$

$$m_{j,k} = \sum_{s=k}^{j-1} \rho_s \sum_{m=s+1}^{\infty} \frac{1}{p_m \rho_m}, \; k > j.$$

TABLE 6.7. **Some criteria for classifying irreducible Markov chains with state space $\{0, 1, 2, \cdots\}$**

(1) Positive recurrent if and only if	$u_i = \sum\limits_{j=0}^{\infty} u_j \, p_{j,i}, \quad i = 0, 1, \cdots$ has solution $\{u_i\}$ which is absolutely convergent and not identically equal to 0.
(2) Positive recurrent if and only if	$\sum\limits_{j=0}^{\infty} p_{i,j} \, x_j \leq x_i - 1, \quad i \neq 0$ has non-negative solution $\{x_i\}$ satisfying $\sum\limits_{j=0}^{\infty} p_{0,j} \, x_j < \infty.$
(3) Non-recurrent if and only if	$\sum\limits_{j=1}^{\infty} p_{i,j} \, y_j = y_i, \quad i \neq 0,$ has bounded non-constant solution $\{y_i\}$.
(4) Recurrent if	$\sum\limits_{j=0}^{\infty} p_{i,j} \, y_j \leq y_i, \qquad i \neq 0,$ has unbounded solution $\{y_i\}$; that is, $y_i \to \infty$ as $i \to \infty$.

Let us show that the imbedded Markov chain $\{X_n\}$ is

    (i) positive recurrent   if $\rho < 1$,
    (ii) null recurrent   if $\rho = 1$,
    (iii) non-recurrent   if $\rho > 1$.

Consider first the case that $\rho \leq 1$. Define a sequence $y_j = j$. Now,

$$\sum_{j=0}^{\infty} p_{i,j} \, y_j = \sum_{j=i-1}^{\infty} a_{j-i+1} \, j = i - 1 + \sum_{k=0}^{\infty} k \, a_k$$
$$= i - 1 + \rho = y_i - (1 - \rho) \leq y_i.$$

By criterion (4) of Table 6.7, the chain is recurrent if $\rho \leq 1$. From the above calculation it also follows that, if $\rho < 1$,

$$\sum_{j=0}^{\infty} p_{i,j} \left(\frac{j}{1-\rho} \right) \le \frac{i}{1-\rho} - 1.$$

Since also [defining $x_j = j/(1-\rho)$]

$$(1-\rho) \sum_{j=0}^{\infty} p_{0,j} \, x_j = \sum_{j=0}^{\infty} a_j \, j < \infty,$$

it follows by criterion (2) that the chain is positive recurrent if $\rho < 1$. To prove that the chain is non-recurrent if $\rho > 1$, we use criterion (3). Let $y_j = z^j$ where z is a number, satisfying $0 < z < 1$, to be determined. Then, for $i > 0$,

$$\sum_{j=0}^{\infty} p_{i,j} \, y_j = \sum_{j=i}^{\infty} a_{j-i+1} \, z^j = z^{i-1} \sum_{k=0}^{\infty} a_k \, z^k.$$

The sequence $y_j = z^j$ therefore satisfies the equations

$$\sum_{j=0}^{\infty} p_{i,j} \, y_j = y_i, \quad i \ne 0,$$

if z satisfies

$$A(z) = \sum_{k=0}^{\infty} a_k \, z^k = z. \tag{8.52}$$

By the fundamental theorem of branching processes there exists a number z in the interval $0 < z < 1$ satisfying Eq. 8.52 if $\rho > 1$.

We next determine the long-run distribution $\{\pi_k, \, k = 0, 1, 2, \cdots \}$ which exists in the case that $\rho < 1$ by solving the system of equations

$$\sum_{j=0}^{\infty} \pi_j \, p_{j,k} = \pi_k, \quad k = 0, 1, \cdots,$$

which becomes

$$\pi_k = \pi_0 \, a_k + \sum_{j=1}^{k+1} \pi_j \, a_{k-j+1}, \quad k = 0, 1, 2, \cdots. \tag{8.53}$$

One may solve these equations by means of the generating functions

$$\Pi(z) = \sum_{j=0}^{\infty} \pi_j \, z^j, \quad A(z) = \sum_{j=0}^{\infty} a_j \, z^j.$$

By multiplying both sides of Eq. 8.53 by z^k and summing over k one obtains

$$\Pi(z) = \pi_0 \, A(z) + \frac{1}{z} \{ A(z)\Pi(z) - \pi_0 \, a_0 \} - \frac{\pi_0}{z} \{ A(z) - a_0 \},$$

which may be solved for $\Pi(z)$:

$$\Pi(z) = \pi_0 \frac{(z-1)A(z)}{z - A(z)}.$$ (8.54)

The generating function of the long-run distribution is determined by Eq. 8.54 up to a constant factor π_0. To determine π_0, we take the limit as z tends to 1 of both sides of Eq. 8.54. Now, as $z \to 1$,

$$\frac{z - A(z)}{z-1} = 1 - \frac{1 - A(z)}{1-z} \to 1 - A'(1) = 1 - \rho,$$

$$A(z) \to 1, \quad \Pi(z) \to 1.$$

Therefore,

$$1 = \frac{\pi_0}{1 - \rho},$$

$$\pi_0 = 1 - \rho.$$

Queueing models are further discussed in Section 7–2.

COMPLEMENTS

8A Consider an irreducible Markov chain C with infinitely many states whose transition probability matrix P is *doubly stochastic*; that is for all i and j in C

$$\sum_{i \in C} p_{i,j} = \sum_{j \in C} p_{i,j} = 1.$$

Show that C is not positive recurrent.

8B Show that an irreducible finite Markov chain is positive recurrent.

Hint. Show that $\sum_{k \in C} \pi_k = 1$, and hence Eq. 8.30 holds.

8C Show that two communicating states have the same period.
Hint. Show that it suffices to prove that if $j \to k$ then $d(j)$ divides $d(k)$. Choose M and N so that $p_{j,k}(M)p_{k,j}(N) > 0$. Show that if $p_{j,j}(n) > 0$ then $p_{k,k}(M + n + N) > 0$ and $p_{k,k}(M + 2n + N) > 0$. Consequently, $d(k)$ divides $M + n + N$ and $M + 2n + N$, and $d(k)$ divides any n for which $p_{j,j}(n) > 0$. Therefore, $d(k)$ divides $d(j)$, since $d(j)$ is the greatest common divisor of the set of n for which $p_{j,j}(n) > 0$.

8D Show that to each state k there is an integer M_k such that for all $m \geq M_k$

$$p_{k,k}(m\, d(k)) > 0.$$

by showing that the following lemma holds.

LEMMA

If A is a set of positive integers closed under addition, and d is the greatest common divisor of A, then there exists a positive integer M such that for all integers $m \geq M$, the positive integers md belong to A.

Hint. Form the set S of positive integers which are finite linear combinations

$$b_1 n_1 + b_2 n_2 + \cdots + b_k n_k,$$

where $n_1 \cdots, n_k$ belong to A, and $b_1, \cdots, b_k$ are positive or negative integers. Denote by d' the minimum positive integer in S. Show that d' is a common divisor of all integers in A (if it were not, then there would be an integer n in A and a positive integer k such that $n - kd'$ is a positive integer smaller than d', which is impossible since $n - kd'$ belongs to S, and d' is the smallest positive integer in S). Now, d' may be expressed as

$$d' = a_1 n_1 + a_2 n_2 + \cdots + a_r n_r \tag{1}$$

for some integers $n_1, \cdots, n_r$ in A and positive or negative integers $a_1, \cdots, a_r$. It then follows that d' is equal to the greatest common divisor d of all members of A, since by Eq. 1 any number d'' which divides all integers in A divides d'. Now, rearrange the terms in Eq. 1 so that the terms with positive coefficients are written first. Since A is closed under addition it follows that d may be written

$$d = N_1 - N_2$$

for some integers N_1 and N_2 in A. Let M be a positive integer satisfying $M \geq (N_2)^2 / d$. Now for any integer $m \geq M$, the integer md may be written $md = aN_2 + bd$ where a and b are non-negative integers satisfying $a \geq N_2$ and $b < N_2$. Consequently, md belongs to A.

8E Show that if j and k communicate, and if $p_{j,k}(n) > 0$ and $p_{j,k}(n') > 0$ then $d(k)$ divides $\mid n - n' \mid$. In words, a state can be reached with positive probability only at times which differ by multiples of the period of the state.

8F *Decomposition of an irreducible periodic communicating class into subclasses.* Show that an irreducible Markov chain with period d may be split into d disjoint subclasses $C(0), C(1), \cdots, C(d-1)$, ordered so that one may pass in one step from a state of $C(r)$ only to a state of $C(r+1)$, where we define $r + 1 = 0$ if $r = d - 1$. Show that if we consider the chain only at times $d, 2d, \cdots$, then we get a new chain, whose transition probability matrix is P^d, in which each subclass $C(r)$ forms a closed aperiodic class.
Hint. Choose a state j in C. Show that to every k in C there exists an integer $r(k)$ such that $0 \leq r(k) \leq d - 1$, and $p_{j,k}(n) > 0$ implies $n \equiv r(k)$ (mod d). Define $C(r)$ to be the set of all states k such that $r(k) = r$.

EXERCISES

In Exercises 8.1 to 8.10, a Markov chain is described by its state space S and transition probability matrix.

(i) Find T, the set of non-recurrent states.
(ii) For all states j in S and k not in T, find $f_{j,k}$
(iii) For all states j in T, find the mean time m_j to absorption in the set of recurrent states, given that the chain started in j.

(iv) For all recurrent states j and k which communicate find the mean first passage time $m_{j,k}$.

(v) For all states j and k in S, find

$$\lim_{n \to \infty} \frac{1}{n} \sum_{m=1}^{n} p_{j,k}(m).$$

(vi) For all states j and k in S, state whether or not the limit

$$\lim_{n \to \infty} p_{j,k}(n) \text{ exists.}$$

In Exercises 8.1 to 8.8, the state space $S = \{0, 1, 2, 3, 4\}$.

8.1 $P = \begin{bmatrix} 1 & 0 & 0 & 0 & 0 \\ \frac{1}{3} & 0 & \frac{2}{3} & 0 & 0 \\ 0 & \frac{1}{3} & 0 & \frac{2}{3} & 0 \\ 0 & 0 & \frac{1}{3} & 0 & \frac{2}{3} \\ 0 & 0 & 0 & 0 & 1 \end{bmatrix}$
  $\quad$ **8.2** $P = \begin{bmatrix} 1 & 0 & 0 & 0 & 0 \\ \frac{1}{6} & \frac{1}{2} & \frac{1}{3} & 0 & 0 \\ 0 & \frac{1}{6} & \frac{1}{2} & \frac{1}{3} & 0 \\ 0 & 0 & \frac{1}{6} & \frac{1}{2} & \frac{1}{3} \\ 0 & 0 & 0 & 0 & 1 \end{bmatrix}$

8.3 $P = \begin{bmatrix} 1 & 0 & 0 & 0 & 0 \\ \frac{1}{3} & 0 & \frac{2}{3} & 0 & 0 \\ 0 & \frac{1}{3} & 0 & \frac{2}{3} & 0 \\ 0 & 0 & \frac{1}{3} & 0 & \frac{2}{3} \\ 0 & 0 & 0 & \frac{1}{3} & \frac{2}{3} \end{bmatrix}$
  $\quad$ **8.4** $P = \begin{bmatrix} \frac{1}{3} & \frac{2}{3} & 0 & 0 & 0 \\ \frac{1}{3} & 0 & \frac{2}{3} & 0 & 0 \\ 0 & \frac{1}{3} & 0 & \frac{2}{3} & 0 \\ 0 & 0 & \frac{1}{3} & 0 & \frac{2}{3} \\ 0 & 0 & 0 & 0 & 1 \end{bmatrix}$

8.5 $P = \begin{bmatrix} 1 & 0 & 0 & 0 & 0 \\ \frac{1}{2} & 0 & \frac{1}{2} & 0 & 0 \\ 0 & \frac{1}{4} & \frac{1}{2} & \frac{1}{4} & 0 \\ 0 & 0 & \frac{1}{8} & \frac{3}{4} & \frac{1}{8} \\ 0 & 0 & 0 & 0 & 1 \end{bmatrix}$
  $\quad$ **8.6** $P = \begin{bmatrix} \frac{1}{2} & \frac{1}{2} & 0 & 0 & 0 \\ \frac{1}{2} & 0 & \frac{1}{2} & 0 & 0 \\ 0 & \frac{1}{4} & \frac{1}{2} & \frac{1}{4} & 0 \\ 0 & 0 & \frac{1}{8} & \frac{3}{4} & \frac{1}{8} \\ 0 & 0 & 0 & 0 & 1 \end{bmatrix}$

8.7 $P = \begin{bmatrix} \frac{1}{3} & \frac{2}{3} & 0 & 0 & 0 \\ \frac{1}{3} & 0 & \frac{2}{3} & 0 & 0 \\ 0 & \frac{1}{3} & 0 & \frac{2}{3} & 0 \\ 0 & 0 & \frac{1}{3} & 0 & \frac{2}{3} \\ 0 & 0 & 0 & \frac{1}{3} & \frac{2}{3} \end{bmatrix}$
  $\quad$ **8.8** $P = \begin{bmatrix} 0 & \frac{2}{3} & 0 & 0 & \frac{1}{3} \\ \frac{1}{3} & 0 & \frac{2}{3} & 0 & 0 \\ 0 & \frac{1}{3} & 0 & \frac{2}{3} & 0 \\ 0 & 0 & \frac{1}{3} & 0 & \frac{2}{3} \\ \frac{2}{3} & 0 & 0 & \frac{1}{3} & 0 \end{bmatrix}$

8.9 $S = \{1, 2, 3, 4, 5, 6\}$
  $\qquad\qquad$ **8.10** $S = \{0, 1, 2, 3, 4, 5, 6\}$

$$P = \begin{bmatrix} 1 & 0 & 0 & 0 & 0 & 0 \\ 0 & 1 & 0 & 0 & 0 & 0 \\ \frac{1}{4} & 0 & \frac{1}{2} & 0 & 0 & \frac{1}{4} \\ 0 & \frac{1}{4} & 0 & \frac{1}{2} & 0 & \frac{1}{4} \\ 0 & 0 & 0 & 0 & 0 & 1 \\ \frac{1}{16} & \frac{1}{16} & \frac{1}{4} & \frac{1}{4} & \frac{1}{8} & \frac{1}{4} \end{bmatrix}$$

$$P = \begin{bmatrix} \frac{1}{3} & \frac{1}{3} & \frac{1}{3} & 0 & 0 & 0 & 0 \\ \frac{1}{3} & \frac{1}{3} & \frac{1}{3} & 0 & 0 & 0 & 0 \\ \frac{1}{3} & \frac{1}{3} & \frac{1}{3} & 0 & 0 & 0 & 0 \\ 0 & \frac{1}{4} & 0 & \frac{1}{2} & 0 & 0 & \frac{1}{4} \\ 0 & 0 & \frac{1}{4} & 0 & \frac{1}{2} & 0 & \frac{1}{4} \\ 0 & 0 & 0 & 0 & 0 & 0 & 1 \\ 0 & \frac{1}{16} & \frac{1}{16} & \frac{1}{4} & \frac{1}{4} & \frac{1}{8} & \frac{1}{4} \end{bmatrix}$$

In Exercises 8.11–8.13 consider an irreducible Markov chain with state space $S = \{0, 1, \cdots\}$. State whether the chain is positive recurrent, null recurrent, or non-recurrent. If it is positive recurrent, find its stationary distribution and state whether or not it possesses a long-run distribution.

8.11 The chain has transition probabilities, for $k = 0, 1, \cdots$,

$$p_{k,0} = \frac{k+1}{k+2} \quad \text{and} \quad p_{k,k+1} = \frac{1}{k+2}.$$

8.12 The chain has transition probabilities, for $k = 0, 1, \cdots$,

$$p_{k,0} = \frac{1}{k+2} \quad \text{and} \quad p_{k,k+1} = \frac{k+1}{k+2}.$$

8.13 $\{X_n\}$ is the imbedded Markov chain of a $GI/M/1$ queue and

$$\frac{\text{mean service time}}{\text{mean inter-arrival time}} < 1.$$

Hint. Look for a solution of the form $\pi_j = c \, x^j$.

**6-9 LIMIT THEOREMS
FOR OCCUPATION TIMES**

The stationary distribution $\{\pi_k, \, k \in C\}$ of an irreducible positive recurrent Markov chain with state space C may also be interpreted as the limit, as n tends to infinity, of the average occupation time of the state k in the first n transitions; more precisely,

$$P\left[\lim_{n \to \infty} \frac{1}{n} N_k(n) = \pi_k \mid X_0 = j\right] = 1 \tag{9.1}$$

where $\pi_k = 1/m_{k,k}$. It is not surprising that Eq. 9.1 holds since intuitively the probability π_k (that the chain is in state k) represents the relative frequency of trials on which the chain is in state k, which is represented by $\lim_{n \to \infty} (1/n) N_k(n)$. It should be noted that Eq. 9.1 implies that

$$E\left[\frac{1}{n} N_k(n) \mid X_0 = j\right] = \frac{1}{n} \sum_{n=1}^{n} p_{j,k}(n) \to \pi_k, \tag{9.2}$$

as $n \to \infty$, so that by proving Eq. 9.1 one obtains a probabilistic proof of Eq. 8.4.

In order to prove that Eq 9.1 holds, let us consider an irreducible positive recurrent Markov chain which at time 0 is in state j, and consider a sequence of random variables $T_1, T_2, \cdots$ defined as follows: T_1 is the first passage time from j to k (that is, T_1 is the number of transitions until the chain visits k for the first time), T_2 is the recurrence time to return to k (that is, T_2 is the number of transitions the chain makes from the time it visits k for the first time until the time it visits k for the second time), and, for $n > 1$, T_n is the number of transitions the chain makes from the time it visits k for the $(n - 1)$st time until the time it visits k for the nth time.

Alternately, the sequence $\{T_n\}$ may be defined in terms of the occupation times $N_k(n)$ as follows. For $n = 1, 2, \cdots$, let

$$W_n = \text{minimum of the set of } w \geq 1 \text{ such that } N_k(w) = n. \quad (9.3)$$

In words, W_n is the number of transitions the chain makes until it visits state k for the nth time. We call W_n the *waiting time* until the nth visit to k. It is easily verified that

$$\begin{aligned} T_1 &= W_1, \\ T_n &= W_n - W_{n-1} \qquad \text{for } n \geq 2. \end{aligned} \quad (9.4)$$

We call T_n the inter-arrival time between the $(n - 1)$st and nth visits to k.

THEOREM 9A

In an irreducible positive recurrent Markov chain, the sequence of inter-arrival times $T_1, T_2, \cdots$ between visits to k are independent random variables. The random variables $T_2, T_3, \cdots$ are identically distributed with distribution given in terms of the first passage probabilities of k,

$$P[T_n = t] = f_{k,k}(t), \qquad t = 1, 2, \cdots. \quad (9.5)$$

Given that at time 0 the chain was in state j

$$P[T_1 = t] = f_{j,k}(t), \qquad t = 1, 2, \cdots. \quad (9.6)$$

Proof. That Eq. 9.6 holds follows from the fact that since $X_0 = j$

$$P[T_1 = t] = P[X_t = k, X_\nu \neq k \text{ for } \nu = 1, \cdots, t - 1 \mid X_0 = j] = f_{j,k}(t).$$

To prove Eq. 9.5, we first note that for $n = 1, 2, \cdots$

$$\begin{aligned} P[T_{n+1} = t \mid W_n = w] = P[X_{w+t} = k, \\ X_{w+\nu} \neq k, \nu = 1, \cdots, t - 1 \mid X_w = k] = f_{k,k}(t). \quad (9.7) \end{aligned}$$

Consequently,

$$P[T_{n+1} = t] = \sum_{w=1}^{\infty} P[T_{n+1} = t \mid W_n = w] \, P[W_n = w] = f_{k,k}(t).$$

To prove that for any integer n, $T_1, T_2, \cdots T_n$ are independent, we show that (for any integers $t_1, \cdots, t_n$)

$$P[T_1 = t_1, \ T_2 = t_2, \ \cdots, \ T_n = t_n] = P[T_1 = t_1] \ P[T_2 = t_2] \ \cdots P[T_n = t_n].$$
$$(9.8)$$

The left-hand side of Eq. 9.8 is equal to

$$P[T_1 = t_1] \ P[T_2 = t_2 \mid T_1 = t_1] \ \cdots P[T_n = t_n \mid T_1 = t_1, \ \cdots, \ T_{n-1} = t_{n-1}].$$

For $\nu = 1, 2, \cdots$ it follows by Eq. 9.7 that

$$P[T_{\nu+1} = t_{\nu+1} \mid T_1 = t_1, \ \cdots, \ T_\nu = t_\nu] = P[T_{\nu+1} = t_{\nu+1} \mid W_\nu = t_1 + \cdots + t_\nu]$$
$$= f_{k,k}(t_{\nu+1}) = P[T_{\nu+1} = t_{\nu+1}].$$

The proof of Eq. 9.8 is complete.

Since the waiting time W_n until the nth visit to k may be represented as the sum of n independent random variables $T_1, \cdots, T_n$, it follows from the classical limit theorems of probability theory that one may obtain limit theorems for the sequence of waiting times $\{W_n\}$.

THEOREM 9B

Let k be a recurrent state with finite mean recurrence time $m_{k,k}$. Let j be a state which communicates with k. Let W_n be defined by Eq. 9.3. Then

$$P\left[\lim_{n \to \infty} \frac{1}{n} W_n = m_{k,k} \mid X_0 = j\right] = 1. \qquad (9.9)$$

Further, if the recurrence time of k has a finite second moment $m_{k,k}^{(2)}$, and therefore, a finite variance

$$\sigma_{k,k}^2 = m_{k,k}^{(2)} - \{m_{k,k}\}^2, \qquad (9.10)$$

then W_n is asymptotically normal in the sense that for any real number x

$$\lim_{n \to \infty} P\left[\frac{W_n - n \, m_{k,k}}{\sqrt{n} \, \sigma_{k,k}} \leq x\right] = \Phi(x) = \frac{1}{\sqrt{2\pi}} \int_{-\infty}^{x} e^{-(1/2)v^2} \, dy. \qquad (9.11)$$

It is beyond the scope of this book to prove Eqs. 9.9 and 9.11. They are immediate consequences of the Strong Law of Large Numbers and the Central Limit Theorem for the consecutive sums of independent identically distributed random variables. For an elementary discussion of these theorems, see *Mod Prob*, Chapter 10. For a more advanced discussion see Loève (1960).

From Eq. 9.9 one may obtain Eq. 9.1 by using the following basic relationship between the sequence $\{W_n\}$ of waiting times and the sequence $\{N_k(n)\}$ of occupation times:

$$N_k(w) < n \quad \text{if and only if } W_n > w \qquad (9.12)$$

for any integers n and w. To prove Eq. 9.12 use the definitions of the concepts involved.

Proof of Eq. 9.1: Let us write $[x]$ to denote the largest integer smaller than or equal to x, and write m to denote $m_{k,k}$. Then

$$\frac{N_k(w)}{w} - \frac{1}{m} < \epsilon$$

if

$$N_k(w) < \left[w \left\{ \epsilon + \frac{1}{m} \right\} \right]$$

if

$$W_{[w\{ \epsilon + 1/m \}]} > w$$

if

$$\frac{1}{\left[w \left\{ \epsilon + \frac{1}{m} \right\} \right]} W_{[w\{ \epsilon + 1/m \}]} - m > \frac{w}{\left[w \left\{ \epsilon + \frac{1}{m} \right\} \right]} - m$$

$$> \frac{w}{w \left\{ \epsilon + \frac{1}{m} \right\}} - m = \frac{-\epsilon}{\epsilon + \frac{1}{m}}.$$

From Eq. 9.9 it follows that for a given $\epsilon > 0$ and $\delta > 0$, the last of these assertions holds for all w greater than some integer N (depending on ϵ and $\delta > 0$) with probability exceeding $1 - \delta$ (see *Mod Prob*, p. 416). Consequently, to each $\epsilon > 0$ and $\delta > 0$ there exists an N such that

$$P\left[\frac{1}{w} N_k(w) - \frac{1}{m} < \epsilon \text{ for all } w > N \right] > 1 - \delta.$$

Similarly, one may show that to each ϵ and δ there exists an N such that

$$P\left[\frac{1}{w} N_k(w) - \frac{1}{m} > - \epsilon \text{ for all } w > N \right] > 1 - \delta.$$

The proof of Eq. 9.1 is now complete.

In view of Eq. 9.1, the average occupation time $(1/n)N_k(n)$ provides an estimate of π_k which is consistent. In order to proceed further with questions of statistical inference on Markov chains, it is necessary to know the distribution of $N_k(n)$.

THEOREM 9C

Asymptotic normality of occupation times. Let k be a recurrent state whose recurrence time has finite mean $m_{k,k}$ and finite variance $\sigma_{k,k}{}^2$. Then for every real number x

$$\lim_{n \to \infty} P\left[\frac{N_k(n) - (n/m_{k,k})}{\sqrt{n\{\sigma_{k,k}^2/(m_{k,k})^3\}}} \le x\right] = \Phi(x). \tag{9.13}$$

In words Eq. 9.13 says that $N_k(n)$ approximately obeys a normal probability law with mean

$$E[N_k(n)] = n\frac{1}{m_{k,k}} \tag{9.14}$$

and variance

$$\text{Var}[N_k(n)] = n\frac{\sigma_{k,k}^2}{(m_{k,k})^3}. \tag{9.15}$$

It may be shown that Eqs. 9.14 and 9.15 hold precisely in the sense that

$$\lim_{n \to \infty} \frac{1}{n} E[N_k(n)] = \frac{1}{m_{k,k}}, \tag{9.16}$$

$$\lim_{n \to \infty} \frac{1}{n} \text{Var}[N_k(n)] = \frac{\sigma_{k,k}^2}{(m_{k,k})^3}. \tag{9.17}$$

Proof of Eq. 9.13. From Eq. 9.12 it follows that

$$P[N_k(w) < n] = P[W_n > w]. \tag{9.18}$$

Let $m = m_{k,k}$, $\sigma^2 = \sigma_{k,k}^2$ and let

$$n(w) = \left[\frac{w}{m} + x\sqrt{\frac{\sigma^2 w}{m^3}}\right]$$

where x is a fixed real number.

It may be verified that

$$\lim_{w \to \infty} \frac{n(w) - (w/m)}{\sqrt{\sigma^2 w/m^3}} = x, \quad \lim_{w \to \infty} \frac{w - mn(w)}{\sqrt{\sigma^2 n(w)}} = -x.$$

From the fact that the sequence of waiting times $\{W_n\}$ satisfy the Central Limit Theorem it follows that

$$\lim_{w \to \infty} P\left[\frac{W_{n(w)} - mn(w)}{\sqrt{\sigma^2 n(w)}} > \frac{w - mn(w)}{\sqrt{\sigma^2 n(w)}}\right]$$

$$= \lim_{n \to \infty} P\left[\frac{W_n - nm}{\sqrt{n\sigma^2}} > -x\right] = 1 - \Phi(-x) = \Phi(x). \tag{9.19}$$

From Eqs. 9.18 and 9.19 it follows that

$$\lim_{w \to \infty} P\left[\frac{N_k(w) - (w/m)}{\sqrt{w\sigma^2/m^3}} < x\right] = \lim_{w \to \infty} P\left[\frac{N_k(w) - (w/m)}{\sqrt{w\sigma^2/m^3}} < \frac{n(w) - (w/m)}{\sqrt{w\sigma^2/m^3}}\right]$$

$$= \lim_{w \to \infty} P\left[\frac{W_{n(w)} - mn(w)}{\sqrt{\sigma^2 n(w)}} > \frac{w - mn(w)}{\sqrt{\sigma^2 n(w)}}\right]$$

$$= \Phi(x).$$

It should be pointed out that the results of this section are special cases of limit theorems for renewal counting processes (see Section 5–3).

6-10 LIMIT THEOREMS FOR TRANSITION PROBABILITIES OF A FINITE MARKOV CHAIN

In this section we show that for an irreducible aperiodic recurrent Markov chain with finite state space C

$$\lim_{n \to \infty} p_{j,k}(n) = \frac{1}{m_{k,k}} \quad \text{for all } j, k \text{ in } C. \tag{10.1}$$

For the proof of this assertion in the case that C is infinite, see Feller (1957), p. 306 or Chung, (1960), p. 27.

In order to prove Eq. 10.1 in the case of a finite Markov chain it suffices to prove that for each k in C there is a real number π_k such that

$$\lim_{n \to \infty} p_{j,k}(n) = \pi_k. \tag{10.2}$$

By the alternative proof of Eq. 8.35, it follows that the limit π_k is necessarily equal to $1/m_{k,k}$. To prove Eq. 10.2, we proceed as follows. Let

$$M_k(n) = \max_{j \in C} p_{j,k}(n), \quad m_k(n) = \min_{j \in C} p_{j,k}(n) \tag{10.3}$$

denote respectively the largest and smallest entries in the kth column of the transition probability matrix $P(n)$. Since

$$p_{j,k}(n+1) = \sum_{i \in C} p_{j,i} \, p_{i,k}(n) \le M_k(n) \sum_{i \in C} p_{j,i} = M_k(n), \tag{10.4}$$

it follows that

$$M_k(n+1) \le M_k(n), \quad m_k(n+1) \ge m_k(n), \tag{10.5}$$

so that the sequence $\{M_k(n)\}$ is monotone non-increasing while the sequence $\{m_k(n)\}$ is monotone non-decreasing. Because these sequences are monotone it follows that there exist real numbers M_k and m_k such that

$$\lim_{n \to \infty} M_k(n) = M_k, \quad \lim_{n \to \infty} m_k(n) = m_k. \tag{10.6}$$

If it is shown that $m_k = M_k$ then it follows that Eq. 10.2 holds, with $\pi_k = m_k = M_k$.

To prove that $M_k = m_k$, let us consider the sequence of differences

$$d_k(n) = M_k(n) - m_k(n)$$

and show that

$$\lim_{n \to \infty} d_k(n) = 0. \tag{10.7}$$

From Eq. 10.5 it follows that

$$0 \le d_k(n+1) \le d_k(n),$$

so that $\{d_k(n)\}$ is a monotone non-increasing sequence. In order to prove that Eq. 10.7 holds, it suffices to show that there is a subsequence $\{d_k(nN), n = 1, 2, \cdots\}$ which satisfies

$$\lim_{n \to \infty} d_k(nN) = 0; \tag{10.8}$$

here N is a fixed integer.

We now use the fact that C is a finite irreducible aperiodic chain. It then follows that there exists an integer N such that for all states j and k,

$$p_{j,k}(N) > 0.$$

Let

$$c = \min_{j,k} p_{j,k}(N).$$

By assumption, $0 < c < \frac{1}{2}$. We shall show that for any $n = 1, 2, \cdots$

$$d_k((n+1)N) \le (1 - 2c)d_k(nN). \tag{10.9}$$

From Eq. 10.9 it follows that

$$d_k(nN) \le (1 - 2c)^n d_k(N), \tag{10.9'}$$

which tends to 0 as n tends to ∞.

We now prove Eq. 10.9. Let n be an integer. Then for any state i

$$p_{i,k}((n+1)N) = \sum_{r \in C} p_{i,r}(N)\, p_{r,k}(nN). \tag{10.10}$$

In particular, choose i so that

$$p_{i,k}((n+1)N) = M_k((n+1)N),$$

Next, choose q so that

$$p_{q,k}(nN) = m_k(nN).$$

Then from Eq. 10.10 it follows that

$$M_k((n+1)N) = p_{i,q}(N)p_{q,k}(nN) + \sum_{r \neq q} p_{i,r}(N) \, p_{r,k}(nN)$$

$$\leq p_{i,q}(N) \, m_k(nN) + M_k(nN) \sum_{r \neq q} p_{i,r}(N)$$

$$= p_{i,q}(N) \, m_k(nN) + M_k(nN) \, \{1 - p_{i,q}(N)\}$$

$$= M_k(nN) - \{M_k(nN) - m_k(nN)\} \, p_{i,q}(N)$$

$$\leq M_k(nN) - \{M_k(nN) - m_k(nN)\} \, c.$$

We have thus shown that

$$M_k((n+1)N) \leq M_k(nN) - \{M_k(nN) - m_k(nN)\} \, c. \qquad (10.11)$$

By choosing i so that $p_{i,k}((n+1)N) = m_k((n+1)N)$ and choosing q so that $p_{q,k}(nN) = M_k(nN)$ one may similarly show that

$$m_k((n+1)N) \geq m_k(nN) + \{M_k(nN) - m_k(nN)\}c. \qquad (10.12)$$

Subtracting Eq. 10.12 from Eq. 10.11, it follows that

$$M_k((n+1)N) - m_k((n+1)N) \leq \{M_k(nN) - m_k(nN)\} \{1 - 2c\},$$

which proves Eq. 10.9. The proof of Eq. 10.2 is now complete.

Geometric ergodicity. Having determined that the transition probabilities possess limits, it is next of interest to determine the rate of convergence.

An aperiodic irreducible Markov chain is said to be *geometrically ergodic* if for each pair of states j and k there exist numbers $M_{j,k}$ and $\rho_{j,k}$ such that

$$0 \leq M_{j,k} < \infty, 0 \leq \rho_{j,k} < 1 \qquad (10.13)$$

and for $n = 1, 2, \cdots$

$$| \, p_{j,k}(n) - \pi_k \, | \leq M_{j,k}(\rho_{j,k})^n. \qquad (10.14)$$

Conditions for geometric ergodicity have been extensively investigated by Kendall (1959), who shows that Eq. 10.14 holds if and only if for some state k in the Markov chain and for some finite real number $c > 1$

$$\sum_{n=1}^{\infty} c^n f_{k,k}(n) < \infty. \qquad (10.15)$$

Kendall (1960) shows how using Eq. 10.15 one may investigate the geometric ergodicity of the imbedded Markov chains of various queueing models.

A proof of Eq. 10.15 is beyond the scope of this book. However, let us show that a *finite* aperiodic irreducible Markov chain is geometrically ergodic. Since by definition

$$m_k(n) \leq p_{j,k}(n) \leq M_k(n)$$

and by Eq. 10.5

$$m_k(n) \leq \pi_k \leq M_k(n)$$

it follows that

$$|\, p_{j,k}(n) - \pi_k \,| \leq M_k(n) - m_k(n).$$

From Eq. 10.9' it follows that

$$M_k(n) - m_k(n) \leq M_0 \rho^n$$

where

$$\rho = (1 - 2c)^{1/N}$$
$$M_0 = d_k(N)(1 - 2c)^{-1}.$$

The proof that a finite aperiodic irreducible Markov chain is geometrically ergodic is now complete.

APPENDIX:
THE INTERCHANGE OF LIMITING PROCESSES

The mathematical theory of probability is founded on the modern theory of integration. This book seeks to develop some of the major concepts and techniques of the theory of stochastic processes without requiring that the reader possess an advanced mathematical background. This does not mean that only heuristic arguments are employed. On the contrary, mathematically sound proofs are given for many of the theorems stated in this book. In order to justify explicitly some of the steps one must appeal to various basic theorems of modern integration theory, especially the dominated convergence theorem, Fatou's lemma, and Fubini's theorem. In this appendix these theorems are stated for infinite series since this is the form used in this chapter. It has been the author's experience that even students familiar with these theorems from a study of modern integration theory achieve a better appreciation of them by seeing them stated for infinite series rather than for integrals over abstract measure spaces.

For $n = 1, 2, \cdots$ let $a_n(t)$ be a function on a discrete set T. One may re-label the members of T, so that we may assume that $T = \{1, 2, \cdots\}$.

The dominated convergence theorem. Suppose that

$$\lim_{n \to \infty} a_n(t) \quad \text{exists for each } t \text{ in } T. \tag{1}$$

Let $B(t)$ be a function on T such that

$$| a_n(t) | \leq B(t) \quad \text{for each } t \text{ in } T \text{ and } n = 1, 2, \cdots, \tag{2}$$

$$\sum_{t \in T} B(t) < \infty; \tag{3}$$

then

$$\lim_{n \to \infty} \sum_{t \in T} a_n(t) = \sum_{t \in T} \left\{ \lim_{n \to \infty} a_n(t) \right\}. \tag{4}$$

Proof. Let $a(t) = \lim_{n \to \infty} a_n(t)$. From Eq. 2 it follows that $| a(t) | \leq B(t)$. Therefore, $\sum_{t=1}^{\infty} a(t)$ converges. Now, for any integer M

$$\left| \sum_{t=1}^{\infty} a_n(t) - \sum_{t=1}^{\infty} a(t) \right| \leq \sum_{t=1}^{M} | a_n(t) - a(t) | + \sum_{t=M+1}^{\infty} \{ | a_n(t) | + | a(t) | \}.$$

Further,

$$\lim_{n \to \infty} \sum_{t=1}^{M} | a_n(t) - a(t) | = \sum_{t=1}^{M} \{ \lim_{n \to \infty} | a_n(t) - a(t) | \} = 0,$$

$$\sum_{t=M+1}^{\infty} \{ | a_n(t) | + | a(t) | \} \leq 2 \sum_{t=M+1}^{\infty} B(t).$$

Therefore, for any integer M

$$\lim_{n \to \infty} \sup \left| \sum_{t=1}^{\infty} a_n(t) - \sum_{t=1}^{\infty} a(t) \right| \leq 2 \sum_{t=M+1}^{\infty} B(t). \tag{5}$$

The right-hand side of Eq. 5 is the remainder term of a convergent series, and therefore tends to 0 as M tends to ∞. Therefore, taking the limit of Eq. 5 as M tends to ∞ it follows that

$$\lim_{n \to \infty} \sup \left| \sum_{t=1}^{\infty} a_n(t) - \sum_{t=1}^{\infty} a(t) \right| = 0.$$

The proof of Eq. 4 is now complete.

In the case that a dominating function $B(t)$ is not available, one may still be able to replace Eq. 4 by an inequality.

Fatou's lemma. Suppose that Eq. 1 holds, and that

$$a_n(t) \geq 0 \qquad \text{for all } t \text{ in } T \text{ and } n = 1, 2, \cdots; \tag{6}$$

then

$$\sum_{t \in T} \{\lim_{n \to \infty} a_n(t)\} \leq \lim_{n \to \infty} \sum_{t \in T} a_n(t). \tag{7}$$

Proof. For any integer M

$$\sum_{t=1}^{M} \lim_{n \to \infty} a_n(t) = \lim_{n \to \infty} \sum_{t=1}^{M} a_n(t) \leq \lim_{n \to \infty} \sum_{t=1}^{\infty} a_n(t). \tag{8}$$

Taking the limit of Eq. 8 as M tends to ∞ one obtains Eq. 7.

We next state (without proof) a theorem, often referred to as Fubini's Theorem, concerned with the conditions under which one may interchange orders of summation.

Fubini's theorem. Let $a_n(t)$ be a function defined for $n = 1, 2, \cdots$ and $t = 1, 2, \cdots$. In order that

$$\sum_{n=1}^{\infty} \sum_{t=1}^{\infty} a_n(t) = \sum_{t=1}^{\infty} \sum_{n=1}^{\infty} a_n(t) \tag{9}$$

it is sufficient that at least one of the following conditions be satisfied:

(i) $a_n(t) \geq 0$ for all n and t,

(ii) $\sum_{n=1}^{\infty} \sum_{t=1}^{\infty} |a_n(t)| < \infty,$

(iii) $\sum_{t=1}^{\infty} \sum_{n=1}^{\infty} |a_n(t)| < \infty.$

Markov chains:
continuous parameter

THIS CHAPTER discusses the basic ideas and applications of the theory of continuous parameter Markov chains with emphasis on birth and death processes.

7-1 LIMIT THEOREMS FOR TRANSITION PROBABILITIES OF A CONTINUOUS PARAMETER MARKOV CHAIN

Let $\{N(t), t \geq 0\}$ be a Markov chain with state space S and homogeneous transition probability function,

$$p_{j,k}(t) = P[N(t+s) = k \mid N(s) = j], \qquad (1.1)$$

assumed to be continuous at $t = 0$,

$$\lim_{t \to 0} p_{j,k}(t) = \delta_{j,k} = \begin{cases} 1 & \text{if} \quad j = k \\ 0 & \text{if} \quad j \neq k. \end{cases} \qquad (1.2)$$

The transition probability function satisfies the Chapman-Kolmogorov equation: for any states j and k, and positive numbers s and t

$$p_{j,k}(s+t) = \sum_{\text{states } h} p_{j,h}(s) p_{h,k}(t). \qquad (1.3)$$

A pair of states j and k are said to *communicate* if there are times t_1 and t_2 such that $p_{j,k}(t_1) > 0$ and $p_{k,j}(t_2) > 0$. The Markov chain is irreducible if all pairs of states in the chain communicate.

It may be shown that for each pair of states j and k, $p_{j,k}(t)$ is uniformly continuous as a function of $t > 0$, and further is either always zero or is always positive. Consequently, in an irreducible Markov chain, $p_{j,k}(t) > 0$ for all $t > 0$ and all states j and k.

In an irreducible Markov chain with homogeneous transition probability function $p_{j,k}(t)$ the limits

$$\lim_{t \to \infty} p_{j,k}(t) = \pi_k, \quad k \in S, \tag{1.4}$$

always exist and are independent of the initial state of the chain. For a finite Markov chain Eq. 1.4 may be proved by the methods used in Section 6–10. For the proof in general, the reader is referred to Chung (1960), p. 178.

As in the case of discrete parameter irreducible Markov chains, the limits $\{\pi_k, \ k \in S\}$ either vanish identically,

$$\pi_k = 0 \qquad \text{for all } k \text{ in } S, \tag{1.5}$$

or are all positive and form a probability distribution,

$$\pi_k > 0 \quad \text{for all } k \text{ in } S, \quad \sum_{k \in S} \pi_k = 1. \tag{1.6}$$

If Eq. 1.6 holds, the irreducible chain is said to be *positive recurrent*. It should be noted that if Eq. 1.6 holds then $\{\pi_k, k \in S\}$ is both a long-run distribution and a stationary distribution for the Markov chain (as defined in Section 6–8).

For a discrete parameter Markov chain with one-step transition probabilities $\{p_{j,k}\}$, in order to determine whether an irreducible Markov chain possesses a long-run (or a stationary) distribution, it suffices to determine whether or not the system of equations

$$\pi_k = \sum_{j \in S} \pi_j \, p_{j,k}, \quad k \in S \tag{1.7}$$

possesses an absolutely convergent non-null solution. If such a solution exists, then (normalized so as to sum to one) it is the long-run distribution.

For a continuous parameter Markov chain, the role of the one-step transition probabilities is played by the *transition intensities* which are defined in terms of the derivatives at 0 of the transition probability functions. Assume that for every state k

$$q_k = -\frac{d}{dt} p_{k,k}(0) = \lim_{t \to 0} \frac{1 - p_{k,k}(t)}{t} \tag{1.8}$$

exists and is finite, while for every pair of states j and k, $j \neq k$,

$$q_{j,k} = \frac{d}{dt} p_{j,k}(0) = \lim_{t \to 0} \frac{p_{j,k}(t)}{t} \qquad (1.9)$$

exists and is finite.

These limits have the following probabilistic interpretation. The probabilities of transition within a time interval of length h are asymptotically proportional to h; the probability $1 - p_{j,j}(h)$ of a transition from a state j to some other state during a time interval of length h is equal to hq_j plus a remainder which, divided by h, tends to 0 (as $h \to 0$), while the probability $p_{j,k}(h)$ of transition from j to k during a time interval of length h is equal to $hq_{j,k}$ plus a remainder which, divided by h, tends to 0 (as $h \to 0$).

We call q_j the *intensity of passage*, given that the Markov chain is in state j. We call $q_{j,k}$ the *intensity of transition* to k, given that the Markov chain is in state j.

Now let s tend to ∞ in the Chapman-Kolmogorov equation 1.3. It follows that the sequence $\{\pi_k, k \in S\}$ satisfies for every $t \ge 0$

$$\pi_k = \sum_{\text{states } h} \pi_h \, p_{h,k}(t), \quad k \in S. \qquad (1.10)$$

If one differentiates Eq. 1.10 with respect to t, and if one formally interchanges the operations of summation and differentiation (which can be justified by sufficiently strong assumptions on the rate of convergence of the limits Eqs. 1.8 and 1.9) it follows that $\{\pi_k, k \in S\}$ satisfies the system of linear equations

$$\pi_k \, q_k = \sum_{h \neq k} \pi_h \, q_{h,k}, \quad k \in S. \qquad (1.11)$$

7-2   BIRTH AND DEATH PROCESSES AND THEIR APPLICATION TO QUEUEING THEORY

A continuous parameter Markov chain $\{N(t), t \ge 0\}$ with state space $\{0, 1, 2, \cdots\}$ and homogeneous transition probabilities is called a *birth and death process* if its transition intensities satisfy the following conditions: if j and k are states such that $|j - k| \ge 2$, then

$$q_{j,k} = 0. \qquad (2.1)$$

In words, a birth and death process is a continuous parameter Markov chain which changes only through transitions from a state to its immediate neighbors. For a birth and death process, it is convenient to introduce quantities λ_j and μ_j defined as follows:

$$\lambda_j = q_{j,j+1} \quad \text{for} \quad j \geq 0,$$
$$\mu_j = q_{j,j-1} \quad \text{for} \quad j \geq 1, \quad (2.2)$$
$$\lambda_j + \mu_j = q_j \quad \text{for} \quad j \geq 0,$$

where we define

$$\mu_0 = 0.$$

More explicitly,

$$\lim_{h \to 0} \frac{p_{n,n+1}(h)}{h} = \lambda_n \qquad \text{for} \quad n \geq 0,$$

$$\lim_{h \to 0} \frac{p_{n,n-1}(h)}{h} = \mu_n \qquad \text{for} \quad n \geq 1, \quad (2.3)$$

$$\lim_{t \to 0} \frac{1 - p_{n,n}(h)}{h} = \lambda_n + \mu_n \quad \text{for} \quad n \geq 0.$$

In words, these equations state that in a very small time interval the population size [represented by $N(t)$] either increases by one, decreases by one, or stays the same. The conditional probability of an increase by 1 (a "birth") may depend on the population size n, and is denoted by λ_n. The conditional probability of a decrease by 1 (a "death") may depend on the population size n, and is denoted by μ_n.

For a birth and death process, the system of equations 1.11 becomes

$$\lambda_0 \pi_0 = \mu_1 \pi_1, \quad (2.4)$$
$$(\lambda_n + \mu_n)\pi_n = \lambda_{n-1} \pi_{n-1} + \mu_{n+1} \pi_{n+1}, \qquad n \geq 1. \quad (2.5)$$

From these equations one may obtain a recursive relationship which may be used to obtain the sequence $\{\pi_n\}$:

$$\mu_n \pi_n = \lambda_{n-1} \pi_{n-1} \qquad \text{for} \quad n \geq 1. \quad (2.6)$$

To prove Eq. 2.6, define $\alpha_n = \mu_n \pi_n - \lambda_{n-1} \pi_{n-1}$. From Eq. 2.5 it follows that $\alpha_n = \alpha_{n+1}$ for $n \geq 1$. Therefore $\alpha_1 = \alpha_2 = \cdots$. But from Eq. 2.4 it follows that $\alpha_1 = 0$. Therefore $\alpha_n = 0$ for $n = 1, 2, \cdots$ and the proof of Eq. 2.6 is complete.

One may give a heuristic explanation of Eq. 2.6. For a small positive quantity h, and for large values of t

$$\pi_{n-1} \lambda_{n-1} h = P[N(t+h) = n \mid N(t) = n - 1] \, P[N(t) = n - 1]$$
$$= P[N(t+h) = n \text{ and } N(t) = n - 1], \quad (2.7)$$
$$\pi_n \mu_n h = P[N(t+h) = n - 1 \mid N(t) = n] \, P[N(t) = n]$$
$$= P[N(t+h) = n - 1 \text{ and } N(t) = n]. \quad (2.8)$$

It seems reasonable that if the size $N(t)$ of a population is in statistical equilibrium then the probability of an increase by 1 in population size during a small time interval should be equal to the probability of a decrease by 1. Consequently, the right-hand sides of Eqs. 2.7 and 2.8 are equal, for all $n \geq 1$, from which one obtains Eq. 2.6.

In the case that $\mu_n > 0$ for all $n \geq 1$, one may obtain from Eq. 2.6 an explicit expression for π_n:

$$\pi_n = \frac{\lambda_{n-1}}{\mu_n} \frac{\lambda_{n-2}}{\mu_{n-1}} \cdots \frac{\lambda_0}{\mu_1} \pi_0, \quad n \geq 1. \tag{2.9}$$

To determine π_0 one uses the normalization condition

$$1 = \pi_0 + \pi_1 + \pi_2 + \cdots = \pi_0 \left\{ 1 + \frac{\lambda_0}{\mu_1} + \frac{\lambda_1 \lambda_0}{\mu_2 \mu_1} + \cdots \right\}. \tag{2.10}$$

It may be shown (see, for example, Karlin and McGregor [1957]) that the birth and death process possesses a long-run distribution if the infinite series in Eq. 2.10 is convergent.

One can give a plausibility argument for the last assertion by approximating a birth and death process by a random walk. Suppose one observes a continuous parameter Markov process $\{N(t), t \geq 0\}$ at a sequence of discrete instants of time, a distance h apart, so that one observes a sequence

$$X_n = N(nh), \quad n = 0, 1, 2, \cdots. \tag{2.11}$$

Then X_n is a discrete parameter Markov chain. If $\{N(t), t \geq 0\}$ is a birth and death process we may regard $\{X_n\}$ as a random walk with transition probability matrix given by Eq. 6.10 of Chapter 6 where $p_0 = \lambda_0 h$, $r_0 = 1 - \lambda_0 h$, and for $n \geq 1$

$$q_n = \mu_n h, \; p_n = \lambda_n h, \; r_n = 1 - (\lambda_n + \mu_n)h. \tag{2.12}$$

Let $\rho_0 = 1$ and, for $m \geq 1$, let

$$\rho_m = \frac{q_1 \cdots q_m}{p_1 \cdots p_m} = \frac{\mu_1 \cdots \mu_m}{\lambda_1 \cdots \lambda_m}.$$

According to Table 6.6, the random walk is positive recurrent if and only if

$$\sum_{m=1}^{\infty} \frac{1}{p_m \rho_m} < \infty,$$

which is equivalent to

$$\sum_{m=1}^{\infty} \frac{1}{p_m \rho_m} = \frac{1}{h} \left\{ \frac{1}{\mu_1} + \frac{\lambda_1}{\mu_1 \mu_2} + \cdots \right\} < \infty,$$

which holds if the infinite series in Eq. 2.10 is convergent.

EXAMPLE 2A

Telephone traffic problems. Consider a telephone exchange. Suppose that subscribers make calls at the instants $\tau_1, \tau_2, \cdots, \tau_n, \cdots$ where $0 < \tau_1 < \tau_2 < \cdots < \infty$. It is often reasonable to assume that the successive inter-arrival times $T_1 = \tau_1$, $T_2 = \tau_2 - \tau_1$, $T_n = \tau_n - \tau_{n-1}, \cdots$, are independent exponentially distributed random variables with mean $1/\lambda$. The duration of the conversation initiated (and therefore the length of time that a channel is held) by the call arriving at time τ_n is a random variable, denoted by S_n, and called the *holding time* (or service time) of the nth call. The successive service times (or durations) $S_1, S_2, \cdots$ may often be assumed to be independent random variables exponentially distributed with mean $1/\mu$ (in a study of the duration of 1837 local telephone calls in New Jersey, Molina [1927] found their durations to be well fitted by an exponential distribution).

An incoming call is connected (gives rise to a conversation) if there is a free channel to handle it. The number of channels available is either a finite number M or is ∞. In the case that the telephone exchange possesses only a finite number of channels, two assumptions are possible concerning the policy of calls which arrive when all channels are busy: either they form a waiting system (that is, each new call joins a waiting line and waits until a channel is free) or they do not form a waiting system (a new call leaves immediately if all channels are busy). The assumption of an infinite number of channels, while unrealistic, is worth considering since it provides insight into the correct design of a telephone exchange with a finite number of channels.

Let $N(t)$ denote the number of channels busy at time t in a telephone exchange with an infinite number of channels. In the language of queueing theory, the set of busy channels may be considered a queue, and $N(t)$ is then the length of the queue at time t. Under the assumptions made, $\{N(t), t \geq 0\}$ is a birth and death process with transition intensities (for $n = 0, 1, \cdots$)

$$\lambda_n = \lambda, \quad \mu_n = n\mu, \tag{2.13}$$

where $1/\lambda$ is the mean time between customer arrivals, and $1/\mu$ is the mean service time of a customer. In words, the probability for exactly one channel to join the queue in a small time interval of length h is the same, no matter how many are already in the queue, and is approximately equal to λh. On the other hand, the probability for exactly one channel in the queue to leave, given that there are n channels in the queue, is approximately equal to $n\mu h$. The probability for the size of the queue to change by more than one in a small time interval of length h is negligible compared to h.

We can only sketch proofs of these assertions. The assertion that $\{N(t), t \geq 0\}$ is a Markov chain seems difficult to prove rigorously. However, it is intuitively clear because of the property of the exponential distribution that it is without memory of the past. Thus if we regard the queue at a given time, what has gone on up to that time will have no effect on the probability law of future arrivals and departures from the queue.

The assertion that the transition intensities of the Markov chain $\{N(t), t \geq 0\}$ are of birth and death type and satisfy Eq. 2.13 can be proved by an argument along the following lines. Let I be an interval of length h, and let $o(h)$ denote a quantity depending on h which, divided by h, tends to 0 as h tends to 0. Then

$$p_{n,n+1}(h) = P \text{ [a new call arrives during } I] + o(h)$$
$$= \lambda h + o(h),$$
$$p_{n,n-1}(h) = P \text{ [exactly one of } n \text{ conversations in}$$
$$\text{progress terminates during } I] + o(h)$$
$$= n(1 - e^{-\mu h})e^{-(n-1)\mu h} + o(h)$$
$$= n\mu h + o(h).$$

From these expressions one may infer Eq. 2.13.

In view of Eq. 2.9, the long-run probabilities are given (for $n \geq 1$) by

$$\pi_n = \lim_{t \to \infty} P[N(t) = n] = \frac{1}{n!} \left(\frac{\lambda}{\mu}\right)^n \pi_0. \tag{2.14}$$

We determine π_0 by the condition

$$1 = \sum_{k=0}^{\infty} \pi_k = \pi_0 \sum_{n=0}^{\infty} \frac{1}{n!} \left(\frac{\lambda}{\mu}\right)^n = \pi_0 e^{(\lambda/\mu)}. \tag{2.15}$$

Therefore,

$$\pi_n = e^{-\rho} \frac{\rho^n}{n!} \tag{2.16}$$

in which

$$\rho = \text{mean length of queue} = \frac{\lambda}{\mu} = \frac{\text{mean service time}}{\text{mean inter-arrival time}}. \tag{2.17}$$

In words, $N(t)$ is, in the long run, Poisson distributed with mean given by Eq. 2.17. It was shown in Section 4–5 that this result remains true for a general service time distribution and exponentially distributed inter-arrival times.

EXAMPLE 2B

Erlang's loss formula. Let $N(t)$ be the number of channels busy in a telephone exchange with a finite number M of channels, in which cus-tomers do not wait for service (that is, if all channels are busy then an in-coming call is lost). Under the assumptions made in Example 2A it follows that $\{N(t), t \geq 0\}$ is a birth and death process with transition intensities

$$\lambda_n = \lambda, \quad n = 0, 1, 2, \cdots, M - 1,$$
$$= 0, \quad n \geq M;$$
$$\mu_n = n\mu, \quad n = 1, 2, \cdots, M,$$
$$= 0, \quad n > M.$$

It follows that the long-run probability distribution of $N(t)$ is given by a censored Poisson distribution: for $n = 0, 1, 2, \cdots, M$

$$\pi_n = \lim_{t \to \infty} P[N(t) = n] = \frac{e^{-\rho} \dfrac{\rho^n}{n!}}{\displaystyle\sum_{n=0}^{M} e^{-\rho} \dfrac{\rho^n}{n!}} = \frac{\dfrac{\rho^n}{n!}}{\displaystyle\sum_{n=0}^{M} \left\{ \dfrac{\rho^n}{n!} \right\}} \tag{2.19}$$

in which ρ is given by Eq. 2.17. The formula 2.19 is called Erlang's loss formula after A. K. Erlang who was one of the pioneer workers in queueing theory (see Brockmeyer, Halstroem, Jensen [1948]). It has been extended by Sevastyanov (1957).

EXAMPLE 2C

Stationary probabilities for a finite server queue. As a final appli-cation of the limit theorems for the transition probabilities of a birth and death process, we shall determine the stationary probabilities for queue length and for waiting time for service in a queue with a finite number M of servers, in which customers are served in the order in which they arrive, and customers who arrive when all servers are busy wait for service. Sup-pose that successive inter-arrival times between customers are independent and exponentially distributed with mean $1/\lambda$, and that successive service times are independent and exponentially distributed with mean $1/\mu$. Let $N(t)$ denote the number of customers either being served or waiting for service. It may be shown that $\{N(t), t \geq 0\}$ is a birth and death process with

$$\lambda_n = \lambda, \quad n = 0, 1, \cdots$$

and

$$\mu_n = n\mu, \quad n = 0, 1, \cdots, M - 1$$
$$= M\mu, \quad n \geq M.$$

The stationary probabilities $\pi_n = \lim_{t \to \infty} P[N(t) = n]$ are given (if they exist) by

$$\pi_n = \pi_0 \frac{1}{n!}\left(\frac{\lambda}{\mu}\right)^n, \quad n \leq M$$

$$= \pi_0 \frac{M^M}{M!}\left(\frac{\lambda}{M\mu}\right)^n, \quad n \geq M.$$

We determine π_0 by the condition

$$1 = \sum_{k=0}^{\infty} \pi_k = \pi_0 \left(\sum_{k=0}^{M-1} \frac{1}{n!}\left(\frac{\lambda}{\mu}\right)^n + \frac{M^M}{M!} \sum_{n=M}^{\infty} \left(\frac{\lambda}{M\mu}\right)^n \right). \tag{2.20}$$

Now the infinite series in Eq. 2.20 converges if and only if

$$\rho = \frac{\lambda}{M\mu} < 1. \tag{2.21}$$

If Eq. 2.21 does not hold, a stationary distribution $\{\pi_n\}$ cannot exist. In this case, $\pi_n = 0$ for all n, which means that the waiting line becomes arbitrarily large. If Eq. 2.21 does hold, then the long-run distribution $\{\pi_n\}$ exists and is given by

$$\pi_0 = \frac{1}{\displaystyle\sum_{n=0}^{M-1} \frac{(M\rho)^n}{n!} + \frac{(M\rho)^M}{M!(1 - \rho)}},$$

$$\pi_n = \pi_0 \frac{(M\rho)^n}{n!}, \quad n = 0, \cdots, M,$$

$$\pi_n = \pi_M \rho^{n-M}, \quad n \geq M + 1. \tag{2.22}$$

The case of a single server is of special interest. Then $M = 1$, and the stationary probabilities of queue size are a geometric distribution:

$$\pi_0 = 1 - \rho,$$
$$\pi_n = (1 - \rho)\rho^n, \quad n \geq 1. \tag{2.23}$$

We next find the stationary distribution of the stochastic process $\{W(t), t \geq 0\}$, where $W(t)$ denotes the time that a customer arriving at time t waits to begin service. We shall show that for $y > 0$

$$\lim_{t \to \infty} P[W(t) \geq y] = \pi_M \left(\frac{1}{1-\rho}\right) e^{-(1-\rho)M\mu y}, \tag{2.24}$$

and

$$\lim_{t \to \infty} P[W(t) > 0] = \lim_{t \to \infty} \{1 - P[W(t) = 0]\} = \pi_M \left(\frac{1}{1-\rho}\right). \tag{2.25}$$

In words, for large values of t, the waiting time for service has a mixed distribution; it has a positive probability mass at zero and is otherwise exponentially distributed.

That Eq. 2.25 holds is clear, since

$$P[W(t) > 0] = P[N(t) \geq M] \to \sum_{n=M}^{\infty} \pi_n = \pi_M \left(\frac{1}{1-\rho}\right) \quad \text{as} \quad t \to \infty.$$

To prove Eq. 2.24, we first determine the conditional distribution of $W(t)$, given $N(t) = m$ and t is very large.

If $N(t) = M$, a customer arriving at time t finds M customers in the queue, all being served. His waiting time for service is the minimum of the times it takes each of the latter to be served; this minimum is exponentially distributed with mean $1/M\mu$. Consequently

$$P[W(t) \geq y \mid N(t) = M] = \int_y^{\infty} M\mu e^{-M\mu x} \, dx.$$

If $N(t) = M + n$, for some $n \geq 1$, an arriving customer finds $M + n$ customers already in the queue. His waiting time for service is the sum of $(n+1)$ independent random variables, each representing the time it takes for one of M busy servers to become free, and each exponentially distributed with mean $1/M\mu$. Consequently, given that $N(t) = M + n$, $W(t)$ has a gamma distribution:

$$P[W(t) \geq y \mid N(t) = M + n] = \int_y^{\infty} M\mu \frac{(M\mu x)^n}{n!} e^{-M\mu x} \, dx.$$

Therefore, for $y > 0$,

$$P[W(t) \geq y] = \sum_{n=0}^{\infty} P[N(t) = M + n] \int_y^{\infty} M\mu \frac{(M\mu x)^n}{n!} e^{-M\mu x} \, dx.$$

For large values of t,

$$P[N(t) = M + n] \doteq \pi_M \rho^n.$$

Consequently, for large values of t, and $y > 0$,

$$P[W(t) \geq y] \doteq \pi_M \frac{1}{1-\rho} e^{-(1-\rho)M\mu y}.$$

The proof of Eq. 2.24 is now complete. ·

As a measure of the effectiveness of a service facility, one may adopt the ratio

$$R = \frac{\text{mean time spent by a customer waiting for service}}{\text{mean time spent by a customer being served}} \quad (2.26)$$

called the customer loss ratio, which represents the ratio of the time lost by the customer in the queue to the time usefully spent there.

The numerator in Eq. 2.26 is for large t approximately equal to

$$E[W(t)] = P[W(t) > 0] \int_0^\infty y(1-\rho)M\mu \, e^{-(1-\rho)M\mu y} \, dy$$
$$= \frac{P[W(t) > 0]}{M\mu(1-\rho)},$$

while the denominator in Eq. 2.26 is equal to $1/\mu$. Consequently, the customer loss ratio is given (for large t) by

$$R = \frac{P[W(t) > 0]}{M(1-\rho)} = \frac{\pi_M}{M(1-\rho)^2}. \quad (2.27)$$

One sees that R is a function only of M and ρ. If one graphs R as a function of ρ (for a given value of M) one sees that, as ρ approaches 1, R approaches infinity. The quantity ρ is called the utilization factor of a queue, since $1 - \rho$ is a measure of the fraction of time that the servers are idle (no one waiting for service). In order to reduce the customer loss ratio, management should allow for a substantial amount of idle time in a service facility at which demands for service occur randomly and the time required to render service is random, rather than attempting to attain a utilization factor as close to 1 as possible. For example, in a 4-station service facility with utilization factor 90% the customer loss ratio is 200%. If one more service station is added to the facility, the utilization factor is reduced to 72%, and the customer loss ratio is less than 10%.

EXERCISES

2.1 *Servicing of machines.* Consider M automatic machines serviced by a single repairman. If at time t a machine is working, the probability that it will break down is $\mu h + o(h)$. A machine which breaks down is serviced immediately unless the repairman is servicing another machine, in which case the machines which have broken down form a waiting line for service. The time it takes the repairman to repair a machine is exponentially distributed with mean $1/\lambda$. Let $N(t)$ be the number of machines working at time t, Find the long-run probability distribution of $N(t)$.

2.2 Consider N mice in a cage who have an infinite supply of food. If at time t a mouse is eating, the probability that he ceases eating at time $t + h$ is $\mu h + o(h)$; if at time t he is not eating, the probability that he begins eating before $t + h$ is $\lambda h + o(h)$. The mice eat independently of each other. Let $N(t)$ be the number of mice eating at time t.

(i) Find the long-run probability distribution of $N(t)$.

(ii) Find the long-run probability that more than half the mice are eating if $N = 10$, $\lambda = 60$, $\mu = 30$.

2.3 *Queues with impatient arrivals.* Consider an M-server queue. Suppose that inter-arrival and service times are exponentially distributed with means $1/\lambda$ and $1/\mu$ respectively. Let $N(t)$ denote the queue size at time t, counting both those being served and those waiting. Let β denote the probability that a customer who arrives when the queue size is M or greater will join the queue.

(i) Find $\pi_n = \lim_{t \to \infty} P[N(t) = n]$.

(ii) Find the stationary distribution of the waiting time to service.

In Exercises 2.4 to 2.6 consider an M-server "first come, first served" queue, in which customers wait for service and have exponential inter-arrival times (in hours) with mean $1/\lambda$; and service times (in hours) are exponentially distributed with mean $1/\mu$. In each case, find in the stationary state

(i) the probability that a customer will have to wait for service;

(ii) the mean length of the queue, including those waiting and those being served;

(iii) the mean time that a customer waits for service;

(iv) the conditional mean waiting time of a customer who has waited;

(v) the probability that a customer will have to wait more than 2 minutes before beginning to receive service;

(vi) the mean time that a customer spends in the queue, waiting for and receiving service;

(vii) the ratio of the mean time a customer waits for service to the mean time he spends in service;

(viii) the probability that there will be 2 or more persons in line waiting for service.

2.4 Suppose that $M = 1$ and (a) $\lambda = 18$, $\mu = 20$, (b) $\lambda = 18$, $\mu = 30$.

2.5 Suppose that $\lambda = 18$, $\mu = 20$, and (a) $M = 1$, (b) $M = 2$.

2.6 Suppose that $\lambda = 18$, and (a) $\mu = 20$, $M = 1$, (b) $\mu = 9$, $M = 2$.

2.7 Consider (in the stationary state) a service facility with exponential inter-arrival and service times. Assume that the rate of customer arrivals is fixed. By comparing customer loss ratios, state whether it is preferable to have two servers, or a single server whose service rate is twice that of each server in the two-server queue.

2.8　Consider a taxi station where taxis looking for customers and customers looking for taxis arrive in accord with Poisson processes, with mean rates per minute of 1 and 1.25. A taxi will wait no matter how many are in line. However, an arriving customer waits only if the number of customers already waiting for taxis is 2 or less. Find:

(i) the mean number of taxis waiting for customers;
(ii) the mean number of customers waiting for taxis;
(iii) the mean number of customers who in the course of an hour do not join the waiting line because there were 3 customers already waiting.

7-3　KOLMOGOROV DIFFERENTIAL EQUATIONS FOR THE TRANSITION PROBABILITY FUNCTIONS

To obtain the transition probability functions of a continuous parameter Markov chain one usually solves a system of differential equations for the transition probability functions. We shall derive these differential equations for the case of non-homogeneous Markov chains $\{N(t), t \geq 0\}$ with transition probability functions

$$p_{j,k}(s,t) = P[N(t) = k \mid N(s) = j], \tag{3.1}$$

defined for any states j and k and times $t > s \geq 0$.

We make the following assumptions: for each state j there is a non-negative continuous function $q_j(t)$ defined by the limit

$$\lim_{h \to 0} \frac{1}{h} \{1 - p_{j,j}(t, t + h)\} = q_j(t) \tag{3.2}$$

and for each pair of states j and k (with $j \neq k$) there is a non-negative continuous function $q_{j,k}(t)$ defined by the limit

$$\lim_{h \to 0} \frac{1}{h} p_{j,k}(t, t + h) = q_{j,k}(t). \tag{3.3}$$

These functions have the following probabilistic interpretation. The probabilities of transition within a time interval of length h are asymptotically proportional to h; the probability $1 - p_{j,j}(t + h)$ of a transition from a state j to some other state during the time interval $(t, t + h)$ is equal to $hq_j(t)$ plus a remainder which, divided by h, tends to 0 (as $h \to 0$) while the probability $p_{j,k}(t, t + h)$ of transition from j to k during the time interval $(t, t + h)$ is equal to $hq_{j,k}(t)$ plus a remainder which, divided by h, tends to 0 (as $h \to 0$).

We call $q_j(t)$ the *intensity of passage*, given that the Markov chain is in state j at time t. We call $q_{j,k}(t)$ the *intensity of transition to k*, given that the Markov chain is in state j at time t.

The intensity functions $q_j(t)$ and $q_{j,k}(t)$ are said to be *homogeneous* if they do not depend on t:

$$q_j(t) = q_j, \tag{3.4}$$
$$q_{j,k}(t) = q_{j,k}.$$

The intensity functions of a homogeneous Markov chain are clearly homogeneous.

EXAMPLE 3A

A failure process. Let $\{N(t), t \geq 0\}$ denote the number of components of a certain mechanism that have failed and been replaced in the time interval 0 to t. Assume that $\{N(t), t \geq 0\}$ is a Markov chain with state space $\{0, 1, 2, \cdots\}$. The intensity function $q_n(t)$ then has the following meaning: $q_n(t)h$ is approximately equal to the probability that during the time interval t to $t + h$ one or more components of the mechanism will fail. In general, one would expect $q_n(t)$ to depend on both n and t, since n represents the number of components that have failed previously, and t represents the length of time the mechanism has been operating. A convenient formula for $q_n(t)$ which might be adopted in order to develop the properties of the Markov chain $\{N(t), t \geq 0\}$ is

$$q_n(t) = \frac{a + bn}{c + dt}, \tag{3.5}$$

where $a, b, d \geq 0$ and $c > 0$ are constants to be specified. If it is assumed that $d = 0$, then the intensity function $q_n(t)$ is homogeneous.

Assumptions 3.2 and 3.3 may be conveniently written in matrix form. Let us define a matrix $A(t)$ as follows:

$$a_{j,k}(t) = q_{j,k}(t) \quad \text{if} \quad j \neq k,$$
$$= -q_j(t) \quad \text{if} \quad j = k. \tag{3.6}$$

The identity matrix I is defined as usual by

$$I = \{\delta_{j,k}\}, \ \delta_{j,k} = \begin{cases} 1 & \text{if} \quad j = k \\ 0 & \text{if} \quad j \neq k. \end{cases} \tag{3.7}$$

Then Eqs. 3.2 and 3.3 may be expressed in matrix form as follows:

$$\frac{1}{h} \{P(t,t+h) - I\} \to A(t) \quad \text{as} \quad h \to 0. \tag{3.8}$$

From the assumption 3.8 and the Chapman-Kolmogorov equation one may obtain a system of differential equations for the transition probability functions $p_{j,k}(s,t)$. We first formally derive these equations in matrix form. By the Chapman-Kolmogorov equations we may write

$$P(s,t+h) = P(s,t)\, P(t,t+h)$$
$$\frac{1}{h} \{P(s,t+h) - P(s,t)\} = \frac{1}{h}\, P(s,t)\, \{P(t,t+h) - I\}. \tag{3.9}$$

Now, let h tend to 0 in Eq. 3.9. By Eq. 3.8 it seems plausible that the right-hand side of Eq. 3.9 tends to

$$P(s,t)\, A(t).$$

Next, suppose that the partial derivatives

$$\frac{\partial}{\partial t} P(s,t) = \left\{ \frac{\partial}{\partial t}\, p_{j,k}(s,t) \right\} \tag{3.10}$$

exist. Then the left-hand side of Eq. 3.9 tends to Eq. 3.10. Consequently, for any $t > s \geq 0$, we obtain

$$\frac{\partial}{\partial t} P(s,t) = P(s,t)\, A(t), \tag{3.11}$$

which may be written: for $t > s \geq 0$ and states j and k

$$\frac{\partial}{\partial t} p_{j,k}(s,t) = \sum_i p_{j,i}(s,t)\, a_{i,k}(t)$$
$$= - q_k(t)\, p_{j,k}(s,t) + \sum_{i \neq k} p_{j,i}(s,t)\, q_{i,k}(t). \tag{3.12}$$

On the other hand, we could have written

$$P(s-h,t) = P(s-h,s)\, P(s,t),$$
$$\frac{1}{h} \{P(s,t) - P(s-h,t)\} = \frac{1}{h} \{I - P(s-h,s)\}\, P(s,t). \tag{3.13}$$

Let

$$\frac{\partial}{\partial s} P(s,t) = \left\{ \frac{\partial}{\partial s}\, P_{j,k}(s,t) \right\} \tag{3.14}$$

be the matrix of partial derivatives with respect to the first time variable s. Then letting h tend to 0 in Eq. 3.13 it follows that

$$\frac{\partial}{\partial s}P(s,t) = A(s) P(s,t),$$ (3.15)

which may be written: for $t \geq s > 0$ and states j and k

$$\frac{\partial}{\partial s}p_{j,k}(s,t) = \sum_i a_{j,i}(s) p_{i,k}(s,t)$$

$$= -q_j(s)p_{j,k}(s,t) + \sum_{i \neq j} q_{j,i}(s) p_{i,k}(s,t).$$ (3.16)

The systems 3.12 and 3.16 of differential equations for the transition probability functions of a Markov chain were first derived by Kolmogorov (1931) in a fundamental paper. The system 3.12 is usually called *Kolmogorov's forward equation* since it involves differentiation with respect to the later time t while the system 3.16 is usually called *Kolmogorov's backward equation* since it involves differentiation with respect to the earlier time s.

It should be noted that, despite the appearance of the partial derivative symbol in Eqs. 3.12 and 3.16, these systems of equations are not really partial differential equations. Rather they are ordinary differential equations since in Eq. 3.12 s and j are not variables but fixed parameters while in Eq. 3.16 t and k are not variables but fixed parameters. The parameters appear only in the initial conditions; for Eq. 3.12,

$$\begin{aligned} p_{j,k}(s,t) &= 1 \quad \text{if} \quad t = s, \ k = j \\ &= 0 \quad \text{if} \quad t = s, \ k \neq j, \end{aligned}$$ (3.17)

while for Eq. 3.16,

$$\begin{aligned} p_{j,k}(s,t) &= 1 \quad \text{if} \quad s = t, \ j = k \\ &= 0 \quad \text{if} \quad s = t, \ j \neq k. \end{aligned}$$ (3.18)

In the foregoing development many questions have been left unanswered. Assuming the existence of the intensity functions $q_i(t)$ and $q_{j,k}(t)$, we have formally derived the systems of differential equations 3.12 and 3.16. It is easily shown that the forward equations 3.12 hold if in addition to Eqs. 3.2 and 3.3 one makes the assumption that, for fixed k, the passage to the limit in Eq. 3.3 is uniform with respect to j. It may be shown (see Feller [1957], p. 427) that the backward equations 3.16 hold under Eqs. 3.2 and 3.3 without any additional assumption. For this reason, in the theory of Markov chains the backward equations are regarded as more fundamental than the forward equations. While the forward equations are intuitively easier to comprehend, the backward equations are easier to deal with from a rigorous point of view, since to establish their validity requires less restrictive assumptions.

There are several questions one can raise concerning the intensity functions $q_j(t)$ and $q_{j,k}(t)$. (i) Do intensity functions satisfying Eqs. 3.2 and 3.3 exist for all Markov chains? (ii) What conditions need non-negative functions $q_j(t)$ and $q_{j,k}(t)$ satisfy in order to be the intensity functions of a Markov chain? This question is of particular concern because one usually will describe a Markov chain by means of its intensity functions. In view of the fact that for any state j and time t

$$1 - p_{j,j}(t,t+h) - \sum_{k \neq j} p_{j,k}(t,t+h) = 0 \qquad (3.19)$$

it seems reasonable to *require* that the intensity functions satisfy, for any state j and time t,

$$q_j(t) = \sum_{k \neq j} q_{j,k}(t). \qquad (3.20)$$

Given a family of non-negative continuous functions $q_j(t)$ and $q_{j,k}(t)$ satisfying Eq. 3.20 it may be shown that there exists a family of non-negative functions $p_{j,k}(s,t)$ satisfying the Kolmogorov differential equations 3.12 and 3.16, the Chapman-Kolmogorov equations, and Eqs. 3.2 and 3.3. However, the functions $p_{j,k}(s,t)$ do not necessarily represent probability distributions, since it may happen that

$$\sum_k p_{j,k}(s,t) < 1. \qquad (3.21)$$

It can be shown that Eq. 3.21 happens if there is a positive probability that in the time interval s to t an infinite number of transitions will occur. A Markov process satisfying Eq. 3.21 is said to be *dishonest* or *pathological.*

In this brief outline of the general theory of Markov chains, it has been our aim to indicate some of the questions that need to be considered in order to construct the foundations of the theory of Markov chains. In the sequel our aim is to examine the Markov chains which arise when one assumes that the intensity functions are of various simple forms.

Homogeneous Markov chains. It is easily verified that, if the intensity functions $q_j(t)$ and $q_{j,k}(t)$ are homogeneous, then the corresponding solutions $p_{j,k}(s,t)$ of the Kolmogorov differential equations 3.12 and 3.16 are functions only of the time difference $t - s$. Consequently, the corresponding Markov chain is homogeneous. To find the transition probabilities

$$p_{j,k}(t) = P[X_{t+u} = k \mid X_u = j] \qquad (3.22)$$

one will usually attempt to find the solutions of the forward Kolmogorov differential equations:

$$\frac{\partial}{\partial t} p_{j,k}(t) = - q_k\, p_{j,k}(t) + \sum_{i \neq k} p_{j,i}(t)\, q_{i,k}. \qquad (3.23)$$

7-4 TWO-STATE MARKOV CHAINS AND PURE BIRTH PROCESSES

In this section we show how, for the two-state Markov chain and the pure birth process, one may solve the Kolmogorov differential equations to obtain the transition probability functions.

Two-state Markov chains. Let $\{X(t), t \geq 0\}$ be a homogeneous Markov chain such that for each t the only possible values of $X(t)$ are 0 or 1. (In Section 1–4 it was shown how such processes arise naturally in the study of system reliability and semi-conductor noise.) Let the intensities of passage from 0 and 1 be given respectively by

$$q_0 = \lambda, \quad q_1 = \mu. \qquad (4.1)$$

It then follows that the transition intensities are given by

$$q_{0,1} = \lambda, \quad q_{1,0} = \mu. \qquad (4.2)$$

The Kolmogorov differential equations 3.23 then become

$$\frac{\partial}{\partial t}\, p_{0,0}(t) = - \lambda\, p_{0,0}(t) + \mu\, p_{0,1}(t),$$

$$\frac{\partial}{\partial t}\, p_{0,1}(t) = - \mu\, p_{0,1}(t) + \lambda\, p_{0,0}(t), \qquad (4.3)$$

$$\frac{\partial}{\partial t}\, p_{1,1}(t) = - \mu\, p_{1,1}(t) + \lambda\, p_{1,0}(t),$$

$$\frac{\partial}{\partial t}\, p_{1,0}(t) = - \lambda\, p_{1,0}(t) + \mu\, p_{1,1}(t).$$

Since $p_{0,1}(t) = 1 - p_{0,0}(t)$, the first of these equations may be rewritten

$$\frac{\partial}{\partial t} p_{0,0}(t) = - (\lambda + \mu)\, p_{0,0}(t) + \mu, \quad 0 \leq t < \infty. \qquad (4.4)$$

Now, Eq. 4.4 is an ordinary differential equation of the form [with $g(t) = p_{0,0}(t)$, $\nu = \lambda + \mu$, $h(t) = \mu$]

$$g'(t) = - \nu\, g(t) + h(t), \quad a \leq t < \infty, \qquad (4.5)$$

whose solution may be found by using the following theorem.

THEOREM 4A

General solution of a first order linear inhomogeneous differential equation. If $g(t)$ is the solution of the differential equation

$$g'(t) + \nu\, g(t) = h(t), \ a \leq t \leq b, \tag{4.6}$$

where ν is a real number and $h(t)$ is a continuous function, then

$$g(t) = \int_a^t e^{-\nu(t-s)} h(s)\, ds + g(a)\, e^{-\nu(t-a)}, \ a \leq t \leq b. \tag{4.7}$$

Proof. Define

$$G(t) = e^{\nu t} g(t). \tag{4.8}$$

Then $g(t) = e^{-\nu t}\, G(t)$ and $g'(t) = e^{-\nu t}\, G'(t) - \nu\, e^{-\nu t}\, G(t)$. Consequently, $g'(t) + \nu\, g(t) = e^{-\nu t}\, G'(t)$. From the fact that Eq. 4.6 holds, it follows that $G(t)$ satisfies

$$e^{-\nu t}\, G'(t) = h(t). \tag{4.9}$$

The differential equation 4.9 has the solution

$$G(t) = \int_a^t e^{\nu s}\, h(s)\, ds + G(a). \tag{4.10}$$

Combining Eqs. 4.8 and 4.10, one obtains Eq. 4.7.

From Eqs. 4.4 and 4.7, and the boundary condition $p_{0,0}(0) = 1$, it follows that

$$p_{0,0}(t) = \mu \int_0^t e^{-(\lambda+\mu)(t-s)}\, ds + e^{-(\lambda+\mu)t}. \tag{4.11}$$

Using Eq. 4.11 one may establish the following results.

Let $\{X(t), t \geq 0\}$ be a Markov chain with state space $\{0, 1\}$ and passage intensities given by Eq. 4.1. Then, for any $s, t \geq 0$,

$$
\begin{aligned}
p_{0,0}(t) &= P[X(t+s) = 0 \mid X(s) = 0] = \frac{\mu}{\lambda+\mu} + \frac{\lambda}{\lambda+\mu}\, e^{-(\lambda+\mu)t}, \\
p_{1,0}(t) &= P[X(t+s) = 0 \mid X(s) = 1] = \frac{\mu}{\lambda+\mu}\, (1 - e^{-(\lambda+\mu)t}), \\
p_{0,1}(t) &= P[X(t+s) = 1 \mid X(s) = 0] = \frac{\lambda}{\lambda+\mu}\, (1 - e^{-(\lambda+\mu)t}), \\
p_{1,1}(t) &= P[X(t+s) = 1 \mid X(s) = 1] = \frac{\lambda}{\lambda+\mu} + \frac{\mu}{\lambda+\mu}\, e^{-(\lambda+\mu)t}.
\end{aligned}
\tag{4.12}
$$

Let $p_0 = P[X(0) = 0]$. It follows from Eq. 4.12 that

$$E[X(t)] = \frac{\lambda}{\lambda + \mu} - \left(p_0 - \frac{\mu}{\lambda + \mu}\right) e^{-(\lambda+\mu)t}, \qquad (4.13)$$

$$\text{Cov}[X(s), X(s+t)] = E[X(s)] \{p_{1,1}(t) - E[X(s+t)]\}$$
$$= e^{-(\lambda+\mu)t} \left\{ \frac{\mu}{\lambda + \mu} + \left(\frac{\mu}{\lambda + \mu} - p_0\right) e^{-(\lambda+\mu)s} \right\} \left\{ \frac{\lambda}{\lambda + \mu} + \left(p_0 - \frac{\mu}{\lambda + \mu}\right) e^{-(\lambda+\mu)s} \right\}.$$

For $t \geq 0$, let $\beta(t)$ be the fraction of time during the interval 0 to t that the stochastic process takes the value 1. Then $\beta(t)$ may be represented as

$$\beta(t) = \frac{1}{t} \int_0^t X(u) \, du.$$

From Eq. 4.13 it follows that, as $t \to \infty$,

$$E[\beta(t)] = \frac{1}{t} \int_0^t E[X(u)] \, du \to \frac{\lambda}{\lambda + \mu}, \qquad (4.14)$$

$$t \, \text{Var}[\beta(t)] = \frac{1}{t} \int_0^t \int_0^t \text{Cov}[X(u), X(v)] \, du \, dv \to \frac{2\lambda\mu}{(\lambda + \mu)^3}.$$

It may be shown that, for large t, $\beta(t)$ is approximately normally distributed with asymptotic mean and variance satisfying Eq. 4.14.

It should be noted that Eq. 4.14 is a special case of Theorem 4B of Chapter 1, in which U is exponentially distributed with mean $1/\lambda$ and V is exponentially distributed with mean $1/\mu$.

Pure birth processes. A birth and death process is said to be a *pure birth process* if $\mu_n = 0$ for all n (that is, death is impossible). For a pure birth process, the differential equations for the transition probability functions $p_{m,n}(t)$ become

$$\frac{\partial}{\partial t} p_{m,n}(t) = -\lambda_n \, p_{m,n}(t) + \lambda_{n-1} \, p_{m,n-1}(t) \quad \text{for} \quad n \geq m+1, \qquad (4.15)$$

$$\frac{\partial}{\partial t} p_{m,m}(t) = -\lambda_m p_{m,m}(t). \qquad (4.16)$$

Using Theorem 4A one obtains the following results.

THEOREM 4B

Transition probability functions of a pure birth process. The solution of Eq. 4.16 is

$$p_{m,m}(t) = e^{-\lambda_m t},$$

while the solution of Eq. 4.15 is, for $n \geq m + 1$,

$$p_{m,n}(t) = e^{-\lambda_n t} \int_0^t e^{\lambda_n s} \lambda_{n-1} \, p_{m,n-1}(s) \, ds.$$

EXAMPLE 4A

The birth process with constant birth rate $\lambda_n = \nu$ as the Poisson process with intensity ν. Using Theorem 4B one may show by mathematical induction that the transition probability function of a pure birth process $\{N(t), t \geq 0\}$ with constant birth rate ν (that is, $\lambda_n = \nu$ for all n) is given by

$$p_{m,n}(t) = e^{-\nu t} \frac{(\nu t)^{n-m}}{(n-m)!} \quad \text{for} \quad n \geq m \quad (4.17)$$
$$= 0, \qquad\qquad \text{otherwise.}$$

In words, Eq. 4.17 says that the increase $N(t + s) - N(s)$ in population size in an interval of length t is Poisson distributed with mean νt, no matter what the population size at the beginning of the interval. In symbols, for any time $s < t$, and integers k and m,

$$P[N(t) - N(s) = k \mid N(s) = m] = e^{-\nu(t-s)} \frac{\{\nu(t-s)\}^k}{k!}. \quad (4.18)$$

From Eq. 4.18 it follows that $N(t)$ is Poisson distributed with mean νt.

To show that the pure birth process $N(\cdot)$ with constant birth rate is indeed the Poisson process, we need to show that $N(\cdot)$ has independent increments. Now for any times $0 = t_0 < t_1 < t_2 < \cdots < t_n$ [assuming $N(0) = 0$]

$$P[N(t_1) - N(t_0) = k_1, \cdots, N(t_n) - N(t_{n-1}) = k_n]$$
$$= P[N(t_1) = k_1] \, P[N(t_2) - N(t_1) = k_2 \mid N(t_1) = k_1] \cdots$$
$$\quad P[N(t_n) - N(t_{n-1}) = k_n \mid N(t_1) = k_1, \cdots, N(t_{n-1}) - N(t_{n-2}) = k_{n-1}]$$
$$= P[N(t_1) = k_1] \, P[N(t_2) - N(t_1) = k_2] \cdots P[N(t_n) - N(t_{n-1}) = k_n],$$

which proves that $\{N(t), t \geq 0\}$ has independent increments.

EXAMPLE 4B

The birth process with linear birthrate. Consider a population whose members can (by splitting or otherwise) give birth to new members but cannot die. Assume that the probability is approximately λh that in a short time interval of length h a member will create a new member. More precisely, assume that if $N(t)$ is the size of the population at time t, then $\{N(t), t \geq 0\}$ is a pure birth process with

$$\lambda_n = n\lambda \quad \text{for} \quad n = 0, 1, \cdots. \quad (4.19)$$

Using Theorem 4B, one may show by mathematical induction that, for $n \geq m \geq 1$,

$$p_{m,n}(t) = \binom{n-1}{n-m}(e^{-\lambda t})^m(1 - e^{-\lambda t})^{n-m}. \quad (4.20)$$

In words, Eq. 4.20 says that the increase $N(t + s) - N(s)$ in population size in an interval of length t has a negative binomial distribution with parameters $p = e^{-\lambda t}$ and $r = m$, where m is the value of $N(s)$. Consequently,

$$E[e^{iu\{N(t+s)-N(s)\}} \mid N(s) = m] = \left\{ \frac{e^{-\lambda t}}{1 - (1 - e^{-\lambda t})e^{iu}} \right\}^m \qquad (4.21)$$

$$E[N(t + s) - N(s) \mid N(s) = m] = me^{\lambda t}(1 - e^{-\lambda t}), \qquad (4.22)$$

$$\mathrm{Var}[N(t + s) - N(s) \mid N(s) = m] = me^{2\lambda t}(1 - e^{-\lambda t}). \qquad (4.23)$$

If the initial population size is m, so that $N(0) = m$, then the foregoing conditional expectations (with $s = 0$) are respectively equal to $\varphi_{N(t)-N(0)}(u)$, $E[N(t) - N(0)]$, and $\mathrm{Var}[N(t) - N(0)]$. Consequently, given that $N(0) = m$,

$$E[N(t)] = E[N(t) - N(0)] + m = me^{\lambda t}, \qquad (4.24)$$

$$\mathrm{Var}[N(t)] = \mathrm{Var}[N(t) - N(0)] = me^{\lambda t}(e^{\lambda t} - 1). \qquad (4.25)$$

The birth process with linear birthrate is sometimes called the Yule process or the Furry process. It was essentially used by Yule (1924) in his mathematical theory of evolution and by Furry (1937) as a model for cosmic-ray showers. In Yule's application, $N(t)$ represents the *number of species* in some genus of animals or plants. It is assumed that a species, once originated, does not die out and that if $N(t)$ species exist at time t, then $\lambda N(t) h$ is the probability that a new species will be created, by mutation, in the interval t to $t + h$.

Yule was interested in the number of genera having n species at a given T. Suppose that new genera originate at times $\tau_1 < \tau_2 < \cdots$ which occur in accord with a non-homogeneous Poisson process with mean value function

$$m(t) = N_0 e^{\alpha t}, \qquad (4.26)$$

where N_0 and α are positive constants. Let $X^{(n)}(T)$ be the number of genera which at time T have exactly n species. One may write $X^{(n)}(T)$ as a filtered Poisson process (as defined in Section 4–5) in the form

$$X^{(n)}(T) = \sum_{m=1}^{\infty} W_m^{(n)}(T, \tau_m), \qquad (4.27)$$

where $W_m^{(n)}(T, \tau_m) = 1$ or 0 depending on whether the genus originating at time τ_m does or does not have n species at time T. By Eq. 4.20

$$\begin{aligned} E[W^{(n)}(t,\tau)] &= p_{1,n}(t - \tau) \\ &= e^{-\lambda(t-\tau)} \{1 - e^{-\lambda(t-\tau)}\}^{n-1} \end{aligned} \qquad (4.28)$$

The expected number of genera with n species at time T is given by

$$E[X^{(n)}(T)] = \int_0^T E[W^{(n)}(T,\tau)] \, dm(\tau)$$
$$= \int_\theta^T e^{-\lambda(T-\tau)}\{1 - e^{-\lambda(T-\tau)}\}^{n-1}\{N_0 \, \alpha \, e^{\alpha\tau}\} \, d\tau. \quad (4.29)$$

For very large T, $E[X^{(n)}(T)]$ is, up to a constant factor which does not depend on n, approximately equal to

$$\int_0^1 (1-y)^{n-1} \, y^{\alpha/\lambda} \, dy. \quad (4.30)$$

Therefore,

$$\frac{E[X^{(1)}(T)]}{\sum_{n=1}^{\infty} E[X^{(n)}(T)]} = \frac{\int_0^1 y^{\alpha/\lambda} \, dy}{\int_0^1 y^{(\alpha/\lambda)-1} dy} = \frac{1}{1 + \left(\dfrac{\lambda}{\alpha}\right)}. \quad (4.31)$$

Using Eq. 4.31 one can develop a procedure for estimating the ratio λ/α, which is a measure of the rapidity of appearance of new species as compared with the appearance of new genera, by observations made only at one particular time. At a given time, count the number M of genera in a certain family (group of genera) and the number M_1 of these genera having just one species. If M and M_1 are large, and these genera have been in existence a long time, the ratio M_1/M should be approximately equal to the left-hand side of Eq. 4.31. Consequently, solving the equation

$$\frac{M_1}{M} = \frac{1}{1 + (\lambda/\alpha)}$$

for λ/α, we may estimate it by

$$\frac{\lambda}{\alpha} = \frac{M - M_1}{M_1}. \quad (4.32)$$

Yule applied this procedure to a family of beetles with 627 genera, of which 34.29% had one species. This yields an estimate of λ/α equal to 1.9. The reader is referred to Yule's paper for further treatment of these questions.

EXERCISES

4.1 *Truncated Poisson process.* Consider a pure birth process $\{N(t), t \geq 0\}$ with constant birthrate ν which has a finite maximum size, say M (that is, $\lambda_n = \nu$ or 0 depending on whether $n < M$ or $n \geq M$). For $t > s > 0$, find the conditional probability mass function of $N(t) - N(s)$, given that $N(s) = k$,

4.2 *Pure death process.* The birth and death process with parameters $\lambda_n = 0$ and $\mu_n = n\mu$ for $n = 0, 1, \cdots$ is called a pure death process with linear death rate. Find (i) the transition probability function $p_{m,n}(t)$, (ii) $E[N(t) \mid N(0) = m]$, (iii) $\text{Var}[N(t) \mid N(0) = m]$.

In Exercises 4.3–4.5 let $\{N(t), t \geq 0\}$ be the birth process with linear birthrate considered in Example 4B.

4.3 Show that Eq. 4.20 holds.

4.4 Suppose that $N(0)$ is a random variable, with a geometric distribution with parameter p (that is, $P[N(0) = n] = p(1 - p)^{n-1}$ for $n = 1, 2, \cdots$). Find $E[N(t)]$ and $\text{Var}[N(t)]$.

4.5 Assume that $N(0) = 1$. Find (i) $\text{Cov}[N(s), N(t)]$; (ii) $\varphi_{N(s),N(t),}(u,v)$. *Hint.* Show, and use, the fact that

$$\varphi_{N(s),N(t)}(u,v) = E[e^{i(u+v)N(s)}E[e^{iv(N(t)-N(s))} \mid N(s)]].$$

7-5 NON-HOMOGENEOUS BIRTH AND DEATH PROCESSES

A number of different methods are available for finding transition probability functions of a homogeneous birth and death process (see, in particular, the papers of Reuter and Ledermann [1953] and Karlin and McGregor [1957] and later papers by these authors). In this section we give a brief sketch of the use of probability generating functions for finding the transition probabilities since it appears to be the only method which works as well for the non-homogeneous case as for the homogeneous case. This section should also be regarded as an introduction to the technique of setting up partial differential equations for probability generating functions or characteristic functions. (For further illustrations of this technique and references to its history, see Bartlett [1949], p. 219 ff., Bartlett [1955], and Syski [1960].)

An integer-valued process $\{N(t), t \geq 0\}$ is said to be a non-homogeneous birth and death process if it is a Markov chain, with transition probability function

$$p_{m,n}(s,t) = P[N(t) = n \mid N(s) = m] \tag{5.1}$$

satisfying the following assumptions. There exist non-negative functions

$$\lambda_0(t), \lambda_1(t), \cdots \quad \text{and} \quad \mu_1(t), \mu_2(t), \cdots \tag{5.2}$$

such that the following limits hold, at each t uniformly in n:

$$\lim_{h \to 0} \frac{p_{n,n+1}(t,\, t+h)}{h} = \lambda_n(t) \qquad \text{for} \quad n \geq 0, \qquad (5.3)$$

$$\lim_{h \to 0} \frac{p_{n,n-1}(t,\, t+h)}{h} = \mu_n(t) \qquad \text{for} \quad n \geq 1,$$

$$\lim_{h \to 0} \frac{1 - p_{n,n}(t,\, t+h)}{h} = \lambda_n(t) + \mu_n(t) \quad \text{for} \quad n \geq 0,$$

where we define, for all $t \geq 0$,

$$\mu_0(t) = 0. \qquad (5.4)$$

The birth and death process is said to be (i) *homogeneous* if, for all n, $\lambda_n(t)$ and $\mu_n(t)$ do not depend on t; (ii) a *pure birth* process if

$$\mu_n(t) = 0 \qquad \text{for all } t \text{ and } n, \qquad (5.5)$$

(iii) a *pure death* process if

$$\lambda_n(t) = 0 \qquad \text{for all } t \text{ and } n. \qquad (5.6)$$

In words, Eq. 5.3 states that in a very small time interval the population size, represented by $N(t)$, either increases by one, decreases by one, or stays the same. The conditional probability of an increase by one (a "birth") may depend both on the time t at which we are observing the process and on the population size n at this time; consequently the conditional probability of a birth is denoted by $\lambda_n(t)$ [or, more precisely, $\lambda_n(t) h + o(h)$]. Similarly the conditional probability of a decrease by one (a "death") is denoted by $\mu_n(t)$, since it may depend on both n and t.

From the Kolmogorov differential equations, one can obtain differential equations for the transition probability function $p_{m,n}(s,t)$ of a non-homogeneous birth and death process. In order to solve the resulting equations, it is more convenient to obtain a partial differential equation for the transition probability generating function

$$\psi_{j,s}(z,t) = \sum_{k=0}^{\infty} z^k p_{j,k}(s,t) \qquad (5.7)$$

defined for any state j, times $s < t$, and $|z| \leq 1$.

THEOREM 5A
    Partial differential equation for the transition probability generating function of a birth and death process. For any initial state j, times $s < t$, and $|z| \leq 1$,

$$\frac{\partial}{\partial t}\psi_{j,s}(z,t) = \sum_{k=0}^{\infty} z^k p_{j,k}(s,t)\{(z-1)\lambda_k(t) + (z^{-1}-1)\mu_k(t)\} \qquad (5.8)$$

with boundary condition

$$\psi_{j,s}(z,s) = z^j \quad \text{if} \quad P[N(s)=j]=1. \qquad (5.9)$$

In the particularly important case that for some functions $\lambda(t)$ and $\mu(t)$

$$\lambda_n(t) = n\,\lambda(t), \qquad \mu_n(t) = n\,\mu(t) \qquad (5.10)$$

then

$$\frac{\partial}{\partial t}\psi_{j,s}(z,t) = \frac{\partial}{\partial z}\psi_{j,s}(z,t)\{(z-1)[z\,\lambda(t) - \mu(t)]\}. \qquad (5.11)$$

Proof. We first note that, as $h \to 0$,

$$\frac{1}{h}\{E[z^{N(t+h)-N(t)} \mid N(t)=k] - 1\}$$

$$= \frac{1}{h}\{z\,p_{k,k+1}(t,t+h) + z^{-1}p_{k,k-1}(t,t+h) + p_{k,k}(t,t+h) - 1 + o(h)\}$$

$$\to (z-1)\,\lambda_k(t) + (z^{-1}-1)\,\mu_k(t).$$

Therefore, as $h \to 0$,

$$\frac{1}{h}\{\psi_{j,s}(z,t+h) - \psi_{j,s}(z,t)\}$$

$$= \sum_{k=0}^{\infty} z^k \frac{1}{h}\{E[z^{N(t+h)-N(t)} \mid N(t)=k] - 1\}\,p_{j,k}(s,t)$$

$$\to \sum_{k=0}^{\infty} z^k p_{j,k}(s,t)\,\{(z-1)\,\lambda_k(t) + (z^{-1}-1)\,\mu_k(t)\}.$$

To obtain Eq. 5.11 note that

$$\sum_{k=0}^{\infty} kz^k p_{j,k}(s,t) = z\frac{\partial}{\partial z}\psi_{j,s}(z,t),$$

$$\lambda(t)\,z(z-1) + \mu(t)\,z(z^{-1}-1) = \lambda(t)\,z^2 - [\lambda(t)+\mu(t)]\,z + \mu(t)$$

$$= \{\lambda(t)\,z - \mu(t)\}\,(z-1).$$

The proof of the theorem is now complete.

In order to solve the partial differential equation Eq. 5.11, we use the following theorem.

THEOREM 5B

Suppose that the probability generating function satisfies the partial differential equation

$$\frac{\partial \psi_{j,s}(z,t)}{\partial t} = a(z,t) \frac{\partial \psi_{j,s}(z,t)}{\partial z} \tag{5.12}$$

subject to the boundary condition Eq. 5.9. Let $u(\,\cdot\,,\,\cdot\,)$ be a function such that the solution $z(t)$ of the differential equation

$$\frac{dz}{dt} + a(z,t) = 0 \tag{5.13}$$

satisfies

$$u(z,t) = \text{constant.} \tag{5.14}$$

Define a function $g(\,\cdot\,)$ by

$$g(z) = u(z,s), \tag{5.15}$$

and let $g^{-1}(\,\cdot\,)$ be the inverse function of $g(\,\cdot\,)$ in the sense that

$$g^{-1}(x) = z \quad \text{if} \quad x = g(z). \tag{5.16}$$

Then

$$\psi_{j,s}(z,t) = \{g^{-1}(u(z,t))\}^{j}. \tag{5.17}$$

The proof of Theorem 5B is beyond the scope of this book. It is an application of Lagrange's method for solving a first-order linear partial differential equation. A sketch of the proof of Theorem 5B is given by Syski (1960), p. 696.

EXAMPLE 5A

Pure birth process with linear birthrate. To show how one uses Theorem 5B to solve the partial differential equation 5.11 let us consider the *non-homogeneous linear growth process* which is a pure birth process with $\lambda_n(t) = n\,\lambda(t)$. Then Eq. 5.11 becomes

$$\frac{\partial}{\partial t}\psi_{j,s}(z,t) = \frac{\partial}{\partial z}\psi_{j,s}(z,t)\{z(z-1)\lambda(t)\} \tag{5.18}$$

which is of the form of Eq. 5.12 with

$$a(z,t) = z(z-1)\,\lambda(t).$$

The ordinary differential equation

$$\frac{dz}{dt} + z(z-1)\,\lambda(t) = 0 \qquad (5.19)$$

may be written

$$\frac{dz}{z(z-1)} + \lambda(t)\,dt = 0. \qquad (5.20)$$

Integrating Eq. 5.20 one obtains

$$\log(1 - z^{-1}) + \int \lambda(t)\,dt = \text{constant}.$$

Any solution $z(t)$ of Eq. 5.19 therefore satisfies

$$u(z,t) = \text{constant},$$

if one defines

$$u(z,t) = \log(1 - z^{-1}) + \rho(t), \qquad (5.21)$$

where

$$\rho(t) = \int_0^t \lambda(t')\,dt'. \qquad (5.22)$$

To solve for $g^{-1}(\,\cdot\,)$ defined by Eq. 5.16 we write

$$x = u(z,s) = \log(1 - z^{-1}) + \rho(s),$$

which implies

$$1 - z^{-1} = \exp\{x - \rho(s)\},$$

which implies

$$g^{-1}(x) = z = (1 - \exp\{x - \rho(s)\})^{-1}.$$

Therefore,

$$g^{-1}(u(z,t)) = (1 - \exp\{\log(1 - z^{-1}) + \rho(t) - \rho(s)\})^{-1}$$
$$= (1 - (1 - z^{-1})\exp\{\rho(t) - \rho(s)\})^{-1}.$$

This may be rewritten

$$g^{-1}(u(z,t)) = \frac{z\,e^{-\{\rho(t)-\rho(s)\}}}{1 - z(1 - e^{-\{\rho(t)-\rho(s)\}})}. \qquad (5.23)$$

The transition probability generating function $\psi_{j,s}(z,t)$ is equal to the jth power of Eq. 5.23.

Now the right-hand side of Eq. 5.23 is the probability generating function of a geometric distribution with parameter

$$p = e^{-\{\rho(t)-\rho(s)\}} \qquad (5.24)$$

Consequently,

$$p_{1,n}(s,t) = e^{-\{\rho(t)-\rho(s)\}}(1 - e^{-\{\rho(t)-\rho(s)\}})^{n-1}, \tag{5.25}$$

which generalizes Eq. 4.20. More generally

$$p_{m,n}(s,t) = \binom{n-1}{n-m}p^m(1-p)^{n-m}, \tag{5.26}$$

where p is given by Eq. 5.24.

EXAMPLE 5B

Birth and death process with $\lambda_n(t) = n\,\lambda(t)$, $\mu_n(t) = n\,\mu(t)$. The transition probability generating function satisfies Eq. 5.11. Following the procedure outlined in Theorem 5B, we first need to find the general solution of the ordinary differential equation

$$\frac{dz}{dt} + (z-1)\{z\,\lambda(t) - \mu(t)\} = 0. \tag{5.27}$$

It was noticed by Kendall (1948) that this equation may be solved by introducing the change of variable $s = (z-1)^{-1}$, under which it becomes

$$\frac{ds}{dt} + \{\mu(t) - \lambda(t)\}s = \lambda(t)$$

with general solution (compare Theorem 4A)

$$s\,e^{\rho(t)} - \int_0^t \lambda(\tau)e^{\rho(\tau)}\,d\tau = \text{constant},$$

defining

$$\rho(t) = \int_0^t \{\mu(\tau) - \lambda(\tau)\}\,d\tau.$$

Any solution $z(t)$ of Eq. 5.27 therefore satisfies

$$u(z,t) = \text{constant},$$

where

$$u(z,t) = \frac{1}{z-1}e^{\rho(t)} - \int_0^t \lambda(\tau)\,e^{\rho(\tau)}\,d\tau.$$

To solve for $g^{-1}(\cdot)$ defined by Eq. 5.16 we write

$$x = u(z,s) = \frac{1}{z-1}e^{\rho(s)} - \int_0^s \lambda(\tau)\,e^{\rho(\tau)}\,d\tau,$$

which implies

$$g^{-1}(x) = z = 1 + \left\{ x\, e^{-\rho(s)} + e^{-\rho(s)} \int_0^s \lambda(\tau)\, e^{\rho(\tau)}\, d\tau \right\}^{-1}.$$

Therefore,

$$g^{-1}(u(z,t)) = 1 + \left\{ \frac{1}{z-1}\, e^{\rho(t)-\rho(s)} - e^{-\rho(s)} \int_s^t \lambda(\tau)\, e^{\rho(\tau)}\, d\tau \right\}^{-1}. \quad (5.28)$$

The transition probability generating function $\psi_{j,s}(z,t)$ is equal to the jth power of Eq. 5.28.

In particular,

$$\psi_{1,0}(z,t) = 1 + \left\{ \frac{1}{z-1}\, e^{\rho(t)} - \int_0^t \lambda(\tau)\, e^{\rho(\tau)}\, d\tau \right\}^{-1}. \quad (5.29)$$

The probability that the population has died out at time t is obtained by letting $z = 0$ in Eq. 5.29. Consequently,

$$P[X(t) = 0 \mid X(0) = 1] = \frac{e^{\rho(t)} + \displaystyle\int_0^t \lambda(\tau)\, e^{\rho(\tau)}\, d\tau - 1}{e^{\rho(t)} + \displaystyle\int_0^t \lambda(\tau)\, e^{\rho(\tau)}\, d\tau}$$

$$= \frac{\displaystyle\int_0^t \mu(\tau)\, e^{\rho(\tau)}\, d\tau}{1 + \displaystyle\int_0^t \mu(\tau)\, e^{\rho(\tau)}\, d\tau},$$

since

$$\int_0^t \{\mu(\tau) - \lambda(\tau)\} e^{\rho(\tau)}\, d\tau = e^{\rho(t)} - 1.$$

One sees that the probability of eventual extinction of the population is equal to 1,

$$\lim_{t\to\infty} P[X(t) = 0 \mid X(0) = 1] = 1,$$

if and only if

$$\lim_{t\to\infty} \int_0^t \mu(\tau)\, e^{\rho(\tau)}\, d\tau = \infty.$$

EXERCISES

5.1 *Pure birth process with immigration.* Consider a pure birth process $\{N(t), t \geq 0\}$ with intensity function

$$\lambda_n(t) = \nu(t) + n\lambda(t), \qquad n = 0, 1, 2, \cdots.$$

One can regard $N(t)$ as the size at time t of a population into which individuals immigrate in accord with a Poisson process with intensity function

$\nu(t)$ and then give rise to offspring in accord with a pure birth process with linear birthrate. Show that

$$\psi_{j,s}(z,t) = z^{-\nu(t)/\lambda(t)} \left\{ \frac{zp}{1-zq} \right\}^{j+\{\nu(s)/\lambda(s)\}},$$

where

$$p = e^{-\{\rho(t)-\rho(s)\}}, \qquad q = 1-p,$$

$$\rho(t) = \int_0^t \lambda(u)\, du.$$

Hint. One can use either the theory of filtered Poisson processes, or one can use the fact that the probability generating function satisfies

$$\frac{\partial}{\partial t} \psi_{j,s}(z,t) = z\,(z-1)\,\lambda(t)\frac{\partial}{\partial z}\,\psi_{j,s}\,(z,t) + (z-1)\,\nu(t)\,\psi_{j,s}(z,t).$$

5.2   *Birth and death process with equal linear birth and death rates.* Consider a non-homogeneous birth and death process with

$$\lambda_n(t) = n\,\lambda(t), \qquad \mu_n(t) = n\,\lambda(t).$$

Show (i) that the probability generating function of the population size satisfies

$$\frac{\partial \psi_{j,s}(z,t)}{\partial t} = \lambda(t)\,(z-1)^2 \frac{\partial \psi_{j,s}(z,t)}{\partial z},$$

(ii) that

$$\psi_{1,0}(z,t) = \frac{\dfrac{\rho(t)}{1+\rho(t)} + z\,\dfrac{1-\rho(t)}{1+\rho(t)}}{1 - \dfrac{\rho(t)}{1+\rho(t)}\,z},$$

where $\rho(t) = \int_0^t \lambda(u)\, du$, and

(iii) that

$$p_{1,0}(0,t) = \frac{\rho(t)}{1+\rho(t)} \to 1 \quad \text{as} \quad t \to \infty \quad \text{if} \quad \rho(t) \to \infty.$$

5.3   *Transition probabilities of the birth and death process with linear rates.* Show that for the birth and death process considered in Example 5B it holds for $n \geq 1$ that

$$p_{1,n}(0,t) = e^{\rho(t)}\{1 - p_{1,0}(0,t)\}^2 \{1 - e^{\rho(t)}[1 - p_{1,0}(0,t)]\}^{n-1},$$

where $p_{1,0}(0,t) = P[X(t) = 0 \mid X(0) = 1]$ is given by Eq. 5.30.

References

Anderson, T. W., and D. A. Darling. "Asymptotic theory of certain goodness of fit criteria based on stochastic processes," *Ann. Math. Stat.*, Vol. 23 (1952), pp. 193–212.

Anderson, T. W., and L. A. Goodman. "Statistical inference about Markov chains," *Ann. Math. Stat.*, Vol. 28 (1957), pp. 89–109.

Arrow, K. J., S. Karlin, and H. Scarf. *Studies in the Mathematical Theory of Inventory and Production.* Stanford, Calif.: Stanford University Press, 1958.

Bachelier, L. "Théorie de la spéculation," *Ann. Sci. Norm. Sup.*, Vol. 3 (1900), pp. 21–86.

——. *Calcul des probabilités.* Paris: Gauthier-Villars, 1912.

Bailey, N. T. J. *The Mathematical Theory of Epidemics.* London: Griffin, 1957.

Barnard, G. A. "Time intervals between accidents—a note on Maguire, Pearson, and Wynn's paper," *Biometrika*, Vol. 40 (1953), pp. 212–213.

Bartlett, M. S. "Some evolutionary stochastic processes," *Jour. Royal Stat. Soc.*, B, Vol. 11 (1949), pp. 211–229.

——. *An Introduction to Stochastic Processes.* London: Cambridge University Press, 1955.

——. *Stochastic Population Models in Ecology and Epidemiology.* London: Methuen, 1960.

——. *Essays on Probability Theory and Statistics.* London: Methuen, 1962.

Bell, D. A. *Statistical Methods in Electrical Engineering.* London: Chapman and Hall, 1953.

Bharucha-Reid, A. T. *Elements of the Theory of Markov Processes and Their Applications.* New York: McGraw-Hill, 1960.

Billingsley, P. "Statistical methods in Markov chains," *Ann. Math. Stat.*, Vol. 32 (1961), pp. 12–40.

Birkhoff, G. D. "Proof of the ergodic theorem," *Proc. Nat. Acad. Sci.*, Vol. 17 (1931), pp. 656–660.

Birnbaum, A. "Some procedures for comparing Poisson processes or populations," *Biometrika*, Vol. 45 (1953), pp. 447–449.

——. "Statistical methods for Poisson processes and exponential populations," *Jour. Amer. Stat. Assoc.*, Vol. 49 (1954), pp. 254–266.

Birnbaum, Z. W. "Numerical tabulation of the distribution of Kolmogorov's statistic for finite sample size," *Jour. Amer. Stat. Assoc.*, Vol. 47 (1952), pp. 425–441.

Blackwell, D. "A renewal theorem," *Duke Math. Jour.*, Vol. 15 (1948), pp. 145–151.

Blanc-Pierre, A., and R. Fortet. *Théorie des fonctions aléatoires.* Paris: Masson, 1953.

Brockmeyer, E., H. L. Halstroem, and A. Jensen. *The Life and Works of A. K. Erlang.* Copenhagen: Copenhagen Telephone Company, 1948.

Chandrasekhar, S. "Stochastic problems in physics and astronomy," *Rev. Mod. Phys.*, Vol. 15 (1943), pp. 1–89.

Chapman, D. "A comparative study of several one-sided goodness-of-fit tests," *Ann. Math. Stat.* Vol. 29 (1958), pp. 655–674.

Chung, K. L. *Markov Processes with Stationary Transition Probabilities.* Heidelberg: Springer-Verlag, 1960.

Cox, D. R. "The analysis of non-Markovian stochastic processes by the inclusion of supplementary variables," *Proc. Cambridge Phil. Soc.*, Vol. 51 (1955), pp. 433–441.

Cramér, H. *Mathematical Methods of Statistics.* Princeton, N.J.: Princeton University Press, 1946.

Davenport, W. B., Jr., and W. L. Root. *Random Signals and Noise.* New York: McGraw-Hill, 1958.

Davis, D. J. "An analysis of some failure data," *Jour. Amer. Stat. Assoc.*, Vol. 47 (1952), pp. 113–150.

Donsker, M. D. "Justification and extension of Doob's heuristic approach to the Kolmogorov-Smirnov theorems," *Ann. Math. Stat.*, Vol. 23 (1952), pp. 271–281.

Doob, J. L. "The Brownian movement and stochastic equations," *Ann. Math.*, Vol. 43 (1942), pp. 351–369.

——. "Time series and harmonic analysis," *Proc. Berkeley Symp. Math. Stat. Prob.*, ed. by J. Neyman, University of California Press, 1949(a), pp. 303–343.

——. "Heuristic approach to the Kolmogorov-Smirnov theorems," *Ann. Math. Stat.*, Vol. 20 (1949)(b), pp. 393–402.

——. *Stochastic Processes.* New York: Wiley, 1953.

Einstein, A. "Investigations on the theory of the Brownian movement," New York: Dover, 1956. (Contains translations of Einstein's 1905 papers.)

Epstein, B. "Tests for the validity of the assumption that the underlying distribution of life is exponential, I and II," *Technometrics*, Vol. 2 (1960), pp. 83–102, 167–184.

Evans, R. D. *The Atomic Nucleus.* New York: McGraw-Hill, 1955.

Feller, W. "On the integral equation of renewal theory," *Ann. Math. Stat.*, Vol. 12 (1941), pp. 243–267.

Feller, W. "On a general class of contagious distributions," *Ann. Math. Stat.*, Vol. 14 (1943), pp. 389–400.

———. "On probability problems in the theory of counters," *Courant Anniversary Volume*, 1948, pp. 105–115.

———. *An Introduction to Probability Theory and Its Applications*, 2nd ed. New York: Wiley, 1957.

———. "Non-Markovian processes with the semi-group property," *Ann. Math. Stat.*, Vol. 30 (1959), pp. 1252–1253.

Fortet, R. "Random functions from a Poisson process," *Proc. 2nd Berkeley Symp. Math. Stat. Prob.*, ed. by J. Neyman, University of California Press, 1951, pp. 373–385.

Foster, F. G. "On the stochastic matrices associated with certain queueing problems," *Ann. Math. Stat.*, Vol. 24 (1953), pp. 355–360.

Friedman, B. *Principles and Techniques of Applied Mathematics.* New York: Wiley, 1956.

Furry, W. "On fluctuation phenomena in the passage of high energy electrons through lead," *Phys. Rev.*, Vol. 52 (1937), p. 569.

Galton, F. *Natural Inheritance.* London: Macmillan, 1889.

Gilbert, E. N., and H. O. Pollak. "Amplitude distribution of shot noise," *Bell System Tech. Jour.*, Vol. 39 (1960), pp. 333–350.

Girshick, M. A., H. Rubin, and R. Sitgreaves. "Estimates of bounded relative error in particle counting," *Ann. Math. Stat.*, Vol. 26 (1955), pp. 189–211.

Good, I. J. "The real stable characteristic functions and chaotic acceleration," *Jour. Royal Stat. Soc.*, B, Vol. 23 (1961), pp. 180–183.

Greenwood, M., and G. U. Yule. "An inquiry into the nature of frequency-distributions representative of multiple happenings with particular reference to the occurrence of multiple attacks of disease or of repeated accidents," *Jour. Royal Stat. Soc.*, Vol. 83 (1920), p. 255.

Grenander, U., and M. Rosenblatt. *Statistical Analysis of Stationary Time Series.* New York: Wiley, 1957.

Gumbel, E. J. *Statistics of Extremes.* New York: Columbia University Press, 1958.

Gupta, S. S., and P. A. Groll, "Gamma distribution in acceptance sampling based on life time," *Jour. Amer. Stat. Assoc.*, Vol. 56 (1961), pp. 942–970.

Hagstroem, K. G. "Stochastik, ein neues — und doch ein altes Wort." *Skandinavisk Aktuarietidskrift*, Vol. 23 (1940), pp. 54–57.

Hannan, E. J. *Time Series Analysis.* London: Methuen, 1960.

Hardy, G. H. *Divergent Series.* Oxford: Oxford University Press, 1949.

Harris, T. E. *Branching Processes.* Heidelberg: Springer-Verlag, 1962.

Helstrom, C. W. *Statistical Theory of Signal Detection.* London: Pergamon Press, 1960.

Jenkins, G. M., "General considerations in the analysis of spectra," *Technometrics*, Vol. 3 (1961), pp. 133–166.

Kac, M. "Random walk and the theory of Brownian motion," *Amer. Math. Monthly*, Vol. 54 (1947), p. 369.

———. "On some connections between probability theory and differential and integral equations." *Proc. 2nd Berkeley Symp. Math. Stat. Prob.*, ed. by J. Neyman, University of California Press, 1951, pp. 189–215.

Kac, M. *Probability and Related Topics in Physical Sciences*. New York: Interscience, 1959.

Karlin, S. "On the renewal equation," *Pacific Jour. Math.*, Vol. 5 (1955), pp. 229–257.

Karlin, S., and J. L. McGregor. "The differential equations of birth and death processes and the Stieltjes moment problem," *Trans. Amer. Math. Soc.*, Vol. 85 (1957), pp. 489–546.

———. "The classification of birth and death processes," *Trans. Amer. Math. Soc.*, Vol. 86 (1957), pp. 366–400.

Kemeny, J. G., and J. L. Snell. *Finite Markov Chains*. Princeton, N.J.: Van Nostrand, 1960.

Kendall, D. G. "Geometric ergodicity and the theory of queues," in *Mathematical Methods in the Social Sciences, 1959*, ed. by K. J. Arrow, S. Karlin, and P. Suppes. Stanford: Stanford University Press, 1960.

———. "On the generalized birth and death process," *Ann. Math. Stat.*, Vol. 19 (1948), pp. 1–15.

———. "Some problems in the theory of queues," *Jour. Royal Stat. Soc.*, B, Vol. 13 (1951), pp. 151–185.

———. "Stochastic processes and population growth," *Jour. Royal Stat. Soc.*, B, Vol. 11 (1949), pp. 230–264.

———. "Stochastic processes occurring in the theory of queues and their analysis by means of the imbedded Markov chain," *Ann. Math. Stat.*, Vol. 24 (1953), pp. 338–354.

———. "Unitary dilations of Markov transition operators and the corresponding integral representations for transition probability matrices," in *Probability and Statistics*, ed. by U. Grenander. New York: Wiley, 1959, pp. 139–161.

Khinchin, A. I. (Khintchine, A. Ya.). "Korrelationstheorie der stationären stochastischen Prozesse," *Math. Ann.*, Vol. 109 (1934), pp. 415–458.

———. *Mathematical Foundations of Statistical Mechanics*. New York: Dover, 1949.

———. "On Poisson sequences of chance events," *Theory of Probability and its Applications* (English translation of the Soviet journal), Vol. 1 (1956), pp. 291–297.

Kolmogorov, A. N. "Über die analytischen Methoden in der Wahrscheinlichkeitsrechnung," *Math. Ann.*, Vol. 104 (1931), pp. 415–458.

———. *Foundations of the Theory of Probability*. New York: Chelsea, 1950 (translation of *Grundbegriffe der Wahrscheinlichkeitsrechnung* [1933]).

Korff, S. A. *Electron and Nuclear Counters*, 2nd ed. Princeton, N.J.: Van Nostrand, 1955.

Lanning, J. H., and R. H. Battin. *Random Processes in Automatic Control*. New York: McGraw-Hill, 1956.

Lawson, J. L., and G. E. Uhlenbeck. *Threshold Signals*. New York: McGraw-Hill, 1950.

Ledermann, W., and G. E. H. Reuter. "Spectral theory for the differential equation of simple birth and death processes," *Phil. Trans.*, A, Vol. 246 (1954), pp. 321–369.

Lin, C. C. "On the motion of a pendulum in a turbulent fluid," *Quarterly App. Math.*, Vol. 1 (1943), pp. 43–48.

Loève, M. *Probability Theory*, 2nd ed. Princeton, N.J.: Van Nostrand, 1960.

Lomax, K. S. "Business failures: another example of the analysis of failure data," *Jour. Amer. Stat. Assoc.*, Vol. 49 (1954), pp. 847–852.

Longuet-Higgins, M. S. "On the intervals between successive zeros of a random function," *Proc. Royal Soc.*, A, Vol. 246 (1958), pp. 99–118.

Lotka, A. J. "The extinction of families," *Jour. Wash. Acad. Sci.*, Vol. 21 (1931), pp. 377, 453.

MacFadden, J. A. "The axis-crossing intervals of random functions II," *I.R.E. Trans. on Information Theory*, Vol. IT—4 (1958), pp. 14–24.

Machlup, S. "Noise in semi-conductors: spectrum of a two parameter random signal," *Jour. App. Phys.*, Vol. 25 (1954), pp. 341–343.

Maguire, B. A., E. S. Pearson, and A. H. A. Wynn. "The time intervals between industrial accidents," *Biometrika*, Vol. 39 (1952), pp. 168–180.

———. "Further notes on the analysis of accident data," *Biometrika*, Vol. 40 (1953), pp. 213–216.

Malakhov, A. N. "Shape of the spectral line of a generator with fluctuating frequency," *Soviet Physics*, J E T P (English Translation), Vol. 3 (1956), pp. 653–656.

Mandelbrot, B. "The Pareto-Lévy law and the distribution of income," *Internat. Econ. Rev.*, Vol. 1 (1960), pp. 79 ff.

Middleton, D. *Statistical Communication Theory*. New York: McGraw-Hill, 1960.

Miller, G. A. "Finite Markov processes in psychology," *Psychometrika*, Vol. 17 (1952), pp. 149–167.

Molina, E. C. "Telephone trunking problems," *Bell System Tech. Jour.*, Vol. 6 (1927), p. 463.

Montroll, E. W. "Markoff chains, Wiener integrals, and Quantum Theory," *Commun. Pure App. Math.*, Vol. 5 (1952), pp. 415–453.

Moran, P. A. P. *The Theory of Storage*. London: Methuen, 1959.

Moyal, J. E. "Stochastic processes and statistical physics," *Jour. Royal Stat. Soc.*, B, Vol. 11 (1949), pp. 150–210.

Neyman, J., and E. L. Scott. "The distribution of galaxies," in *The Universe* (a Scientific American book). New York: Simon and Schuster, 1957, pp. 99–111.

———. "Statistical approach to problems of cosmology," *Jour. Royal Stat. Soc.*, B, Vol. 20 (1958), pp. 1–43.

———. "Stochastic models of population dynamics," *Science*, Vol. 130 (1959), pp. 303–308.

Olkin, I., and J. W. Pratt. "A multivariate Tchebycheff inequality," *Ann. Math. Stat.*, Vol. 29 (1958), pp. 201–211.

Osborne, M. F. M. "Reply to comments on Brownian motion in the stock market," *Oper. Res.*, Vol. 7 (1959), pp. 807–811.

Owen, A. R. G. "The theory of genetical recombination. I. Long-chromosome arms." *Proc. Royal Soc.*, B, Vol. 136 (1949), p. 67.

Parzen, E. *Modern Probability Theory and Its Applications*. New York: Wiley, 1960.

———. "Conditions that a stochastic process be ergodic," *Ann. Math. Stat.*, Vol. 29 (1958), pp. 299–301.

Parzen, E. "Mathematical considerations in the estimation of spectra," *Techno-metrics.* Vol. 3 (1961)(a), pp. 167–190.
———. "An approach to time series analysis," *Ann. Math. Stat.*, Vol. 32 (1961)(b), pp. 951–989.
———. "Spectral analysis of asymptotically stationary time series," *Bulletin of the International Statistical Institute*, 33rd Session, Paris, 1961(c).
Perrin, J. *Atoms.* London: Constable, 1916.
Prais, S. J. "Measuring social mobility," *Jour. Royal Stat. Soc.*, A, Vol. 118 (1955), pp. 56–66.
Pyke, R. "On renewal processes related to type I and type II counter models," *Ann. Math. Stat.*, Vol. 29 (1958), pp. 737–754.
Rényi, A. "On the asymptotic distribution of the sum of a random number of independent random variables," *Acta Math., Acad. Scient. Hung.*, Vol. 8 (1957), pp. 193–199.
Reuter, G. E. H., and W. Ledermann. "On the differential equations for the transition probabilities of Markov processes with enumerably many states," *Proc. Cambridge Phil. Soc.*, Vol. 49 (1953), pp. 247–262.
Rice, S. O. "Mathematical analysis of random noise," *Bell System Tech. Jour.*, Vol. 23 (1944), pp. 282–332; Vol. 24 (1945), pp. 46–156.
Satterthwaite, F. E. "Generalized Poisson distributions," *Ann. Math. Stat.*, Vol. 13 (1942), pp. 410–417.
Schottky, W. "Über spontane Stromschwankungen in verschiedenen Elektrizitäts-leitern," *Ann. der Physik*, Vol. 57 (1918), pp. 541–567.
Sevastyanov, B. A. "An ergodic theorem for Markov processes and its application to telephone systems with refusals," *Theory of Probability and its Applications* (English translation of the Soviet journal), Vol. 2 (1957), pp. 104–112,
Skellam, J. G. "Studies in statistical ecology," *Biometrika*, Vol. 39 (1952), pp. 346–362.
Slepian, D. "First passage problem for a particular Gaussian process," *Ann. Math. Stat.*, Vol. 32 (1961), pp. 610–612.
Steffensen, J. F., "Om Sandsynligheden for at Afkommet uddør," *Mat. Tidsskr.*, B, Vol. 19 (1930), pp. 19–23.
Smith, W. L. "Renewal theory and its ramifications," *Jour. Royal Stat. Soc.*, B, Vol. 20 (1958), pp. 243–302 (with discussion).
———. "On renewal theory, counter problems, and quasi-Poisson processes," *Proc. Cambridge Phil. Soc.*, Vol. 53 (1957), pp. 175–193.
Smoluchowski, M. V. "Drei Vorträge über Diffusion, Brownsche Bewegung and Koagulation von Kolloidteilchen," *Physik. Zeit.*, Vol. 17 (1916), pp. 557–585.
Syski, R. *Introduction to Congestion Theory in Telephone Systems.* Edinburgh: Oliver and Boyd, 1960.
Takacs, L. "On a probability problem in the theory of counters," *Ann. Math. Stat.*, Vol. 29 (1958), pp. 1257–1263.
———. *Stochastic Processes. Problems and Solutions.* London: Methuen, 1960.
ter Haar, D. *Elements of Statistical Mechanics.* New York: Rinehart, 1954.
Titchmarsh, E. C. *Introduction to the Theory of Fourier Integrals.* Oxford: Oxford University Press, 1948.

Tukey, J. W. "An introduction to the measurement of spectra," in *Probability and Statistics*, ed. by U. Grenander. New York: Wiley, 1959, pp. 300–330.

Uhlenbeck, G. E., and L. S. Ornstein. "On the theory of Brownian motion," *Phys. Rev.*, Vol. 36 (1930), pp. 823–841.

von Neumann, J. "Proof of the quasi-ergodic hypothesis," *Proc. Nat. Acad. Sci.*, Vol. 18 (1932), p. 70.

———. "Physical applications of the ergodic hypothesis," *Proc. Nat. Acad. Sci.*, Vol. 18 (1932), pp. 263–266.

Wang, Ming Chen, and G. E. Uhlenbeck. "On the theory of the Brownian motion II," *Rev. Mod. Phys.*, Vol. 17 (1945), pp. 323–342.

Wax, N. *Noise and Stochastic Processes*. New York: Dover, 1954.

Whittaker, J. M. "The shot effect for showers," *Proc. Cambridge Phil. Soc.*, Vol. 33 (1937), pp. 451–458.

Whittle, P. "Continuous generalizations of Tchebichev's inequality," *Theory of Probability and its Applications (Teoriya Veroyatnostei i ee Primeneniya)*, Vol. 3 (1958), pp. 386–394.

Wiener, N. "Differential space," *J. Math. Phys. Mass. Inst. Tech.*, Vol. 2 (1923), pp. 131–174.

———. "Generalized harmonic analysis," *Acta. Math.*, Vol. 55 (1930), p. 117.

Wiener, N., and A. Siegel. "A new form of the statistical postulate for quantum mechanics," *Phys. Rev.*, Vol. 91 (1953), pp. 1551–1560.

Wold, H. *A Study in the Analysis of Stationary Time Series*, 2nd ed. (with an appendix by Peter Whittle). Uppsala, Sweden, 1954.

Wonham, W. M., and A. T. Fuller. "Probability densities of the smoothed random telegraph signal." *Jour. of Electronics and Control*, Vol. 4 (1958), pp. 567–576.

Yule, G. U. "A mathematical theory of evolution, based on the conclusions of Dr. J. C. Willis, F. R. S.," *Phil. Trans. Royal Soc. London. B*, Vol. 213 (1924), pp. 21–87.

Author index

Subject index

316

272

324   SUBJECT INDEX

273   7   1867